THE MINICOMPUTER
IN THE LABORATORY

THE MINICOMPUTER IN THE LABORATORY:

With Examples Using the PDP-11

Second Edition

JAMES W. COOPER

Vice President, Software Development
Bruker Instruments, Inc.
Billerica, Massachusetts

A WILEY-INTERSCIENCE PUBLICATION

JOHN WILEY & SONS

NEW YORK □ CHICHESTER □ BRISBANE □ TORONTO □ SINGAPORE

Library of Congress Cataloging in Publication Data:

Cooper, James William, 1943–
  The minicomputer in the laboratory.

  "A Wiley-Interscience publication."
  Includes index.
  1. Minicomputers—Programming. 2. PDP-11 (Computer)—
Programming. 3. Assembler language (Computer program
language) I. Title.

QA76.6.C653 1982   001.64′2   82-8490
ISBN 0-471-09012-3   AACR2

Printed in the United States of America

10 9 8 7 6 5 4 3 2 1

 **PREFACE** □

Since the publication of the first edition, there has been a tremendous increase in the availability of small, sophisticated minicomputers such as the PDP-11, especially in LSI forms. In fact, there are currently versions costing only a few thousand dollars, with much greater capability than that represented by the $25,000 machine at Tufts of a few years ago. This trend will undoubtedly continue in the years to come.

Thus, in this edition, I have practically eliminated any mention of paper tape equipment in favor of the increasingly common floppy diskette units used on most modern systems. In addition, I have chosen to cover a few more sophisticated topics, since both the scientists and the machines are more sophisticated.

My one attempt at humor in the first edition was a reference to the mathematician Lobachevski as the alleged inventor of Polish notation. This was a jest that was meant to refer to the Tom Lehrer song, "Lobachevski," in which this fictionalized mathematician discovered one theorem after another by plagiarizing everyone's work. Needless to say, I was corrected, most notably by a scientist in Warsaw, and I have thus noted in this edition that Polish notation was invented by Lukasiewicz.

The rapid growth of the minicomputer as a teaching and research tool has made it increasingly important that the laboratory scientist have some knowledge of the minicomputer both as a "black box" and as a versatile research tool. Many scientists have some familiarity with high-level languages such as FORTRAN, BASIC or Pascal, but little assembly language experience. Although each computer has its own assembly language, the mastery of one such fundamental language usually leads to extremely fast mastery of other languages as needed.

The objectives of this book are to provide a basic introduction to minicomputer architecture, to teach the PDP-11 assembly language in detail, and to discuss the various types of programming techniques or *algorithms* appropriate to the laboratory computer.

The PDP-11 computer was selected because of its great popularity and because of the author's experience in teaching a course in laboratory computing for chemists at Tufts University. However, many of the algorithms discussed in this text are applicable to any number of other computers and languages.

v

In learning about laboratory computing, it is absolutely essential that the student write a large number of programs. For this reason, various exercises are provided, many with answers. These exercises or ones of similar difficulty should be worked out before proceeding to new concepts, as it is impossible to learn anything about the minicomputer without a great deal of hands-on experience.

JAMES W. COOPER

Billerica, Massachusetts
November 1982

□□□□□□ ACKNOWLEDGMENTS □

Since the publication of the first edition, I have learned many things that I have included in this edition from correspondents, co-workers, and students. These people include Professor Terry Haas and Dr. Charles Thibault at Tufts University, Dr. Harold Boll, Dr. Elaine Braun-Keller, and George Pawle at Bruker Instruments, and Janus Zalewski of the Warsaw Institute of Nuclear Research.

Spectra for the stacked plot in Chapter 17 were obtained by the applications group at Bruker, and much of the manuscript was written using word processing programs on Bruker's Aspect-2000 computer system. The cooperation of Bruker Instruments in the final production of the manuscript is gratefully acknowledged.

The infrared spectra in Chapter 21 were obtained with the cooperation of Dr. J. Gronholz and the IR applications group of Bruker-Karlsruhe.

Lastly, I wish to acknowledge the continuing patience and understanding of my wife Vicki while I worked on this project.

J.W.C.

□□□□□□□□□□ CONTENTS □

Contents

□□□□□□□□□□□□ **PART ONE**□

INTRODUCTION TO THE PDP-11 COMPUTER

□□□□□□□□□CHAPTER ONE□

INTRODUCTION TO THE COMPUTER

Computer memories are the most fundamental part of the computer system. The computer's *main memory* consists of some substance or device that will allow storage of the 1's and 0's of *binary information*. Historically, this has been a set of small magnetic *cores*, or doughnuts, each the size of a period, with wires strung through them. These cores can be magnetized into one of two magnetic states, represented by 1's or 0's. Today, we see more and more semiconductor memory, where each unit acts more like a set of small capacitors that are either charged or not charged and that can be examined electronically to determine their state.

The principal advantage of these memories is their much lower manufacturing cost, since as much as 16,384 to 262,144 individual bits can be stored in a chip the size of a postage stamp. Semiconductor memory has the disadvantage that it will only maintain information while power is applied; thus several types of backup are usually supplied for such systems.

Core memory, on the other hand, has much more permanence, but it substantially more expensive and much slower in response.

Read-Only Memory

One common way around the problem of the volatility of semiconductor memory is the inclusion of some common programs in read-only memory for starting up the computer when a power failure or system failure has occurred. These small start-up programs are often referred to as *bootstrap* programs and usually contain information on how to read in more complex programs to run the computer system.

Computer Programming

In early computers, the logic and arithmetic functions were grouped in a section called the *central processor*, and they included such simple abilities as addition and subtraction, multiplication and division, fetching and storing numbers, and testing for positive, negative, and zero numbers. These functions could be accessed by the use of a patch panel, much like an old-fashioned telephone switchboard. In these computers, programming amounted to patching the various operations to the desired addresses, and thus it was not only a logical but a mechanical skill.

Although a program for adding or multiplying columns of numbers stored in memory was relatively easy to patch, the capability to perform complex data manipulations such as Fourier transforms, requiring both complex interconnections and more advanced hardware, was still far in the future.

The major breakthrough in computer design is attributed to von Neumann, who suggested that the numbers stored in memory be considered both as data and as *instructions*. This was accomplished by allocating some variable portion of memory as a program and pointing an instruction decoder circuit to those memory addresses that were to be interpreted as instructions.

For example, a simple 4-bit computer might contain the ability to fetch numbers, add them, multiply them and store them, and the binary numbers needed to accomplish these operations might be

0001	Fetch
0010	Add
0100	Store
1000	Multiply
0000	Stop

A simple program for adding two numbers together and multiplying by a third might appear as follows:

0001	Fetch a number from memory.
0010	Add a number to it.
1000	Multiply by a third number.
0000	Stop, displaying result.

This simplified example assumes that the numbers used for addition and multiplication are stored in memory in some predetermined order and that the computer somehow knows which ones they are. In a more practical computer system, we would allocate additional bits to form the *address* of the data being referred to. This address is just the sequential binary number describing the location of the number in memory. Addresses usually start at 0 and go as high as the total number of words in memory.

In our simple little computer, we will consider addresses 5, 6, 7, and 8 as the ones referred to by our program. The program can then be rewritten with 4 bits used for each instruction and 4 more bits used to refer to the actual address.

Address	Instruction	Address	Meaning
0000	0001	0101	Fetch a number from address 0101 (5).
0001	0010	0110	Add contents of address 0110 to it.
0010	1000	0111	Multiply by contents of 0111.
0011	0100	1000	Store result in address 1000 (8).
0100	0000	0000	Stop.
0101	0010	1101	Number to be fetched.
0110	1011	1001	Number to be added.
0111	0000	0010	Number to be multiplied.
1000	0000	0000	Location where result is stored.

In no time at all, this becomes quite cumbersome, and for this reason one of the first programs written for any computer is the one that translates simple alphabetic symbols into binary numbers. Such a program is called an *assembler*. An assembler performs two main functions:

1. It translates computer instructions into binary.
2. It allows variables to be referred to by name instead of address.

Thus the three numbers referred to above might be named NUM, ADDEND, and MULPLYR, instead of being called addresses 5, 6, and 7. We could then code the preceding program in assembly language as

```
            FET NUM          ;Fetch contents of NUM
            ADD ADDEND       ;Add a number to it
            MUL MULPLYR      ;Multiply by a third
            STO RESULT       ;And store the result
            STOP             ;That's all

NUM:        55               ;Number to start with
ADDEND:     31               ;Number to add
MULPLYR:    2                ;Number to multiply by
RESULT:     0                ;Place to store result
```

Assembly languages are the mainstay of minicomputers because even though their instructions are tediously elementary, they provide the maximum speed and versatility for the computer system. We will study the PDP-11's assembly language in this book because it will enable us to understand fully the minicomputer's great power to solve laboratory problems. We will regard the assembly language as going "halfway" toward meeting the computer. The binary machine language is clearly the computer's home ground, and higher mathematical languages are clearly closer to human's home ground.

You are probably more familiar with such high-level languages as Pascal[1] and FORTRAN, which allow us to write the operations in the preceding programs as a simple single expression:

RESULT := (NUM + ADDEND)*MULPLYER;

It is possible to write programs to translate such expressions into machine languages such as that shown in the preceding example. Such programs are called *compilers* and are fairly complicated programming projects. We can distinguish compilers from assemblers by the rule of thumb that an assembler produces one binary machine instruction for each assembly language statement, while a compiler may produce many machine instructions from each statement.

For the most part, programs for data acquisition and processing have been written in assembly language. There are several reasons why this has been the case. First, most common high-level languages have no statements for the efficient production of data acquisition, display, or instrument control routines. Some of these features have been added as special subroutines to some manufacturer's versions of BASIC or FORTRAN, but they are usually only for fairly simple, slow experiments. Second, most special-purpose languages for data acquisition have almost no facilities for processing the data, so that even if acquisition were possible, the complex needs of modern data processing would require other languages for the calculations. Third, hybrid languages claiming to have facilities for both are usually limited in scope by the designer's conception of laboratory needs. Finally, it turns out that rapid acquisition and display routines require the speed of assembly language. This is the most persuasive reason for taking it up in the following chapters.

Reference

1. J. W. Cooper, *Introduction to Pascal for Scientists*, Wiley-Interscience, New York, 1981.

CHAPTER TWO

BASIC NUMERICAL CONCEPTS

Computer Words

The memory of a digital computer is made up of binary digits or *bits* combined into logical groups called *words*. Typical word lengths in small computers include 8, 12, 16, 20, 24, and 32 bits. The PDP-11 has a 16-bit word, one of the most common word lengths in use.

A computer word can be used to contain one number, one instruction, or one or more characters for a message to be printed out. Since each word in the PDP-11 contains 16 bits, it can be used to represent numbers from 0 to $2^{16} - 1$. Numbers are represented by setting one or more of the bits of the computer word to 1. The rightmost bit represents 2^0 or 1, and the leftmost bit 2^{15} or 32,756. If all bits are ones, the number is $2^{16} - 1$ or 65,535. The 16-bit computer word with its associated powers of 2 for each bit is illustrated in Figure 2.1. The bits are numbered from right to left to remind us that each bit represents that power of 2.

Number Systems

When the minicomputer is first manufactured and its memory boards inserted, it "knows" nothing unless a specific program is inserted in its read-only memory (ROM). Since it understands no languages or commands to read or print at the terminal, it is only fitting that the programmer go halfway in learning to talk to it in concepts that it can understand.

19	18	17	16	15	14	13	12	11	10	9	8	7	6	5	4	3	2	1	0
524,288	262,144	131,072	65,536	32,768	16,384	8,192	4,096	2,048	1,024	512	256	128	64	32	16	8	4	2	1

Figure 2.1 Bit numbering scheme and associated powers of 2.

These concepts include a few that may be unfamiliar to the average scientist.

Binary Numbers

Since a digital computer is a binary device, all decisions and numeric operations take place in binary. You can count in binary as follows: 0, 1, 10, 11, 100, 101, 110, 111, 1000, 1001, 1010. A given binary number can be converted to decimal (base-10) by simply considering each digit as a multiplier of that power of 2. Thus the number 1010 can be considered as

$$
\begin{aligned}
1 \times 2^3 &= 1 \times 8 = 8 \\
0 \times 2^2 &= 0 \times 4 = 0 \\
1 \times 2^1 &= 1 \times 2 = 2 \\
0 \times 2^0 &= 0 \times 1 = \underline{0} \\
& \qquad\qquad\qquad\;\; 10
\end{aligned}
$$

It is not necessary, however, to convert all numbers from binary to decimal to perform operations with them. Let us examine the addition of numbers in binary.

1	1	10	10111	0111010110111011
+0	+1	+10	+10001	+0101101010100110
1	10	100	101000	1101000001100001

As you can see from these examples, the manipulation of even small binary numbers becomes quite complicated, and the manipulation of 16-bit binary numbers boggles the mind. For this reason, it is customary to use a shorthand method for representing binary numbers. This method contains as much information, is readily convertable to binary, and is much easier for the human mind to assimilate. This shorthand method is called the *octal* or *base-8* number system.

The Octal Number System

The octal number system is used because it looks almost like the decimal number system; thus it can be grasped easily by new students. Recall that our conventional decimal number system is a base-10 number system. This means that the first number that cannot be represented in one digit is 10. In octal, the first number that cannot be represented in one digit is 8. Let us compare octal, decimal, and binary in the following chart.

Decimal	Octal	Binary
0	0	000
1	1	001
2	2	010
3	3	011
4	4	100
5	5	101
6	6	110
7	7	111
8	10	1 000
9	11	1 001
10	12	1 010

As you can see, octal and decimal are the same for the first eight digits: 0–7. The octal system requires two digits to represent 8_{10} as 10_8. Note also that in our list of binary numbers in the third column, we have grouped the binary digits into groups of three, even adding leading 0's where necessary. This is done so that conversion between binary and octal can be obvious.

To convert between binary and octal, we simply group the binary digits into groups of three starting at the right. If there are not enough digits when we get to the left end, we simply add leading 0's. Then conversion to octal is done by inspection: writing down an octal digit for each binary group of three. For example, we write the binary number 1101011010011101 in groups of three:

1	101	011	010	011	101
1	5	3	2	3	5

and write the octal digits under each group. There is no carrying between groups, and we find that the number 153235_8 is a much more compact representation of the binary string 1101011010011101.

Conversely, we can convert octal to binary by simply writing down the three-digit binary number for each octal digit. The octal number 75346 is converted to binary as follows:

```
 7    5    3    4    6    1
111  101  011  100  110  001
```

You should learn the table of the eight octal–binary conversions so that you can do this automatically.

Octal Arithmetic

The only rule necessary to learn to add in base-8 or octal is that any number that would be written as an 8 in decimal becomes an octal 10 (or zero with 1 to carry). The corollary is that any number larger than 8 can be converted by simply subtracting 8 from that sum and putting down the remainder. There is then 1 to carry as before. Briefly,

$$7+1 = 10$$
and $6+3 = 11$
and so forth.

For example,

```
  2 2      2 4 6
  6 3      1 5 3
_____      _____
1 0 5      4 2 1
```

Looking at the previous complex example, we see that we can perform the binary addition much more quickly if we convert to octal, do octal addition, and reconvert to binary:

```
10111  =  010 111  =  2 7
10001  =  010 001  =  2 1
_____     _____     ___
                      5 0  =  101 000
```

More importantly, 16-bit numbers can be handled much more easily:

```
0  111  010  110  111  011  =  0  7  2  6  7  3
0  101  101  010  100  110     0  5  5  2  4  6
                               _____
                               1  5  0  1  4  1
                            =  1  101  000  001  100  001
```

At this point, we can introduce an important corollary to our basic rule for octal addition. Add each pair of digits in your head in decimal. If the sum is greater than 7, subtract 8. The remainder is the digit to be put down and 1 is the carry. For example,

$$\begin{array}{r} 7 \\ +5 \\ \hline 12_{10} \end{array} \qquad\qquad \begin{array}{r} 7 \\ +5 \\ \hline 14_8 \end{array}$$

$12 - 8 = 4$, so the sum is   14

Conversion Between Octal and Decimal

Conversion between octal and decimal is performed automatically by the MACRO assembler and can also be accomplished by looking in the table at the back of this text. However, the general principle is the same as that for conversion between binary and decimal. Each digit in an octal number is a multiplier times the appropriate power of 8. Thus the number 246_8 means

$$\begin{array}{rcccl} 2 \times 8^2 & = & 2 \times 64 & = & 128 \\ +4 \times 8^1 & = & 4 \times 8 & = & 32 \\ +6 \times 8^0 & = & 6 \times 1 & = & \underline{6} \\ & & & & 166_{10} \end{array}$$

Conversion Between Decimal and Octal

Conversion between decimal and octal is accomplished by dividing by successively smaller powers of 8. To convert 2453_{10} to octal, we start by dividing by 8^3 or 512. Then we divide by 8_2 or 64, by 8 and by 1.

$$\begin{array}{lll} 2453/512 & = & 4 \text{ with } 405 \text{ remainder} \\ 405/64 & = & 6 \text{ with } 21 \text{ remainder} \\ 21/8 & = & 2 \text{ with } 5 \text{ remainder} \end{array}$$

and the ones column will thus contain the 5. The converted number, then, is

$$2453_{10} = 4625_8$$

A table of octal–decimal conversions is given in Appendix V.

Exercises

2.1   Convert the following binary numbers to octal.
     010   0010101110111000
     101101   0111101000110110
     00010101   0101101110001010

2.2 Convert the following octal numbers to binary.
     223   1264
     11707   65643
     2106463   3006557

2.3 Convert the following octal numbers to decimal.
     7777   10000
     144   256
     4076   12346

2.4 Convert the following decimal numbers to octal.
     4096   524,289
     100   16,383
     512   300

2.5 Perform the following octal additions.
     2467   12   10543   304566
     1234   23   21615   134652

2.6 Perform the following additions in binary. Convert to octal, and
     compare your results.
     110 101 111 001 100   10 100 001 101 111 101 100
     001 010 000 111 101   01 111 010 001 011 111 011

Important Numerical Concepts for PDP-11 Programming

Complements. The term *complement*, or, more fully, *one's complement*, is extremely useful in referring to binary arithmetic. The one's
complement of a number is obtained by simply changing all the 1's to 0's
and all the 0's to 1's. Thus 010 and 101 are one's complements. The binary
number

     1 010 101 111 000 100   is the complement of
     0 101 010 000 111 011   and vice versa.

It must also be true that the sum of any two binary numbers that are
complements of each other must be all 1's, since a 1 must always line up
with a 0 during addition. Thus the following complements sum to produce
all 1's.

     000   1010   010 101 000 110
     111   0101   101 010 111 001
     111   1111   111 111 111 111

This fact suggests a way of determining the complements of *octal* numbers without converting them to binary. Since the sum of two binary complements must be all 1's, the sum of two octal numbers that are complements must be all 7's. Remember that $111_2 = 7_8$.

It is obvious, then, that we can determine the complement of an octal number by simply subtracting it from the octal number representing a binary number that is all 1's.

The complement of 542_8 can be determined as follows:

$$\begin{array}{r} 777 \\ -542 \\ \hline 235 \end{array} = 010 \quad 011 \quad 101$$

In the case of a 16-bit number, the complement is determined by subtracting from that number which represents 16 bits of 1's: 177777. The first digit is only a 1 because the leftmost bit is left over after dividing the 16 bits into groups of three starting at the right. Thus the 16-bit complement of 456 is determined by subtracting it from 177777:

$$\begin{array}{r} 177777 \\ -456 \\ \hline 177321 \end{array}$$

Two's Complements.   The two's complement is closely related to the one's complement and is particularly useful in modern minicomputers. The two's complement of a number is defined simply as the *one's complement plus 1*.

Thus the two's complement of 1101 is the one's complement 0010 plus 1, or 0011. Longer numbers can be determined directly in octal by taking the one's complement in octal and adding 1:

$$\text{two's complement of } 123 = \begin{array}{r} 777 \\ -123 \\ \hline 654 \end{array} + 1 = 655$$

Using 16-bit numbers, we find the two's complement of 5326 by

$$\begin{array}{r} 177777 \\ -5326 \\ \hline 172451 \end{array} + 1 = 172452$$

These two operations can, of course, be combined into a single step. Instead of subtracting the number from 177777, we can subtract it from 177777 + 1, which we will write in whatever form is most useful for this subtraction. To form the two's complement of 1256, we subtract

```
17777"8"
- 125  6
17652  2,  or just 176522
```

To form the two's complement of 123560, we subtract

```
 1777"8"0
-1235  6  0
 0542  2  0
```

Two's complement arithmetic is of great importance because the PDP-11, as well as most other minicomputers, utilizes the two's complement of a number as its *negative*. The total range of unsigned numbers that can be represented in the PDP-11 is $0\text{-}177777_8$. We arbitrarily divide this range in half and call half of those numbers positive and the other half negative. The range of signed numbers looks like this:

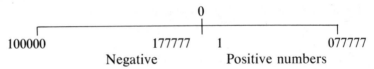

Close examination of the numbers in the negative range reveals that they all have one thing in common: The leftmost bit, bit 15, is set to 1. Conversely, all positive numbers, including 0, have bit 15 set to 0. Consequently, bit 15 is called the *sign bit* and can be tested independently to allow a decision as to whether a particular number is negative.

This division of numbers into positive and negative ranges using bit 15 is not as arbitrary as it seems since arithmetic manipulations can be performed in a consistent manner using this convention. Let us start adding 1's to the large number 177775:

```
 177775
 177776
 177777
*000000     At this point, overflow occurs, and there is a carry to the
 000001     imaginary seventeenth bit.
 000002
 000003
```

The same sort of thing would happen if we started adding 1's to -3. The sequence would be -3, -2, -1, 0, 1, 2, 3. In fact, as far as the computer is concerned this is exactly what we *have* done. The two's complement of 3 is 177775_8 and thus represents -3 to the computer. As an additional proof, we add 3 to 177775:

$$
\begin{array}{r}
177775 \\
+3 \\
\hline
(1)000000
\end{array}
$$

getting 0 as expected. The seventeenth bit, if it existed, would be set to 1 as indicated by the (1). Clearly, since we get 0 by adding -3 to $+3$, the choice of two's complements for negative numbers is not an arbitrary one.

Subtraction. The PDP-11 performs subtraction using the method we are taught in algebra, that is, by changing the sign and then adding. The sign is changed in the computer, of course, by taking the two's complement of the number. Thus if we wished the computer to subtract 5 from 7, the operation would be

$$
\begin{array}{rr}
7 = & 7 \\
-5 = & 177773 \\
\hline
& 000002
\end{array}
$$

In other words, $7 - 5 = 2$, as expected.

The Logical AND

One of the operations fundamental to computer operations is the logical AND. The result of a logical AND between 2 bits is such that the result is only 1 if both of the bits are 1. Otherwise, the result of the AND is 0. This can be represented in a *truth table* as follows:

	0	1
0	0	0
1	0	1

The AND operation can also be represented symbolically by the schematic symbol for the AND gate shown in Figure 2.2.

Here, the two input bits are represented on the left side of the gate, and the resulting value is on the right side of the gate. AND gates, like other fundamental digital building blocks, can be made up of transistors and are readily available with several in a single integrated circuit chip.

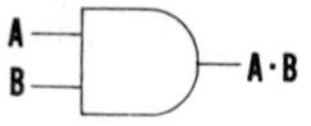

Figure 2.2 The AND gate.

The AND function can be thought of as a masking operation allowing us to examine certain bits of a number without considering the remaining bits, since the AND will only let the bits that are 1's through the mask. Thus, if we wish to examine bit 4 of a number, we AND that number with the binary value 10000_2.

1	101	110	111	011	010	156332
0	000	000	000	010	000	000020
0	000	000	000	010	000	000020

A nonzero result indicates that bit 4 was set. If we wish to examine an entire octal digit, we can set 3 bits of the AND mask:

156732	1	101	110	111	011	010
007000	0	000	111	000	000	000
006000	0	000	110	000	000	000

The Inclusive OR

Another function easily performed electronically by the digital computer is the inclusive OR. This operation returns a 1 if either or both of the bits are 1 and returns a 0 only if both of the bits are 0. The truth table is

	0	1
0	0	1
1	1	1

This can be represented schematically as shown in Figure 2.3.

The inclusive OR is used to set bits in a computer word, whether they were previously set or not. Using the OR to compare two words produces a word that shows which bits are set in either of the two words. The following example shows the inclusive OR between some octal digits:

000	0	010	2	100	4	101	5
111	7	110	6	011	3	001	1
111	7	110	6	111	7	101	5

Figure 2.3 The OR gate.

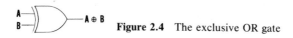

Figure 2.4   The exclusive OR gate

The Exclusive OR

The exclusive OR or XOR is rather like the OR minus the AND. Bits are set in the result only if the input bits are *different*. If the bits are the same, either 1's or 0's, the result is 0. The truth table is

	0	1
0	0	1
1	1	0

The schematic symbol is shown in Figure 2.4.

The XOR is shown for the following octal digits:

000	0	010	2	100	4	101	5
111	7	110	6	111	7	100	4
111	7	100	4	011	3	001	1

Exercises

2.7.   Find the one's complements of the following numbers:
  0 111 011 101 110 011   1 101 010 110 100 100
  0 000 000 101 111 000   1 101 101 111 010 011

2.8.   Find the two's complements of the numbers in Exercise 2.7.

2.9.   Perform the following subtractions, using two's complement arithmetic. Assume 16-bit results.

56342	067542	105423
− 315	− 134527	− 154321

2.10   Write the positive octal number corresponding to each of the following negative octal numbers:
  177560   156120   153210

2.11 Perform the AND operation between each pair of these octal num-
      bers:

7	5	376	70770
3	2	123	146237

2.12 Perform the inclusive OR between the octal numbers in Exercise
      2.11.

2.13 Perform the exclusive OR (XOR) between the octal numbers in
      Exercise 2.11.

CHAPTER THREE

STARTING THE PDP-11 AND THE RT-11 MONITOR

The PDP-11 consists of a central processor unit, memory, and a collection of peripheral devices that all communicate along the Unibus, which transfers high-speed logic signals between devices. Unlike many other computers, the center of the PDP-11 is the Unibus itself, with the central processor, memory, terminals, and disks all peripheral devices. This is particularly true because the devices, the processor, and the memory all have actual addresses.

The PDP-11 numbers memory locations sequentially in 8-bit *bytes*, so that each byte has a separate address. Thus each memory *word* has two addresses, one for each byte. Addresses (and their contents) are generally given in *octal*, and in this form the first 4K (4096 words) has addresses from 0 to 17777_8, representing 4K of words or 8K bytes. This byte addressing scheme was developed for the PDP-11 because most characters can be represented in 8 bits and a major task of minicomputers is intercomputer communication.

The byte addresses run sequentially through memory with even numbers corresponding to the lower bytes of words and odd addresses to the upper bytes of words (Figure 3.1).

The PDP-11 memory consists of blocks of memory on single circuit boards. These blocks may be 4K, 8K, or 16K of words, and a simple PDP-11 may contain up to 28K of usable memory. The last 4K of memory from 28K to 32K is reserved for device addresses that can be referred to from within programs just as if they were memory addresses. These device addresses refer to the central processor status and to all of the

WORD   BYTES

WORD	BYTE odd	BYTE even
0	1	0
2	3	2
4	5	4
6	7	6

Figure 3.1 Word and byte addressing in the PDP-11.

peripheral devices. Newer PDP-11 models can be expanded to up to 124K using a technique called *memory management*, with the block of memory from 125K to 128K reserved for the device addresses.

In addition to the special meaning reserved for the last 4K of PDP-11 memory, the first 400_8 addresses of memory have special uses as interrupt and trap vector addresses, which will be discussed in Chapter 10. Programs usually are designed to begin at address 1000, with addresses 400–776 reserved for temporary storage use as a hardware *stack*. At the upper end of memory, a program is sometimes resident for reading in paper tapes or starting a bootstrap loader for disks in machines that lack a hardware bootstrap feature. The remaining memory can be used for any combination of programs and data that might be useful in various calculations. The layout of memory is shown in Figure 3.2. Today most PDP-11's contain either 24K or 32K, where that last 4K of memory addresses cannot be used, since the device addresses supplant it.

Use of the PDP-11 Console

PDP-11 consoles vary widely in design. Some have a number of switches to represent all of the binary digits of the 16-bit word, and others have a calculator-like keypad for entry of addresses and values to be deposited in memory. Further, more recent PDP-11's have some programs in ROMs for starting up programs and running octal debuggers.

All PDP-11's have a power keyswitch to turn them on, and a position called Panel Lock in which the remaining keys are disabled for protection

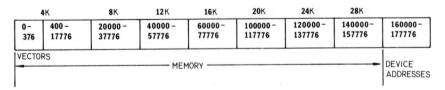

4K	8K	12K	16K	20K	24K	28K	
0– 376	400– 17776	20000– 37776	40000– 57776	60000– 77776	100000– 117776	120000– 137776	140000– 157776	160000– 177776

VECTORS
←——————————————————— MEMORY ———————————————————→ DEVICE ADDRESSES

Figure 3.2 Layout of addressing in the PDP-11.

from curious visitors. Many also have a display of the Address Register and Data Register so that you can see what the computer is doing (when running) or what the computer is going to do (when stopped). When the computer is halted, the switches for selecting the address (Load Add) and examining (Exam) and depositing (Deposit) numbers are activated. In older computers, it is necessary to use these switches to deposit primitive "bootstrap" programs in memory to get the computer to read in paper tapes or start up its disk or diskette drive monitor. In more recent PDP-11's, the switch CNTRL is held down and the Boot switch pressed to start up the floppy disk or hard disk drive monitor.

The front panel of the PDP-11 consists of the following elements:

Power keyswitch
Address display
Data display
Switch Register or keypad
Control switches
   Load Address
   Examine
   Continue
   Enable/Halt
   Start
   Deposit
Status lights
   Run
   Processor
   Bus
   Console
   User
   Virtual

The power keyswitch turns the power on to the computer and can be turned to a Panel Lock position, in which the console switches are disabled to prevent tampering.

The Address display allows display of an 18-bit address. Computers without Memory Management, having only a 28K maximum of memory do not use the upper 2 bits of the address register. Device addresses, however, automatically set these 2 bits, as the processor always recognizes device addresses as being between 124K and 128K. If, for example, the address of general register R0 is toggled into the Switch Register and loaded using the Load Address as 177700, the Address Register will automatically display 777700.

The Data Register displays the contents of memory locations or other bus addresses and can be read only when the processor is stopped. During computer operation, all data passing along the bus are displayed briefly, making interpretation of the display impossible. The Data Register displays the contents of R0 when the processor halts, and it can then be used to display the contents of any memory location or bus address.

The Switch Register allows manual communication between the user and the computer. The value 1 is represented by a switch in the up position, and 0 by a switch in the down position. It is used to toggle in data that are to be deposited in memory, to toggle in addresses that are to be examined, and to start up programs when the computer is halted. It also can be read by programs while they are running and used as a 16-bit sense register in this way.

Status Lights

The Status lights indicate what the computer is doing at present. The Run light shows that the processor clock is running and that the processor can be started by pressing the Start or Continue switch. The Run clock can be turned off by pressing Start while the Halt switch is down, but this is of little value except for maintenance work.

The Console light indicates that the processor is stopped, even though the processor clock may be running, meaning that the processor is in the Console mode. In this mode the Load Address, Examine, Continue, Start, and Deposit switches can be used. When the Console light is off, only the Halt switch can be used, to bring the processor to a stop and thus to the Console mode.

The remaining lamps—Bus, User, and Virtual—are of almost no use except for maintenance. The Bus lamp lights whenever the Unibus is being used, which is almost all the time. The User and Virtual lights indicate which mode the processor is operating in. Neither mode is within the scope of this book.

Console Switches

All control of the PDP-11 is accomplished through the console switches. They are used to start and stop programs and examine and alter memory contents.

Enable/Halt.   The Enable/Halt switch is not spring loaded and stays in either the up or the down position. In the up position the processor is enabled, and programs can run if the processor is started. In the down position the processor is halted and programs are prevented from running. This is the so-called Console mode, in which the remaining control switches are active. When a program is stopped by pressing the Halt switch, the contents of R0 are displayed in the Data Register lights.

Load Address. This switch transfers the contents of the Switch Register into the Address Register. The Switch Register is then shown displayed in the address lights. More importantly, this address can then be used to examine or change the contents of the location of memory of the bus having that address.

Examine. Depressing Examine causes the contents of the location whose address is in the Address Register to be displayed in the Data Register. The contents of any memory location or device address can be examined in this way. The Examine switch automatically increments the Address Register if depressed a second time, thus causing the display of (address reg + 2). Thus a sequence of memory locations can be examined by pressing only the Examine switch.

Deposit. The Deposit switch can be used to change the contents of the memory location currently in the Address Register. The contents of the Switch Register are transferred to the location whose address is in the Address Register. If the Deposit switch is raised a second time, the Address Register is incremented by 2, allowing data to be deposited in successive memory locations without resetting the Address Register. If, however, the location has been examined, it can be changed by changing the Switch Register to the new value and raising the Deposit switch. This sequence of operations does *not* increment the Address Register; it is incremented only if either the Examine or Deposit switch is operated twice *in succession*.

Start. This switch causes the computer to begin executing instructions at the address in the Address Register. For example, to start a program at address 1000, simply set the Switch Register to 1000, press Load Add, and then press Start. If the Start button is depressed while the Halt switch is still down, this will *turn off the Run clock* and the computer *cannot then be started*. The Run clock can be turned back on by depressing Continue while the Halt switch is down.

Continue. The Continue switch causes the processor to continue executing instructions at the address following the one in which it halted, whether the halt was caused by depressing the Halt switch or by a HALT instruction in the program. If the Halt switch is still down when the

Continue switch is pressed, only one instruction will be executed. This is useful in determining what is happening in a program that does not seem to work properly, but the display in the Data Register during this single-stepping procedure is not the same as during program execution and thus cannot be used as a debugging aid. The Address Register display is the same, however.

The LSI-11 Console

The LSI-11 console has a set of switches similar to those of the PDP-11, except that a keypad like that on a pocket calculator replaces the Switch Register. The switches are labeled:

```
DIS AD  EXAM   DEP          HLT/SS
LAD                         CONT
LSR                         BOOT
CLR     INIT   CNTRL   START
```

The EXAM, DEP, HLT/SS, CONT, and START switches have the same functions described earlier. The CLR switch is simply for correcting an error in a number entered using the keypad. The CNTRL switch is protection so that the bootstrap function can only be performed by pressing two keys. To bootstrap the system, hold down the CNTRL switch while pressing BOOT.

The remaining switches are used primarily for maintenance, and their functions can be found in the LSI-11 manual.

The RT-11 Monitor

The PDP-11 is provided with a program for reading programs and data from disks, diskettes and magnetic tape called RT-11, a "real time" monitor. This program maintains directories on the various disks and allows you to read in and start such programs with simple commands. RT-11 supervises all input and output and can thus be used by the programmer to simplify his or her programming.

The RT-11 system is designed so that the programmer can start it up with a few instructions in a bootstrap ROM or deposited in memory, which in turn reads in more instructions until the entire program is resident in the computer's memory. The single job RT-11 monitor has a 1½K resident portion and another section that swaps in and out with your data as needed. A more complex monitor, termed the Foreground/Background monitor is also available, but will not be described here.

Starting Up RT-11 in Hardware (ROM) Bootstrap Systems

To start up RT-11 from diskette, be sure that the floppy disk marked RT-11 System Disk is in the left-hand drive and that the system is Halted. Then hold down the CNTRL button on the console (*not* the CTRL key on the terminal) and press BOOT. The ROM program should read in the RT-11 monitor and type out

    RT-11SJ   Vxxxx

where Vxxxx is a version number. The dot printed on the next line is the RT-11 "prompt" character, indicating that it is ready to accept commands from the user.

Starting RT-11 from a System Without a ROM Bootstrap

In older PDP-11's without the ROM bootstrap, it is necessary to deposit the bootstrap instructions in memory before starting the monitor. For systems containing paper tape readers, the paper tape bootstrap is much shorter than the floppy disk bootstrap, it is easier to toggle in the paper tape bootstrap (Appendix I) and then read in the floppy disk bootstrap from paper tape.

In such systems, it is also easy to modify RT-11 so that the bootstrap remains permanently resident in memory just above RT-11, rather than allowing RT-11 to take over all memory in the computer. Then if a program crashes but does not wipe out all memory, you can usually re-boot by starting at the bootstrap address above RT-11.

In the 24K PDP-11 at Tufts University, we have reserved addresses above 134000 for the RT-11 bootstrapper, a program that moves the bootstrap to its starting address at 1000 and starts it. There is also room for the shorter debugger ODT-11 at 135000 and for the paper tape loader program at 137500 and the paper tape bootstrap at 137744. Directions for modifying RT-11 to reside below this address are given in Appendix II.

To start the RT-11 bootstrapper,

1. Depress Halt and raise it.
2. Set the Switch Register to 134000 (1   011   100   000   000   000).

3. Press Load Add.
4. Press Start.

The Bootstrapper should move the 31-instruction bootstrap to location 1000 and start it. RT-11 should then start up and print out its name, as described earlier.

Filenames in RT-11

All program and data manipulation in RT-11 is done through *filenames,* which refer to specific data on disks or diskettes. Here the term *disk* means a rigid rotating magnetic surface that has binary data recorded in concentric tracks. These tracks are further divided mechanically into *sectors.* The term *diskette* is used for any flexible rotating magnetic medium, often referred to as a *floppy* disk. Disks have higher information density and transfer rates. Diskettes have slower transfer rates and smaller capacities, but are cheaper and can be mailed easily in a stiffened envelope and filed in ordinary file cabinets.

The information on disks and diskettes is stored in the form of *files.* Each file may be a program, data, or simply a set of characters in some text, program, or manual. These types of files are usually distinguished by the sort of filename they are given. Although you can give a file any sort of name, RT-11 usage suggests that you might well follow the following conventions.

1. Each file can be named with a *filename* of 1–6 characters and an *extension* of 1–3 characters. In general, these names must be made up of letters and numbers, and the first character must be a letter.
2. The extension is separated from the filename by a dot (.).
3. Files of different type or content may have the same name but differ in extension. Thus we might name an assembly language file ADD10.MAC, meaning that it is to be assembled by MACRO, and name the actual executable file ADD10.SAV.
3. The most common extensions used in RT-11 are
    - .MAC   MACRO assembly language source file
    - .BAK   Backup copy of any ASCII source file
    - .OBJ   Object file produced by MACRO to be linked by the LINK program
    - .SAV   Saved copy of any memory region
              Usually a file that can be RUN

.SYS   System file or device handler
.LST   Listing file
.MAP   Linker map file ·
.BAD   File having bad blocks
.FOR   FORTRAN source file

Devices in RT-11

Each file is located on some *device*. The most common devices are disk, diskette and magnetic tape. These devices are divided further into the *system device* and other devices. The system devices is usually disk unit 0 or diskette unit 0. Each device has a 2-character name that *must* be followed by a colon in RT-11 parlance. The most common devices are

DK: Default storage device
DKn: Unit n of same type as device DK
DLn: RL01 disk
DMn: RK06, RK07 disk
DPn: RP02, RP03 disk
DSn: RJS03, RJS04 disk
DXn: RX01 diskette (single sided)
DYn: RX02 diskette (double sided)
RF: RF11 fixed disk
RKn: RK05 disk
SY: The system device, regardless of type

Other devices that might be used in RT-11 include

LP: Line printer
PC: High-speed paper tape reader-punch
TT: Console terminal

If you give a device name in an RT-11 command *without* the colon, RT-11 will look for or create a *file* having that name.

The Directory

To find out what files are stored in your disk or diskette, you can give the DIR (DIRrectory) command, and RT-11 will print out a list of all the files on that device. The command DIR by itself prints out all the files

on the default device. After RT-11 is first started, the default device is
SY:, the system disk(ette), but this can be changed with the ASSign
command.

You can also get the directory of any device with the command
DIR dev:, where dev: is the name of any of the preceding devices that
have directories associated with them. Thus you can get the directory of
DX1: by typing

   DIR   DX1:

You can restrict the command further, by specifying the filename to
look for,

   DIR   DX1:FFT

or by using the asterisk wild-card name discussed later.

Naming Specific RT-11 Files

When you want to access a file on a disk or diskette, it is usually to
perform one of the following operations:

1. Run a program.
2. Edit a program text file.
3. Assemble a program.
4. Delete a file.
5. Copy or list a file.

The operations require that you refer to the file by device and name. You
can refer to any file on the system device, which is that containing the
monitor and device handlers, by calling that file SY:filename. You can
refer to any file on another device by prefixing that device name to the
filename:

   RUN   DX1:FFT

RT-11 makes it even easier to keep track of your files in a two-disk
or two-diskette system by allowing you to run any file on the system
device with the command R. Programs on other devices must be started

with the RUN command, and must include the device name:

    R TECO

but

    RUN DY1:PLOT

In addition, you can specify the *default* device for your commands with the ASSIGN command, which allows you to assign any device as the default one to be used when none is specified. For example, you might always want files on DX1: to be referred to in RUN commands. This is accomplished by the command

    ASSIGN DK: DX1:

Then the default device DK: is assumed to be DX1:, and the command

    RUN PEAKS

is then equivalent to

    RUN DX1:PEAKS

RT-11 System Files

The System device in an RT-11 system is the one containing the monitor, bootstrap and a number of utility programs that you will need for program development. Some of the files you should have on your system disk or diskette include

DxMNSJ.SYS	The single job monitor
Dx.SYS	The diskette device handler
PC.SYS PR.SYS	The paper tape device handlers
TT.SYS	The terminal device handler
MACRO.SAV	The assembler
TECO. SAV	The editor
LINK.SAV	the Linker

PATCH.SAV	A file patching program
PIP.SAV	A file handler program called by the monitor
LIBR.SAV	A librarian program
DUP.SAV	A file duplicator utility
DIR.SAV	The directory handler program
CREF.SAV	A cross-reference program called by MACRO
FPMP11.OBJ	The floating point routines
ODT.OBJ	The expanded debugger
SYSMAC.SML	The system Macro library

RT-11 Commands

The commands that we will use most in RT-11 are

ASSIGN dev: dev:	Assigns a logical name to a physical device
COPY file file	Copies a file into another
DELETE file	Delete a specified file
DIR device	List out the directory for that device
EDIT file	Edit a file.
EXE file	Assemble, Link and Run a file
	If extension is .FOR, compile, link, and run
GET file	Loads a memory image (SAV) file into memory
LINK file	Converts relocatable modules into an executable program
MACRO file	Assemble a file using MACRO
PRINT file	Lists a file on the line printer
RENAME file	Renames a file
R file	Runs a file on the system device
RUN file	Runs a file on any device
TYPE file	Lists out a file on the terminal

The Asterisk "Wild-Card" Character

You can also refer to a group of files having the same root name or the same extension using the asterisk to replace either the name or the extension. In a directory consisting of

ADD10.MAC
ADD10.BAK
UNPK.BAK
UNPK.OBJ

    ADD10.OBJ
    UNPK.MAC

the command

    DEL *.BAK

would delete ADD10.BAK and UNPK.BAK, and the command

    REN ADD10.* ADDER.*

would rename

        ADD10.MAC to   ADDER.MAC
        ADD10.BAK to   ADDER.BAK
    and ADD10.OBJ to   ADDER.OBJ

Default Extensions

Many programs assume that any file you refer to will have a given *default*
extension. You can always override this assumption by specifically typing
in the extension as part of the command, but if you type no extension,
the program will look for a specific extension. The command default
extensions are

    .MAC   File to be assembled by MACRO
    .OBJ   Files to be linked by LINK, produced by MACRO
    .SAV   Files to be RUN, produced by LINK

RT-11 Command Entry

All RT-11 commands are one line in length. They usually consist of a
command followed by one or more filenames. Commands are not executed until you type a Return. Until the time you type the Return, you
can edit the command line with

    DEL or Rubout   Deletes one character
                     each time it is struck.
    CTRL/U   Deletes the entire line.

Thus you can correct even complicated commands lines fairly easily.

CHAPTER FOUR

INTRODUCTION TO THE PDP-11 INSTRUCTION SET

All computers are built around a particular architectural concept that determines their efficiency and suitability for certain tasks, as well as the actual manner in which particular elementary operations are performed. These architectural concepts are usually summarized in a brief description of the computer hardware and of its instruction set. The instruction sets for various computers may vary widely in method, but they all allow operations such as addition, subtraction, multiplication, division, moving of data, and testing for various logical conditions such as greater than, equal to, or less than 0, and addition overflow. The PDP-11 instruction set performs virtually all of its operations through a set of eight general registers, usually numbered R0 through R7. Although they all can be used to perform the same sort of operations, two of them have special functions: R6 is known as the Stack Pointer (SP), and R7 as the Program Counter (PC). These registers are used to contain the data, the address of the data, the address of the address of the data, or the offset, which is added to another number to produce the end address of the data. These different modes of accessing the actual 16-bit data words are known as *addressing modes*. Most minicomputers have at least three addressing modes, and some have more. Since every computer instruction must set aside a particular number of bits for the various addressing modes and more bits for the instruction itself, there is usually a trade-off between the number of different instructions and the number of addressing modes.

The PDP-11 is one of the computers with a large number of addressing modes and a small number of instructions. It has only about 54 instructions in its repertoire, but it has eight addressing modes through which data can be accessed by many of these instructions. Whether this is a more or less powerful approach than that adopted by manufacturers offering fewer addressing capabilities and more instructions is a moot point. The total effective power of the PDP-11 is more or less the same as in computers of other types.

The eight general registers of the PDP-11 are central processor high-speed registers which can be accessed rapidly and modified within instructions. Although they are not memory locations, they do have bus device addresses and can be accessed as such from the computer's Switch Register.

Register	32K Add	128K Add
R0	177700	777700
R1	177701	777701
R2	177702	777702
R3	177703	777703
R4	177704	777704
R5	177705	777705
(SP) R6	177706	777706
(PC) R7	177707	777707

The first column above shows the register name, the second its bus address in machines having only 28K of memory maximum, and the third the bus address in machines having 124K maximum with Memory Management. Note that the register addresses are only one address apart. If you examine them from the Switch Register, the Examine key will only advance the Address Register by 1. The registers cannot be addressed by these addresses within a program, however.

The Processor Status Register

One other extremely important register in the PDP-11 is the Processor Status (PS) Register. This word contains the result of the previous operation in 4 bits called the *condition codes*. It also contains information regarding the processor interrupt priority and 4 bits used by the Memory Management hardware. They are assigned as shown in Figure 4.1. The

15	14	13	12	11	10	9	8	7	6	5	4	3	2	1	0
CURRENT MODE		PREVIOUS MODE		←———— UNUSED ————→				←———— PRIORITY ——→			T	N	Z	V	C

Figure 4.1 Processor Status word.

most important part of the Status Register is the condition codes in bits 0–3.

Bit 0   C-bit   Set to 1 if the last operation resulted in a carry-out of bit 15. This could occur if two numbers were added whose sum could not be represented in 16 bits. Detection of carry-out is important in maintaining control of the size of numbers and in writing double-precision code. For example,

$$\begin{array}{r} 123451 \\ + 103022 \\ \hline 026473 \end{array}$$

carry = 1

sets C-bit to 1. If no carry, C-bit is always 0.

Bit 1   V-bit   Set if arithmetic overflow occurs during the preceding instruction. Overflow is defined as addition of two operands of the same sign that produces a result of the opposite sign. For example,

$$\begin{array}{r} 12345 \text{ positive} \\ + 76543 \text{ positive} \\ \hline 111110 \text{ negative number results } (-66670) \end{array}$$

Bit 2   Z-bit   Set to 1 if the result of the previous operation was 0. Set to 0 otherwise. Note that there is a possibility of confusion here: Finding a *zero* result sets the Z-bit to a 1.

Bit 3   N-bit   Set if the result of the previous operation was negative. Remember that a negative number always has bit 15 set. Thus bit 15 of the previous result is copied into the N-bit.

Bit 4   T-bit   This bit is used principally in debugging hardware and software. It causes an automatic trap instruction to be executed at the end of the current instruction. This causes a jump to a location whose address is set in location 14 and a new PS word to be loaded from location 16.

Bits 5–7   Priority   These bits determine the priority of the processor with regard to interrupts. The processor is initially given a zero priority, which means that any device can interrupt it. This can be changed by the program by loading the PS or by interrupt and trap instructions. Most of the lower-priced PDP-11's have implemented only five of the eight possible priority levels. Levels 7, 6, 5, and 4 are set by setting bits 5–7 of the PS. Levels 3, 2, 1, and 0 are considered equivalent in these machines. When a device requests an interrupt, it is granted only if the processor is running at a priority *lower* than the priority of request.

Bits 12–15   These bits are used for the Memory Management option, which is discussed in detail in Chapter 6 of the PDP-11 *Processor Handbook*. Its use is beyond the scope of this text.

Single-Operand Instructions

The PDP-11 instruction set consists of single-operand instructions, double-operand instructions, and instructions that have no operand at all. An operand is simply the end memory location or register that is referenced by the instruction. The single-operand instructions reserve the lower 6 bits, so that bits 0–2 represent the register number referenced in the instruction and bits 3–5 the addressing mode, each of which can take on values from 0 to 7. The bit assignments of this format are illustrated in Figure 4.2.

15	14	13	12	11	10	9	8	7	6	5	4	3	2	1	0
BYTE	←————————— OPERATION CODE ——————————→									DESTINATION ADDRESSING MODE			DESTINATION REGISTER		

Figure 4.2   Single-operand instruction bit assignments.

The basic single-operand instructions are as follows:

CLR	b050ar	CLRB	Clear the operand—set it to 0.
COM	b051ar	COMB	Complement the operand—take the one's complement.
INC	b052ar	INCB	Increment the operand—add 1 to it.
DEC	b053ar	DECB	Decrement the operand—subtract 1 from it.
NEG	b054ar	NEGB	Negate the operand—take the two's complement.
TST	b057ar	TSTB	Test the status of the operand, and set condition codes.

This table shows CLR, COM, INC, DEC, NEG, and TST as instructions that operate on full 16-bit words. For these instructions bit 15, represented by a *b* in the octal code, is a 0. If this bit is set in the instructions, they become *byte instructions*, operating on 8-bit bytes and represented by the second list of names: CLRB, COMB, INCB, DECB, NEGB, and TSTB. The function of each of these is described succinctly alongside. The exact use and function of the TST and TSTB instructions will be described later.

Double-Operand Instructions

There is also a set of very powerful two-operand instructions for the PDP-11. These reserve the lower 12 bits for the addressing modes and register assignments of both a source and a destination operand. Bits 9–11 and 6–8 specify the addressing mode and register used for the source operand, and bits 3–5 and 0–2 the addressing mode of the destination operand. In all cases these instructions leave the source unchanged and change the destination operand. The bit assignments for the dual-operand instructions are shown in Figure 4.3.

The four general dual-operand instructions are as follows:

MOV	b1arar	MOVB	Move source operand to destination.
CMP	b2arar	CMPB	Compare source to destination and set condition codes.
ADD	06arar		Add source to destination.
SUB	16arar		Subtract source from destination.

Note that there are byte versions of both the MOV and the CMP instructions but that ADD and SUB operate on words only. It should also be

	15	14	13	12	11	10	9	8	7	6	5	4	3	2	1	0
BYTE	OPERATION CODE				SOURCE ADDRESSING MODE			SOURCE REGISTER			DESTINATION ADDRESSING MODE			DESTINATION REGISTER		

Figure 4.3   Double-operand instruction bit assignments.

carefully noted that the SUBtract instruction subtracts the source from the destination and the CMP instruction subtracts the destination from the source. This is easily rationalized in reading the instruction out loud:

SUB A, B

means subtract A *from* B, whereas

CMP A, B

means compare A *with* B. If after we compare A with B we ask the question, "Was A greater?" we know that, to answer this question correctly, we must subtract B from A to arrive at a result that is greater than 0. In other words, if A is greater than B, then (A − B) will be greater than 0.

The MOVB instruction may move data from memory locations or to or from registers. If the destination of a MOVB instruction is a register, the sign of the byte (bit 7) is copied or extended into bits 8–15. This is unique to the MOVB instruction and occurs only when the *destination* is a register.

A Simple Program

We are now in the position of being able to write our first simple program. For the present we will use only the register addressing mode, which is mode 0, symbolized by having all three addressing mode bits equal to 0. We will use the following steps for writing any program in any language.

1. Write a concise English statement of the purpose of the program.
2. Sketch a flowchart of the problem.
3. Code the program from the flowchart, commenting it thoroughly.
4. Test and debug the program.

It should be emphasized that step 4 is really the most important. No program can be said to be "working" until it has had thorough testing, using a suitable hand-calculated case. Debugging complex programs may actually take longer than steps 1–3 combined, and in planning programming projects large amounts of time should be allocated to this stage.

Clear English Statement. We are going to write a program to add together three numbers that are stored in registers R1–R3. The result will be held in R0 and displayed when the program halts.

Flowchart of Program. The flowchart shown in Figure 4.4 is, of course, extremely simpleminded, as there are no decision points or other complex logical steps.

Code for Program. This program can be coded as

```
MOV R1, R0    ;PUT FIRST NUMBER IN R0
ADD R2, R0    ;ADD 2ND TO IT
ADD R3, R0    ;ADD 3RD TO THAT
HALT          ;AND STOP
```

This coding shows some of the elementary rules for syntax of the PDP-11 assembly language. Each instruction appears on a separate line, and dual-operand instructions have the two operands separated by commas. Most importantly, a semicolon precedes the comments. Comments are the most important part of an assembly language program. Because so much can happen in an instruction and because the logic can become rather abstruse, the comments tell the reader of the program exactly what the programmer had in mind without the former having to read the logic of the code and thoroughly understand it.

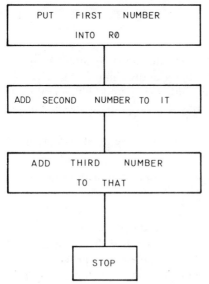

Figure 4.4 Simple flowchart to add three numbers together.

Comments.   The necessity for comments cannot be overemphasized. They are printed out on assembly language listings by the assembly program and are useful in program debugging. They are an absolute necessity in programs that are to be used over any period of time, either by the same author or by a succession of authors, so that the actual workings of the program can be deciphered at any time. The usual counterarguments to including comments in programs run something like these:

1. They are too hard to type.
2. This program is only for my own use.
3. I understand how it works.
4. I'll put in the comments later when the program is working.

Experience by the set of all programmers over all time has shown the fallaciousness of all such arguments. It is never as hard to type comments as it is to figure out what is wrong with a program without any comments. Moreover, a program is never for only one person's use unless that person is stranded on a desert island with his computer system. The comments aid the author in later uses or modifications of his program and help him to transmit his work to others for their uses and modifications. Finally, a program is never "working." There are always changes that would be desirable, and the upshot is that the comments are never added if they are not put in at the beginning.

   Despite all these persuasive arguments, it has been found that beginning programmers refuse to comment their programs out of sheer laziness, and this leads to the requirement in computer courses that programs include comments if credit is to be received.

Analysis of the Assembled Program

The code shown on page 38 is in the assembly language of the PDP-11. It is a series of symbolic representations of binary machine instructions such as 010000 for MOV and 160000 for SUB. Since these instructions are symbolic of the actual operations, they are referred to as *mnemonic codes* or just *mnemonics*. These instruction codes are combined with the bits showing the correct register and addressing modes, and the resulting binary number is said to be assembled.

   In addition, an assembler allows you to define other symbols representing addresses or constants, which are easier to use and remember than the actual octal numbers. These defined symbols are stored in a *symbol table* along with the numeric value associated with each symbol. When the symbol is referred to in the program, the assembler looks it up in the symbol table just as it looks up the predefined symbols of the computer's instruction set.

The assembled code is then stored in a file and printed out on a program listing along with the mnemonic instructions to aid you in testing the program.

Example of How the MACRO Assembler Works

Let us consider the simple instruction

ADD R3, R0

Assembly of this instruction is accomplished as follows:

06xxxx	MACRO looks up the code for the instruction ADD and finds 060000.
060xxx	It determines the source addressing mode as register mode and sets bits 9–11 to 000.
0603xx	It finds that the source register is R3 and sets bits 6–8 to 011.
06030x	It determines the destination addressing mode as register mode and sets bits 3–5 to 000.
060300	It determines that the destination register is R0 and sets bits 0–2 to 000.

The assembled instruction is thus 060300 in *octal* or 0 110 000 011 000 000 in binary. You can assemble such an instruction by hand using the PDP-11 Pocket Programming Card or the tables in Appendix I.

If the program is loaded starting at address 0, the assembled code will appear in locations 0–6 as shown below:

```
0        010100   MOV  R1,  R0      ;PUT FIRST NUMBER IN R0
2        060200   ADD  R2,  R0      ;ADD 2ND NUMBER TO THAT
4        060300   ADD  R3,  R0      ;ADD 3RD NUMBER TO THAT
6        000000   HALT              ;AND STOP
```

This code can be toggled into the computer from the Switch Register, the numbers placed in R1–R3 (addresses 777701–777703), and the program run. When it halts, the sum will be displayed in the Data Register, which always shows the contents of R0 when the computer is not running.

Exercises

The program can also be modified so that we can observe each step
as it occurs. We could step the program an address at a time with the
Halt switch down, resulting in a successive display of addresses 0, 2, 4,
and 6 in the Address Register. However, in this single step mode, the
Data Register does not display useful information. If we actually wish to
watch addition occur in R0, we must toggle the program in with stops in
between and press Continue each time to watch the next step occur.

Such a program would look like this:

```
0     010100    MOV R1, R0
2     000000    HALT
4     060200    ADD R2, R0
6     000000    HALT
10    060300    ADD R3, R0
12    000000    HALT
```

Exercises

4.1   If R0 contains 73426 and R1 contains 31203, what will the condition
codes be after the instruction ADD R1, R0 is executed?

4.2   If R4 contains a 0 and the instructions
DEC   R4
INC   R4
are executed, what will the PS and R4 contain?

4.3   If R2 contains 000005 and the instructions below are performed,
what will R2 contain? Careful!
NEG   R2
MOVB   R2, R3
NEG   R2
ADD   R3, R2

4.4   Which condition code bits could be set by the following instruction
CMP   R5, R5

4.5   If R0 contains 176435, what will it contain after the following in-
structions?
COM   R0
DEC   R0

4.6   If R5 contains 176401, what will the condition codes be after the
      following?
      DECB   R5

4.7   What will R5 contain after the following?
      R5/003702
      NEGB   R5

CHAPTER FIVE

THE PDP-11
ADDRESSING MODES

Most of the power of the PDP-11 instruction set lies in its *addressing modes*. An addressing mode is the manner in which the computer accesses data. The eight PDP-11 addressing modes are as follows:

Number	Symbol	Name	Description
0	Rn	Register mode	The data are in the register.
1	(Rn)	Indirect	The address of the data is in the register.
2	(Rn)+	Indirect autoincrement	The address of the data is in the register and is incremented after use.
3	@(Rn)+	Double indirect autoincrement	The address of the address of the data is in the register. Register is incremented after use.
4	−(Rn)	Indirect predecrement	The address in the register is decremented and then used as the address of the data.
5	@−(Rn)	Double indirect predecrement	The number in the register is decremented and then used as the address of the address of the data.

| 6 | X(Rn) | Index mode | The value of X stored in the location following the instruction is added to the register to produce the address of the data. |
| 7 | @X(Rn) | Indexed indirect | The value of X in the location after the instruction is added to the register to produce the address of the address of the data. |

So many addressing modes may seem rather difficult to assimilate at first, but actually they are quite easy to remember once you become familiar with their uses. The most commonly used ones are Rn, (Rn)+, and (Rn), in that order. The others are used much less, except for some special types of addressing using the PC as a general register, which are discussed under PC addressing modes.

Register Mode Rn

In this mode, the register contains the number that is being operated on. Since "the number" may be the contents of either a source or a destination location, the term *operand* is often used to describe the location where the operation takes place. Thus in register mode the register itself contains the operand. Let us suppose that R0 contained 617 and R1 contained 474 prior to a MOV instruction.

		R0	R1
	Before	617	474
MOV R1, R0			
	After	474	474

This instruction moves the number in R1 to R0. Thus the contents of R1 are copied into R0. The source operand (R1) is unchanged, and the destination operand is changed.

Register Indirect Mode

In this mode, the register contains the *address* of the data. If R3 contains 500 and R5 700 and location 500 contains 16703 and 700 contains 12,

		R3	R5	500	700
	Before	500	700	16703	12
MOV (R3),(R5)					
	After	500	700	16703	16703

This instruction says to move the contents of *address* 500 into *address* 700 or, in other words, to move 1703 into address 700. After this instruction, the registers are unchanged and the source location (500) is unchanged, but the destination address 700 *is* changed.

It is also possible to mix addressing modes promiscuously. For example, using the same initial conditions,

		R3	R5	500	700
	Before	500	700	16703	12
MOV (R3),R5					
	After	500	16703	16703	12

This instruction says to move the contents of the address pointed to by R3 into register R5 or to put 16703 into R5. As before, only the destination operand is changed.

Indirect Autoincrement Mode

This mode performs exactly the same indirect addressing as does the indirect mode, but with the added feature that the register is *incremented* after the operation takes place. The increment is by 2 if a word is addressed and by 1 if a byte is addressed. This addressing mode is used primarily in going through lists or tables of numbers and doing the same thing to all of them.

Let us consider the conditions used earlier and look at the difference in the results:

		R3	R5	500	700
	Before	500	700	16703	12
MOV (R3)+,R5					
	After	502	16703	16703	12

As before, the contents of address 500 are moved into R5. Here, however, R3 is incremented by 2 after being used as a pointer, so that it now points to the next sequential word address.

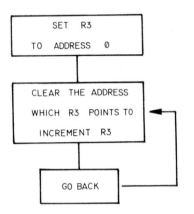

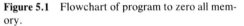

Figure 5.1   Flowchart of program to zero all memory.

We could very easily write a program that zeroes all memory locations in just one or two instructions using this addressing mode. Of course if this program were uncontrolled it could "eat itself," and wipe out the monitor and even the program itself, but since the HLT instruction has the code 000000, the program will eventually stop. Such a program would have a flowchart like that shown in Figure 5.1

This program would be written as follows:

```
005003   CLR R3          ;SET RO TO FIRST ADDRESS TO BE CLEARED
005023   CLR (R3)+       ;CLEAR ADDRESS R3 POINTS TO AND ADD 2 TO R3
000776   BR .-2          ;GO BACK TO PREVIOUS INSTRUCTION
```

This program will work anywhere in memory and will halt when R3 reaches the address of the second instruction, so that addresses from 0 to the position of the program are zeroed.

Suppose that the program is placed at addresses 2050, 2052, and 2054. It will start, and in the first pass clear location 0, in the second, location 2, and so forth. This will continue until R3 contains 2052. Then the CLR (R3)+ instruction will zero address 2052 and increment R3 to 2054. When the BR instruction jumps back to 2052, the instruction will not be 5023 but 0, which means "halt." The program will stop.

Double Indirect Autoincrement Mode @(R3)+

In this mode, the register contains the address of the address of the operand; after accessing the operand, the register is incremented by 2 for word addressing or by 1 for byte addressing. At first, double indirect addressing sounds quite obscure, and indeed it is seldom used, but it is of substantial use when tables of addresses are being used. Then a register can point to an entry in the table, and the contents of the final destination can be referred to through this double indirect scheme. For example, suppose R2 contained 760 and R3 contained 4274. Then we have

	R2	R3	760	4274	2356	1166
	760	4274	2356	1166	13	17600
MOV R2, @(R3)+						
	760	4276	2356	1166	13	760

Note that, as usual, only the destination operands are changed. Originally, R3 pointed to 4274 and 4274 pointed to 1166. After the instruction is executed, the contents of R2, 760, are put into address 1166, and the contents of R3 incremented by 2 from 4274 to 4276.

Predecrement Indirect Mode

This mode decrements the register by 2 for word addressing or by 1 for byte addressing and uses the decremented value as an address pointer to the data. The autoincrement and predecrement modes form a symmetrical pair for moving up and down a list with the least attendant complexity. For example, we could add another element to a list by giving the instruction

    MOV R3, (R0)+   ;ADD R3 TO A LIST

and could later access this same value and restore it to R3 by

    MOV -(R0), R3   ;DECREMENT R0 AND PUT RESULT IN R3

which removes the last value from the list. We could, of course, accomplish this same pair of operations by reversing the addressing modes, if we assume that our data list starts at a high-numbered address and moves downward:

    MOV R3, -(R0)   ;ADD CONTENTS OF R3 TO LIST
    MOV (R0)+,R3   ;REMOVE LAST ELEMENT FROM LIST & PUT IN R3

Note particularly how the assembly language symbolism reminds us of the predecrement and post-increment convention:

    -(Rn)   ;PREDECREMENT
    (Rn)+   ;POST-INCREMENT

If these symbols are given in the wrong order, the assembler will flag
them as errors.

Predecrement Double Indirect Mode @ − (Rn)

In this mode, the register is decremented by 1 or 2 before being used as
the address of the address of the data. For example,

	R4	R5	15716	12606
	2132	15720	12606	41623
MOV R4,@ − (R5)				
	2132	15716	12606	2132

This instruction decrements R5 by 2 so that it points to 15716. The address
12606 is then pointed to by 15716, and the contents of R4, 2132, is placed
in address 12606.

Index Mode X(Rn)

In the indexed mode, the address of the operand is calculated by adding
a value in a register to a number in the word following the instruction.
Since this is a pointer address, which is calculated, we could regard this
as an indirect mode. For example, if we write

    MOV 12(R2), R0

the constant 12 in the next location will be added with the current contents
of R2 to form the address of the number to move. If R2 contained 7462,
then address 7474 would be referenced

	R0	R2	7474
	16	7462	472
MOV 12(R2),R0			
	472	7462	472

This instruction is assembled as the two words

    016200   MOV 12(R0), R2   ;MOVE INSTRUCTION
       12   ;INDEX CONSTANT

and similarly, the instruction MOV 1532(R1), −5(R2) becomes

```
016162   MOV 1532(R1), −5(R2)
1532                         ;FIRST INDEX CONSTANT
777773                       ;2ND INDEX CONSTANT

xxxx     ;Next instruction to be executed
```

The indexed mode is used principally for accessing data in the middle of lists. For example, when we write a program, we might know where it begins but not which element we need. The element in question can be determined by a calculation within the program that puts its offset from the start of a table into the register to be indexed.

Similarly, the program might know which element it wants, but not where the table begins. Then the constant offset is placed as the index and the table start placed in the register during program execution

MOV 4(R5), R1   ;GET 3RD WORD IN LIST

puts the third word of a table starting at R5 into R1.

Indexed Indirect Mode @X(Rn)

This mode is really a double indirect mode. The constant in the address following the instruction is added to the contents of the register to form the address of the address of the operand. This mode is seldom used except with the PC addressing modes described later in this chapter.

Examples of Addressing Modes

In each of the following examples, these initial conditions are assumed. Changed values are marked with an asterisk.

R1	R2	1000	1002	1004	4000	6000	7000
1002	0	4000	6000	7000	16264	3	177776

	R1	R2
MOV R1,R2	1002	1002*
MOV (R1),R2	1002	6000*
MOV (R1)+,R2	1004*	6000*
MOV @(R1)+,R2	1004*	3*
MOV −(R1),R2	1000*	4000*
MOV @−(R1),R2	1000*	16264
MOV 2(R1),R2	1002	7000*
MOV @2(R1),R2	1002	177776*

Exercises

5.1   Fill in the table below for the final values of R3 and R4 given the
      following initial conditions:

           R3(4562)
           R4(17630)
      4560/1342   1342/1515
      4562/2616   2616/000013
      4564/17760   17760/3626
      17626/13042   13042/5662
      17630/1402   1402/1544
      17632/7636   7636/6064

MOV R3, R4	R3( ), R4( )
MOV (R3), R4	R3( ), R4( )
MOV (R3)+, R4	R3( ), R4( )
MOV @(R3)+, R4	R3( ), R4( )
MOV −(R3), R4	R3( ), R4( )
MOV @−(R3), R4	R3( ), R4( )
MOV −2(R3), R4	R3( ), R4( )
MOV @−2(R3), R4	R3( ), R4( )

5.2   What would be the octal code for the following instruction?

                      MOV 2(R3), @−2(R4)

      Given the same initial conditions as in Exercise 5.1, what would be
      the result?

5.3   Given the same initial conditions as in Exercise 5.1, predict the
      result of each of the following:

      (a)   ADD (R3), @−(R4)
      (b)   MOV R4, (R3)
      (c)   MOV R4, (R3)+
      (d)   DEC −(R4)
      (e)   CLR (R4)+
      (f)   CLR R3
      (g)   SUB (R4), (R4)+

The Program Counter Addressing Modes

Of the eight general registers, two have special functions that prevent
them from being used for random data manipulation. However, these two
registers do perform powerful functions related to the central processor,

and the ability to address them through any of the eight addressing modes leads to extremely useful instructions.

Register 6 (R6) is known as the Stack Pointer, or SP. It contains the address of a temporary list of constants and variables that may be used either by the programmer or by the hardware of the central processor. It will be discussed in detail in the sections dealing with subroutines, interrupts, and trap instructions.

Register 7 (R7) is most commonly known as the Program Counter, or PC. It acts as a traffic cop, always pointing to the address of the *next* instruction to be executed. If the computer is started from a Halt, as when the Absolute Loader is run, the Load Add key loads the Switch Register into the PC, and Start begins execution of instructions at that address. When a program is running, the next instruction to be executed is usually at the address of the current instruction plus 2. This changes only when a branch or jump instruction is used to transfer control to some part of the program other than that which immediately follows. In this case the PC contains the address of the instruction to be executed next, regardless of where it lies in memory.

Since we can address the PC as one of the general registers, we can actually increase the addressing power of the PDP-11 in four cases. The other four modes have little value or cause program crash conditions if they are used.

PC Mode 2: Immediate Mode #n

One of the most useful addressing modes is one that will put a constant into a register or memory location. This is an important part of program initialization, as it always is necessary to set counters, pointers, constants, and multipliers before going into the program proper.

The immediate mode is symbolized by the number sign (#), and to move the number 2600 into register 1, we would write

MOV #2600, R1

Now the question arises, Where does the value 2600 go in the instruction? It clearly does not go within the single instruction word because the source bits would only allow a 6-bit integer and it takes 11 bits to represent 2600_8. The next most logical place to put the value would be in the address following the instruction. We would then want to address it so that the number would be accessed as a constant within the instruction and not as an instruction per se. At the time we are executing the instruction MOV, the PC is pointing to the address following the instruction. We

want to use the PC to get the value in that address and then increment
the PC by 2 again so that the next instruction will come from the MOV
address plus 4. This can be done with the register indirect autoincrement
mode, or mode 2. Thus

    MOV #2600, R1   becomes MOV(PC) +, R1   or   012701
                                 2600   2600

Assembler programs that translate assembly language code into binary
automatically perform this conversion when they detect the number symbol (#), forming a two-word instruction for the immediate mode.

PC Mode 3: Absolute Mode @#n

The PC mode 3 is just one indirect level deeper than PC mode 2. The
absolute address to be accessed is stored in the location following the
instruction. Thus

    MOV R1, @#34   becomes MOV R1, @(PC) +   or   010137
                                       34   34

This addressing mode is used mostly when a specific location rather than
a relative one is to be loaded with value. Such locations are usually those
below location 400, which are used as interrupt and trap vector locations.
It is, of course, perfectly possible to mix these addressing modes and
arrive at an instruction like

    MOV #340, @#36   which is translated as
                          MOV (PC) +, @(PC) +   or   012737
                               340   340
                               36   36

Note that in this case the PC is advanced twice. When the instruction is
first started, the PC is pointing to the 340 in the second location; after the
(PC) + source instruction mode is executed, the PC points to the third
location, where the 36 is stored. The @(PC) + addressing mode of the
destination section accesses the 36 and again increments the PC, leaving
it pointing to the third location following the initial instruction code.

PC Mode 6: Relative Mode A

PC modes 2 and 3 actually allow most of the useful addressing capabilities
that we might like to have in a computer. However, they have the disadvantage that every direct reference to a memory location is a reference

to a specific absolute address. It would be nice to be able to write programs that could easily be relocated, and this could be done if we could address a memory location in a relative manner. Relative addressing means *relative to the PC*, and this mode is accomplished by using the index mode of the PDP-11. Recall that the index mode adds the value of an index constant to the contents of a register to determine the effective address. If the index constant is added to the PC, we have generated a PC-relative addressing mode such as

500 MOV 74(PC), R1	which is assembled as	500 016701
.		502 74
.		. .
.		. .
.		. .
600 2602		600 2602

When this instruction is executed, the PC will point to 504 and the value 74 found there will be added to 504 to produce the address 600. The contents of address 600 are moved to register R1. The index mode automatically adds 2 to the PC, so that the next instruction comes from address 504.

We could have accomplished the same thing using the instruction

500 MOV @#600, R1	which becomes	500 013701
		600

but this instruction would always address location 600, whereas the relative mode will address the location which is 74 locations away from the PC, no matter where the program is loaded.

Use of Labels

The assembler program will assume that it is to use the relative addressing mode (PC mode 6) if we write

MOV 600, R1	;MOVE CONTENTS OF ADDRESS 600 INTO
	;R1
or	
MOV 74(PC), R1	;MOVE CONTENTS OF LOCATION 74
	;WORDS DISTANT INTO R1

However, this sort of coding leaves much of the work for the programmer. He or she either must know the absolute address of the location to be addressed or must know how far it is from the current location. Since PDP-11 instructions can, as we have seen, take one, two, or three locations, it is difficult to know just how far a particular constant will be at assembly time. Furthermore, every time we make a program change by inserting or deleting an instruction, we must change these absolute addresses or offsets. Clearly this is much too complicated.

The PDP-11 assemblers and those of most other computers provide the ability to refer to any memory location by name instead of by actual address. These names, or *labels*, are any designations that the programmer wishes to use, which in some convenient way remind the programmer and the program's readers exactly what a particular location is used for.

When an address is to be labeled, the name of the label is written first on the line, followed by a colon and then by the contents of the address:

```
LABEL:  7
TEMP:   0
A64:     -437
START:  MOV #1, I
```

Each of the preceding examples shows an address that is labeled, followed by its contents. The assembler will scan the text for such labels and then create a *symbol table* of them and their absolute addresses. The assembler will then begin assembling the code, substituting the absolute address for each label it encounters:

```
MOV LABEL, R1     would become    MOV X(PC), R1
                                  X    ;distance between
                                       ;LABEL and PC
```

Similarly we could write

```
              CLR TEMP
              ADD LABEL, A64
              MOV TEMP, @POINT
POINT:  1234
TEMP:   0
A64:    4062
LABEL:  12706
```

where in each case the label refers to an address that is assembled as relative mode addressing. Furthermore, the label can be referred to in any place where an address might be used: in relative mode, in relative indirect mode, in a table of addresses, or in any calculation involving an address.

Remember that a *label* is a symbol for a number. It provides a mnemonic way of representing a constant that might otherwise be hard to remember. Therefore, in addition to representing addresses, you can define a label for any constant. The MACRO assembler allows you to define any constant symbolically with an equals sign:

```
CR = 15            ;CARRIAGE RETURN
START = 1000
BUFSIZ = 256.
```

and then use these labels as constants within the program:

```
. = START             ;STARTING ADDRESS
MOV #CR, R0           ;PUT CONSTANT FOR CARRIAGE RETURN
                      ;IN R0
ADD #BUFSIZ, R1       ;ADD BUFFER SIZE TO R1
```

In summary, you use the *equals* sign to define a constant of any size or meaning, and you use the colon to define a label having the value of the current address during assembly.

A label must obey the following rules:

1. It must contain 1–6 characters.
2. The first character must be alphabetic.
3. It may not contain the special characters : @ , + − ( ).
4. It is defined using a colon (:) or an equals sign.
5. It is referred to by name without a colon wherever reference to the address in question is made.

PC Mode 7: Relative Indirect Mode @A

If we wanted to obtain the contents of an address pointed to by another address, where *memory* rather than a register is used as a pointer, we could write

```
500   MOV @600, R1   ;GET CONT OF ADDRESS POINTED TO BY
          :           ;ADDRESS 600
          :
600          2476     ;POINTS TO ADDRESS 2476
```

Like PC mode 6, this is cumbersome when we do not know the absolute address where this pointer will be located at assembly time. Again, we can use a label to represent the address, and this label will take on the value 600 during assembly:

```
MOV @ POINT, R1   ;GET CONTENTS OF ADDRESS
       :           ;POINTED TO BY "POINT"
       :
POINT: 2476                  ;POINTER ADDRESS
```

Note that, although we can use an address as an indirect address pointer, just as MOV(R0), R1 uses a register, we cannot increment or decrement the contents of that address the way we can a register. Thus we can write

MOV (R0) +, R1

but not

MOV (POINT) +, R1

where POINT is defined as a memory address. Thus the principal reason for using an address rather than a register for a pointer is simply that all of the registers are already in use. This could be the case in a subroutine or an interrupt service routine, as we will see later.

In summary, the four PC addressing modes are as follows:

PC mode 2: Immediate mode—MOV #300, R1
PC mode 3: Absolute mode—MOV #300, @#36
PC mode 6: Relative mode—MOV #300, POINT
PC mode 7: Relative indirect mode—MOV @POINT, R1

Branch Instructions

Another powerful type of instructions in the PDP-11 instruction set are the branch instructions. These instructions test the condition code bits of the Processor Status word (N, Z, V, C) and branch accordingly. The branch may be in only a limited range from -128 to $+127$ words from the branch instruction, but this range satisfies many program looping criteria and allows the powerful ability to test for a condition and branch to a new location in the same instruction.

The branch instructions are as follows:

Branch			Conditions for Branch
400	BR	Branch unconditionally	Any
1000	BNE	Branch if not equal	$Z = 0$
1400	BEQ	Branch if equal	$Z = 1$
100000	BPL	Branch if plus	$N = 0$
100400	BMI	Branch if minus	$N = 1$
102000	BVC	Branch if overflow clear	$V = 0$
102400	BVS	Branch if overflow set	$V = 1$
103000	BCC	Branch if carry clear	$C = 0$
103400	BCS	Branch if carry is set	$C = 1$

Signed Conditional Branch

002000	BGE	Branch if greater than or equal	N xor $V = 0$
002400	BLT	Branch if less than	N xor $V = 1$
003000	BGT	Branch if greater than	Z or (N xor V) $= 0$
003400	BLE	Branch if less than or equal	Z or (N xor V) $= 1$

where *or* and *xor* mean the inclusive and exclusive OR, respectively.

The preceding branch instructions test the condition codes and either branch or do not branch, depending on the states of the codes. These codes are set by the last instruction that tests, compares, or operates on some data or operand. The branch instructions do not change the condition codes, so that two or more branches can be performed on the basis of the same conditions. For example,

```
TST   (R1)    ;CHECK THE VALUE POINTED TO BY R1
BEQ   A1      ;GO TO LOCATION A1 IF ZERO
BPL   A2      ;GO TO LOCN A2 IF PLUS
BMI   A3      ;GO TO A3 IF VALUE IS MINUS
```

Many of these branch instructions assume that operations occur on signed numbers. It is sometimes useful to treat the computer word as if it were a 16-bit positive number. In this case, the tests for whether a number is greater or less than another are somewhat different. These instructions are called the unsigned conditional branches.

Unsigned Conditional Branch			Conditions for Branch
101000	BHI	Branch on higher	$C = 0$ and $N = 0$
103000	BHIS	Branch on higher or same	$C = 0$ (Same as BCC)
103400	BLO	Branch if lower	$C = 1$ (Same as BCS)
101400	BLOS	Branch if lower or same	C or $V = 1$

All of the branch instructions, then, use the upper 8 bits for the instruction code. The lower 8 bits of the instruction define where the branch is to go if it is satisfied. These 8 bits are treated as a signed two's complement number. If bit 7 is set, the distance is the two's complement of the distance in words *back* from where the actual branch instruction is executed. If bit 7 is not set, the offset is a positive number that is the distance in words *forward* from the branch instruction. The offsets are always represented in words, even though word addresses are two locations apart. Thus the processor performs the following operations in calculating the address to branch to if the conditions are satisfied:

1. Multiply the lower byte of the instruction by 2.
2. If bit 7 is on, extend the sign throughout the internal calculation register.
3. Add this value to the PC causing the branch.

It will be helpful to consider some examples of how the branch instruction works. Suppose that we are executing an instruction at location 1242, and we give branch instruction to location 1270:

    1242 BR 1270   ;BRANCH TO LOCATION 1270
                      ;UNCONDITIONALLY

At the time the instruction is executed, the PC will be pointing to 1244, so the offset is given by

$$
\begin{array}{ll}
1270 & \text{The code for BR is} \quad 400 \\
-1244 & \qquad\qquad\qquad\quad +12 \text{ word offset} \\
\overline{\quad 24/2} = 12 \text{ word offset} & \qquad\qquad\qquad\quad \overline{412} \text{ is the instruction code}
\end{array}
$$

Similarly, suppose that we are at 1616 and want to branch to location 1530. At the time the instruction is executed, the PC will point to 1620:

    1616   BNE   1530

The offset is given by

$$
\begin{array}{ll}
1530 & \qquad\qquad BNE = 1000 \text{ instruction code} \\
-1620 & \qquad\qquad\qquad\quad + \ \underline{344} \text{ offset} \\
\overline{177710/2} = 177744, \text{ which to 8 bits} & \qquad\qquad\qquad\quad 1344 \text{ is the final code} \\
\qquad\qquad\quad \text{is 344}
\end{array}
$$

Going back the other way, suppose that we encounter the code 100741 at location 131672:

$$131672/100741$$

This breaks down into the instruction code 100400 in the upper 8 bits and the offset 341 in the lower 8 bits. Since bit 7 is set, the offset is the two's complement of the number of words that the instruction will branch *back*. Converting to a 16-bit number, we extend the sign (all 1's) into the upper byte to give 341 = 177741. Multiplying by 2 since words are two addresses apart, we get $177741 \times 2 = 177702$. To find the address to which the branch occurs, we simply add this offset to the PC at the time the instruction is to be executed.

131672/100741 byte offset is 177702   PC will be 131674
$$\begin{array}{r} 177702 \\ \hline 131576 \text{ is the addressed} \\ \text{location} \end{array}$$

In summary, since only 8 bits are available to specify the offset, branches are limited to a range of -128 to $+127$ words. The offset is stored as the two's complement of the number of words of distance between the current PC and the desired destination in the lower byte, and the upper byte specifies which condition codes are to be tested. Most commonly, branch instructions branch to a *labeled* location and are of the form BR LOOP, where LOOP is defined as a label elsewhere in the program.

The JMP Instruction

To jump outside the range of -128 to $+127$ words, the JMP instruction can be used. It is of the form 0001ar, where, just as in the single-operand instructions, "a" is the addressing mode and "r" the register used. The JMP instruction can be used with any of the eight addressing modes, although the most common way is simply JMP LABEL, which is PC mode 6, the relative mode. This, of course, takes two words to represent, whereas the branch instructions require only one word.

Exercises

Determine the octal machine code values for the following programs, toggle them in (with stops in between, if you wish), and watch them execute.

```
5.4   500      ADD R0, R1
      502      SUB R2, R3
      504      DEC R3
      506      HALT

5.5   1000     MOV R0, R1
      1002     MOV(R0), R2
      1004     MOV (R0)+, R3
      1006     MOV@(R0)+, R4
      1010     MOV -(R0), R5
      1012     HALT
```

5.6 What will the contents of R0 be when the following program stops?
     Try it.

```
MOV #123, R0
NEG R0
SUB #3, R0
HALT
```

5.7 The program in Exercise 5.6 was toggled in and executed. When it
     stopped, R7 contained 510. Explain why, since the program occupies
     only 500–506.

5.8 Toggle in and execute the following program, and at each Halt ex-
     amine the contents of the PS, address 177776.

```
R0/0
R1/1000
500   TST   R0
502   HALT
504   MOV (R1)+, R0
506   HALT
510   ADD (R1)+, R0
512   HALT
514   ADD (R1)+, R0
516   HALT
1000    31726
1002   101241
1004    50123
```

5.9 What will R3 contain when the following program stops? Show all
     steps in octal.

```
R2/516
R3/3216
ADD   R3, R2
SUB   R2, R3
CLR   R2
SUB   R3, R2
INC   R2
COM   R2
SUB   R2, R3
HALT
```

5.10 Write down the assembly language equivalents of the following machine instructions:

500	010504
502	005003
504	062403
506	001376
510	160305
512	000000

5.11 Write down the octal equivalents of the following instructions. The symbol ``. = 500'' means that the first instruction goes at address 500.

```
. = 500
MOV   #500, R6
MOV   #340, @#36
MOV   #1000, R0
MOV   #10, R1
CLR   R2
ADD   (R0), R2
  .
  .
  .
```

5.12 What will the contents of TEMP be when (and if) the following program halts?

```
START:  MOV #6, R0
        SUB R0, TEMP
        BEQ START
        COM TEMP
        HALT
TEMP:   6
```

5.13 Write down the octal equivalents of the following two instructions. When executed, how will they differ?

```
. = 500            . = 500
MOV R0, A          MOV R0, @#506
HALT               HALT
A:  0              0
```

CHAPTER SIX

WRITING
SIMPLE PROGRAMS

This chapter discusses the task of writing a simple program in assembly language and converting it to binary code to run on a PDP-11. We will use the TECO text editor, the MACRO assembler, the LINK linker, and the RT-11 command structure.

Designing Algorithms

When you decide how you want a program to work, you have designed a programming *algorithm*. An algorithm is simply computerese for a way of doing something. Writing programs is always a problem of algorithm design. First, you define the problem in simple terms and then flesh out the logic by sketching a flowchart. Next, you write the actual code, including numerous comments, and finally you test and debug it.

Once you have written the program code on a piece of paper, you must type it into the computer in some way. While this was once done by punching out a paper tape, it is done today through the use of text editing programs that allow you to type in text and store it in a *file* on a disk or diskette. This text is written onto the disk and corrected using a text editor program. Then a program is used to read this file of characters and convert it into a file of binary machine code. This program is called an *assembler*. Finally, the program is converted to an absolute format and may be linked with other program modules using a linking program.

A Program for Adding 10 Numbers

We will start by writing a program to add together 10 numbers stored in a list in memory. If these numbers started at address 2000, we could write a simple program of the form

    ADD @#2000, R0
    ADD @#2002, R0
    etc.

but this would be extremely inefficient use of coding time and of memory. A more efficient way is to set the address of the first number in a list into a register and use this register to *point* to the list. Then addressing the register indirectly will allow us to access an element of a list, and incrementing the register will allow us to go on to the next element. The flowchart for this simple program is shown in Figure 6.1.

To code this program, we simply write down the assembly language equivalents of each element in the flowchart and insert the comments from the flowchart alongside. Be sure to actually type in the comments as you write the program code: They will remind you of how the program works later as you test and debug it.

```
;ADD 10 NUMBERS TOGETHER FROM LIST IN MEMORY

START:  MOV #2000, R1   ;SET POINTER TO TOP OF LIST
        MOV #10., R2     ;SET COUNTER TO 10
        CLR R0           ;SET SUM TO 0

LOOP:   ADD (R1)+, R0    ;ADD CONTENTS OF ADDRESS
                         ;PTD TO BY R1 INTO R0
        DEC R2           ;DECREMENT COUNTER
        BNE LOOP         ;GO BACK IF NOT 0
        HALT             ;AND STOP WITH SUM IN R0

;LIST BEGINS BELOW AT ADDRESS 2000
        .=2000
        :
        :
```

This program begins by putting the address of the beginning of the list in R1 and a counter in R2. These are the address pointer and the counter. Let us assume that the data start at address 2000 and are as follows:

    2000   5
    2002   6
    2004   11.
    2006   2
      etc.

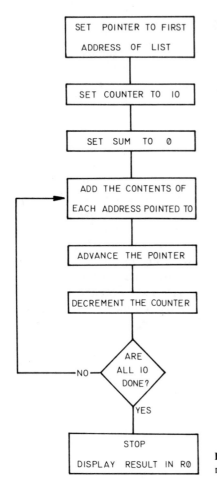

Figure 6.1 Flowchart of program to add 10 numbers together.

When we start the program, the sum is first zeroed by the instruction CLR R0. Then the instruction

ADD (R1)+, R0

is encountered, where

R1 = 1000   and   R0 = 0

This instruction can be read as follows: ADD the contents of the address *pointed to by* R1 into R0 and then increment R1 by 2. After this instruction is executed the first time, the registers have

$R1 = 2002$ and $R0 = 5$

Now R1 contains the address of the second number in the list and is said to be pointing to 2002. Schematically, we could write

R1	Addresses	Contents	R0
	2000	5	0
→	2002	6	5
	2004	11	

Then the instruction DEC R2 is encountered. Since R2 was set as a counter, in this case 10., we are now decrementing it so that it represents one cycle completed. We subtract 1 from 10. to get 9. (or 11_8), and this also sets the condition codes to represent the state of the machine at the end of this instruction. These will be

$$\begin{array}{cccc} N & Z & V & C \\ 0 & 0 & 0 & 0 \end{array}$$

In other words, the number is not negative, not zero, and neither overflow nor carry have occurred.

The instruction BNE LOOP means branch to the address named LOOP unless the Z condition code is 1, or unless the value of the number at the end of the previous instruction was a zero. The counter R2 is not yet 0, so the program returns to the address LOOP to add in yet another number.

At address LOOP, the instruction

ADD (R1)+, R0

is executed again with

$R1 = 2002$ and $R0 = 5$

and after execution

$R1 = 2004$ and $R0 = 5 + 11. = 16.$

Schematically, we can write

R1	Addresses	Contents	R0
	2000	5	0
	2002	6	5
→	2004	11	16

Now R1 points to address 2004, the next entry in the list. If the counter R2 is not yet 0, there will be a branch back to add another number into R0. The instruction DEC R2 makes R2 = 9 − 1 = 8., and a branch indeed occurs. This looping will continue until the tenth number is added into R0, after which the instruction DEC R2 will cause 1 − 1 = 0, and this will set the condition codes as

$$
\begin{array}{cccc}
\text{N} & \text{Z} & \text{V} & \text{C} \\
0 & 1 & 0 & 0
\end{array}
$$

Then the instruction BNE LOOP will find Z = 1 and therefore will *not* branch. Instead, the next instruction will be executed, and the processor will halt with the contents of R0 displayed in the lights of the Data Register.

The assembled listing of this program is shown below.

```
                        ;ADD 10 NUMBERS TOGETHER FROM LIST IN MEMORY
000000                  .ASECT   ;ASSEMBLE AS ABSOLUTE FILE
            001000       .=1000              ;LOAD IT AT THIS ADDRESS
001000      012701      START:  MOV #LIST, R1   ;SET POINTER TO BEGINNING
001004      012702              MOV #10., R2   ;SET COUNTER TO 10
001010      005000              CLR R0          ;CLEAR SUM REGISTER

001012      062100      LOOP:   ADD (R1)+, R0  ;ADD CONTENTS OF ADDRESS
                                               ;   PTD TO BY R1 INTO R0
001014      005302              DEC R2         ;DECREMENT COUNTER
001016      001375              BNE LOOP       ;GO BACK IF NOT DONE
001020      000000              HALT           ;STOP WITH SUM IN R0

                        ;HERE IS THE LIST TO BE SUMMED
001022      000005      LIST:   5
001024      000006              6
001026      000013              11.
001030      000002              2.
001032      000015              13.
001034      000026              22.
001036      000004              4
001040      000011              9.
001042      000002              2
001044      000211              137.
            001000      END START
```

Note that although two of the setup steps require two words each, the actual loop requires only one word per instruction.

The SOB Instruction

Most modern PDP-11's feature the SOB instruction, which has the form

077rnn   SOB reg, offset   ;SUBTRACT 1 AND BRANCH IF NOT 0 where the register is a counter and "nn" is a 6-bit unsigned word offset. This offset is the distance to branch back in *words* if the result is nonzero after the register is decremented. Unlike the branch instruction, the SOB instruction can only branch *back*, and since only 6 bits are used for the offset, the largest jump is 77_8 or 63_{10} words. This instruction is extremely useful in making small loops in program run more efficiently. The instruction

```
SOB R1, LOOP    ;GO BACK UNTIL R1 IS 0
```

effectively replaces

```
DEC R1      ;DECREMENT THE COUNTER
BNE LOOP    ;AND GO BACK IF NOT YET 0
```

where LOOP must be less than 77_8 words distant. Rewriting our program to add 10 numbers, we have

```
;PROGRAM TO ADD 10 NUMBERS TOGETHER
START:  MOV #10., R2     ;SET THE COUNTER
        MOV #2000, R1    ;SET THE POINTER
        CLR R0           ;CLEAR THE SUM

LOOP:   ADD (R1)+, R0    ;ADD EACH INTO THE SUM
        SOB R2, LOOP     ;GO BACK TILL R2 0
        HALT             ;STOP WITH SUM IN R0
```

Programming Without Using R0–R6

Occasionally it is necessary to write a program that does not use the registers as pointers and counters because they are all in use in some other section of the program. This often occurs in interrupt service routines and subroutines called from many points in the main program. One way to write such a program is to use the PC addressing modes with the pointer, counter, and sum in memory locations instead of registers.

```
;PROGRAM TO ADD 10 NUMBERS WITHOUT USING R0-R6

1000    START:  MOV #2000, POINT   ;SET COUNTER IN LOCN POINT
1002            MOV #10., COUNT    ;SET COUNTER IN LOCN COUNT
1014            CLR SUM            ;SET SUM=0

1020    LOOP:   ADD @ POINT, SUM   ;ADD EACH INTO THE SUM
1026            ADD #2, POINT      ;INCREMENT THE POINTER BY 2
1034            DEC COUNT          ;DECREMENT THE COUNTER
1040            BNE LOOP           ;GO BACK TILL DONE
1042            HALT               ;STOP- SUM IS IN 'SUM'

;POINTERS AND VARIABLES
1044    POINT:  0                  ;POINTER
1046    COUNT:  0                  ;COUNTER
1050    SUM:    0                  ;SUM WILL BE HERE
```

This program works exactly like the programs that do use the registers. It does require significantly more space since nearly all the instructions require two or three words. Further, when the program halts, the sum will not be displayed in the Data Register, but will be found in location 1050 only. The instruction

ADD @POINT, SUM

causes the addition of the contents of the address pointed to by address POINT in the same way that the instruction ADD (R1)+, R0 did in the register program. However, it is not possible to increment the pointer location POINT in the same way that we can increment the register R1. Remember that you can automatically increment (and decrement) registers, but not memory locations. Thus the additional instruction

ADD #2, POINT

is needed to complete the task. Note that while ADD (R1)+, R0 is a one-word instruction, the two instructions

ADD @POINT, SUM
ADD #2, POINT

require three words *each*. This is because the instructions each require two offset words on PC mode:

ADD @POINT, SUM means ADD @x(R7), x(R7)
                                  c1
                                  c2

where "c1" and "c2" are two constants that represent the amount to be added to the PC to represent the actual address of POINT and SUM. For example, if POINT is at 1046 and SUM at 1052, and the instruction begins at 1020, the constants will be

```
1020  067767  ADD @POINT, SUM
1022    22              c1        ;1024 + 22 = 1046
1024    24              c2        ;1026 + 24 = 1052

1046          POINT:  0

1052          SUM:    0
```

Note that the offset 22_8 is added to the current PC (1024) to produce the address of POINT (1046) and that the PC is then advanced to 1026 after fetching the next word of the instruction. The 24 is then added to 1026 to give the address of SUM (1052).

The instruction ADD #2, POINT advances the address pointer to the next address and is required because we cannot increment a memory location as we could register R1. The instruction must be an ADD, since the address must be advanced by 2. The INC or INCB instructions would only advance the location by 1.

Proper Initialization

It is sometimes a temptation to write the first program given earlier as

```
            MOV #1000, R1    ;SET POINTER
            MOV #10., R2     ;COUNTER
     LOOP:  ADD (R1)+, R0    ;ADD INTO SUM
            SOB R2, LOOP     ;GO BACK TIL R2 0
            HALT             ;THEN STOP
```

However, note that the sum will be incorrect, since the sum was never zeroed. This is an easy error to make and falls into the category of improper *initialization*. Just as the pointer (R1) and the counter (R2) are initialized, the sum (R0) must be zeroed as well.

A second example of improper initialization is shown in this truncated, incorrect version of the second program.

```
     LOOP:   ADD @ POINT, SUM    ;ADD EACH POINT INTO SUM
             ADD #2, POINT       ;ADVANCE THE POINTER
             DEC COUNT           ;DECREMENT THE COUNTER
             BNE LOOP            ;GO BACK UNTIL DONE
             HALT                ;STOP WHEN ALL ARE DONE
     POINT:  2000                ;POINTER ALREADY SET!
     COUNT:  10,                 ;AND SO IS THE COUNTER!
     SUM:    0                   ;AND THE SUM IS 0
```

This program has no internal initialization *at all*. It appears that it will perform the same task as the earlier ones. In fact, it will the *first* time.

In this program, the pointer and counter are set in the assembly language code. However, after the program has run and halted, the variables will contain the values

POINT: 2024
COUNT: 0
SUM: whatever

Consider what would happen if this program were run again. The instruction ADD INT, SUM would start, not at address 2000, but at address 2024, and the sum would start not at 0 but at the point if left off before. Worst of all, the counter would start at 0 and would be decremented each time to 177777, 177776 . . . , and COUNT would not become 0 again for 65,535 more passes through the loop!

Although initialization is quite simple in this case, it can sometimes be rather involved, and for this reason it should be a standard part of all programs and should occur at the *beginning* of the program or routine.

Exercises

These exercises are to be written out. You may wish to assemble them and try to run them after reading the next chapter on the assembly of programs.

6.1   Write a program to add and subtract alternate numbers in memory which will halt if the sum becomes 0 or after all the memory in your computer has been addressed. In other words, add the first number, subtract the second, add the third, and so forth, testing for 0 after each sum.

6.2   Write a program to put a "ramp" in memory, that is, a 0 in location zero, a 1 in location one, and so forth up to location 17776.

6.3   Write a program to put 15 elements of the Fibonacci series into memory, where each element after the first two is the sum of the previous two:

$$1, 1, 2, 3, 5, 8, \text{etc.}$$

Halt with the fifteenth element displayed in R0.

6.4   Write a program to count the number of negative numbers between addresses 1200 and 4216. Display the result in R0.

6.5   Write a program to count the largest number of adjacent zero words in 4K of memory. Display the result in R0. (Write a flowchart before you start.)

CHAPTER SEVEN

ASSEMBLING PROGRAMS

As we have seen, the conversion between assembly language mnemonics and the actual binary code is merely a matter of looking up the values of the symbols in the table and combining them with values of other symbols. Such operations can easily be performed by a computer program called an *assembler*. In addition, the actual preparation of program text can be facilitated by a text editing program. Although there are a number of editing programs available for the PDP-11, the most versatile and widely used of these is TECO, which stands for Text Editor and Corrector. The assemblers used in a number of operating environments are all invariably called MACRO.

Preparing Programs Using TECO

The TECO text editor can easily be made the standard editor of the RT-11 system and can then be called using the EDIT command or directly using the R TECO command. To call TECO directly for the creation of a new file, you type

```
.R TECO
*EWADD10.MAC$$
```

where the file to be created is called ADD10.MAC, and the "EW" stands for Edit Write. The prompter character for RT-11 is the period (.) and that for TECO and most other system programs is the asterisk (*). These are typed by the RT-11 monitor and by TECO, respectively, and are not entered as part of the commands.

Alternatively, the RT-11 system allows you to condense these commands into a single monitor-level command

```
.EDIT ADD10.MAC/CREATE
*
```

that has exactly the same effect.

Once you are in TECO and have defined the file to be created, you can begin inserting text using the I command. The format of the I (for Insert) command is that all text following the I is inserted into the text of the file, until the first ESC (or ALT MODE) character is found. Thus the command has the form

```
Ixxxxxxx
xxxx
xxxxx$
```

where we use $ to represent the nonprinting ESC character as before. To create our program to add 10 numbers together, we type

```
*I;ADD 10 NUMBERS TOGETHER
.ASECT
.=1000
START:  MOV #LIST, R1   ;STE POINTER
        MOV #10, R2      ;SET COUNTER
        CLR R0           ;SET SUM TO 0

LOOP:   ADD (R1)+, R0    ;ADD EACH NUMBER PTD TO
        DEC R2           ;DECREMENT COUNTER
        BEQ LOOP
        STOP             ;STOP WITH SUM IN R0

        .END START       ;START ADDRESS IS 'START'
```

Then to write this file onto the disk, we exit from TECO with the EX command. This closes the file ADD10.MAC on the disk and returns you to the RT-11 monitor.

```
$$
```

```
*EX$$
.
```

Note that the asterisk prompting character of TECO and the period prompter of RT-11 are typed by those programs, not by the user.

Correcting the Program

The program we wrote earlier has a number of mistakes to be corrected, and we will use TECO to correct them. To reopen the file you just created for changes, give the monitor command

EDIT ADD10.MAC

This starts TECO and will cause the file ADD10.MAC to be read in for corrections and changes. When you are done editing the file in this way and exit back to RT-11 with the EX command, TECO will rename the old ADD10.MAC to ADD10.BAK and create a new, corrected file having the name ADD10.MAC. The great power of TECO lies in its ability to make rapid changes in the text of a program, or indeed of any document that you write in text form.

TECO edits programs through use of a character *pointer*, which you can move around and use in deletions and insertions of text. It is particularly powerful because it allows you to insert and change single characters or groups of characters without retyping the whole line.

When you start an edit of an existing file, the pointer points initially to the first character in the file. You can move it by line or by character using the following commands:

L	Move the pointer down one line.
nL	Move the pointer down *n* lines.
−L	Move the pointer back 1 line.
−nL	Move the pointer back *n* lines.
0L	Move the pointer to the beginning of the current line.
C	Move the pointer forward one character.
R	Move the pointer back one character.
nC, nR	Move the pointer forward or back *n* characters.
nJ	Move the pointer to the *n*th character of the file.
J	Move the pointer to the beginning of the file.
ZJ	Move the pointer to the end of the file.

You can also delete single characters and whole lines with TECO using the commands

D	Delete the next character.
nD	Delete the next *n* characters.
−nD	Delete the previous *n* characters.
K	Delete the current line to the right of the pointer.
nK	Delete the next *n* lines.
−nK	Delete the previous *n* lines.
0K	Delete all of the line to the left of the pointer.

The Search Commands

The most important way of moving the pointer to the correct position for text editing is with the search commands:

Sstring$ Search current buffer for next occurrence of "string."
Nstring$ Search current buffer and all successive buffers for the next occurrence of "string."

The search commands allow you to search for any unique text and then make changes around that point. When a search is completed, the text pointer is pointing to the first character *after* the string you searched for.

As your program files become bigger and bigger, it is useful to divide them into smaller portions that will be read in one after another during an edit. This has the principle advantage that only a small amount of text is kept in memory and thus the speed with which characters can be inserted and deleted is high. You can divide large files into small buffer units by inserting CTRL/L characters (Form Feed) characters every 60 or so lines.

The Search and Replace Commands FS and FN

We could make all our changes with these commands and the I command to insert new characters where we deleted the old ones. For example, we could correct the comment in the line

    START: MOV #LIST, R1   ;STE POINTER

by using the TECO commands:

Search for the characters "; S"

    S;S$

Delete the next 2

    2D

Insert "ET" instead

    IET$

Move the pointer back to the beginning of the line

    0L

and print out the corrected line:

    T$$

This is all strung together in TECO using the single command string:

```
S;S$2DIET$0LT$$
```

and will produce the printout of the corrected line:

```
START:  MOV #LIST, R1     ;SET POINTER
```

However, there is one more powerful set of commands that make this even easier: the search and replace commands:

FSstring1$string2$ Change next occurrence of "string1" to "string2."

FNstring1$string2$ Change next occurrence of "string1" to "string2" in this or any succeeding buffer.

With these commands, you can change any misspelled word to the correct one in just a few keystrokes. Here the search for "STE" and its replacement with "SET" becomes

```
FSSTE$SET$0LT$
```

Other TECO Commands

The TECO commands discussed so far are sufficient for nearly all editing. A complete list of all commands is given in Appendix VI. You should consult the TECO manual for more detailed descriptions of other commands.

Correcting the Program

Here we can correct the mistakes in this program quite easily. Briefly, these mistakes are

1. The #10, should be #10.,
2. The BEQ should be a BNE.
3. The label "LIST" is not defined.
4. The instruction STOP should be HALT.

We will only change the first mistake for now and then observe how we discover the other mistakes. We start by editing the file:

```
EDIT ADD10.MAC
FSSTE$SET$0LT$$
```

and TECO prints
```
START:   MOV #LIST, R1    ;SET POINTER
FS10$10.$0LT$$
```
TECO Prints out
```
         MOV #10., R2      ;SET COUNTER
```

```
EX$$
```

We are now ready to try assembling the program again. We will discuss the rules for representing numbers and special characters and then go on to finish assembling and running this program.

Numbers in the Assembler

The MACRO assembler recognizes numbers in two radixes automatically. Octal numbers are assumed if the digits are not terminated with a decimal point. Decimal numbers always end with a decimal point.

Numbers with fractional parts are termed *floating point* numbers and are discussed in detail in Chapter 11. The MACRO assembler will create a two-word floating point representation of a number if it is preceded with the special directive .FLT2.

Assembler Special Characters

We have already used several characters that the assembler recognizes as having special functions. The complete list of these characters is

:	Defines a label at that location such as LOOP: ADD (R2)+, R1.
=	Defines a label having a specific value such as CR = 15.
%	Used to define a register symbol such as R2 = %2.
#	Used to define immediate addressing mode such as MOV #10., R1.
@	Indirect addressing mode symbol such as MOV #200, @ POINT.
( )	Indirect register mode symbols.
"	defines a pair of ASCII characters.

'   Defines a single ASCII character.
+ −   Used in arithmetic expressions involving symbols
       and in autoincrement and decrement modes.
!   Logical OR between symbols.
&   Logical AND between symbols.
,   Used to separate two operands in a dual-operand
       instruction.
.   Used to represent decimal integers;
       also used to represent the value of the
       current location counter, such as . = 1000.
;   Begins a comment.

The Current Location Counter

The symbol "." is used to represent the value of the current location counter. It is the value of the current address where an instruction is being assembled. It differs from the value of the PC when the instruction is executed in that the PC always contains the address of the *next* instruction to be executed. Further, the value of the current location counter is a concept peculiar to assembly language and can be used to construct various expressions and addressing modes, whereas the value of the PC has meaning when the program is assembled and running.

The most important use of the current location counter is in setting the first address of the code to be assembled. For example, the statement

. = 1000

appears at the beginning of many of our programs and tells the assembler to assemble the first executable instruction following this line at address 1000.

We can also use the location counter to reserve a block of storage, although there are clearer ways of doing this, by writing

. = . + 100

which reserves 100_8 locations by moving the assembly address down by that many locations.

The third use of the location counter is in the construction of small jump and branch instructions without using labels. For example, we could write

```
ADD (R1)+, R0
DEC R2
BNE .-4          ;GO BACK 2 WORDS IF R2 IS NOT 0
```

Since instructions may be one, two, or three locations long, this method can often lead to errors, and the use of labels is strongly recommended.

Assembling the Program

To assemble this program, we use the MACRO assembler, which is invoked by the command

MACRO ADD10/LIST:TT:/NOSHOW:BEX

This command calls the MACRO assembler and assembles the file ADD10.MAC. The .MAC extension is the default extension for files to be assembled by MACRO. The modifying "switch" /LIST:TT: means list the file on the terminal having the device name TT: and the switch /NOSHOW:BEX means list only the first word of two- and three-word instructions. The switch /SH:TTM is optional and is for narrow paper terminals; it limits the output listing to 80 columns.

```
. MAIN.   MACRO V03. 01   PAGE 1


        1                       ; ADD 10 NUMBERS TOGETHER
        2 000000       . ASECT
        3              001000    . =1000
  U     4 001000       012701   START:   MOV #LIST, R1    ; SET POINTER
        5 001004       012702            MOV #10. , R2    ; SET COUNTER
        6 001010       005000            CLR R0           ; SET SUM TO 0
        7
        8 001012       062100   LOOP:    ADD (R1)+, R0    ; ADD EACH NUMBER INTO R0
        9 001014       005302            DEC R2           ; DECREMENT COUNTER
       10 001016       001775            BEQ LOOP
  U    11 001020       000000            STOP             ; STOP WITH SUM IN R0
       12
       13              001000   . END START
. MAIN.   MACRO V03. 01   PAGE 1-1
SYMBOL TABLE

LIST  = ******            START    001000          STOP  = ******
LOOP    001012

 . ABS.   001022       000
          000000       001
ERRORS DETECTED:   2
```

In this assembled listing, we see two lines flagged with the error "U." This means that these lines contain a symbol that is *undefined*. An undefined symbol is one not known to the assembler either as part of its permanent symbol table of instructions or in its table of user-defined symbols.

In line 4, the error is that the symbol LIST is not defined, and in line 11, the error is that the correct instruction is HALT, not STOP. Let's correct these two errors with TECO. First we open the file with the EDIT command:

    EDIT ADD10.MAC

We will fix the STOP error first, since the LIST symbol will be defined as the first address of a list of numbers at the end of the program. So we change STOP to HALT using the FS command:

    FSSTOP$HALT$0LT$$

Teco prints out

    HALT   ;STOP WITH SUM IN R0

Then we insert the list to be summed on the lines after the program. It is *very important* to recognize that constants such as pointers and lists of numbers cannot be interspersed with the actual program, since the numbers will be executed as *instructions* if the computer encounters them in a sequential list of instructions. Thus we will place the list of numbers to be added together at the *end* of the program.

The label LIST will simply be the name of the first location of the list. Since the symbol LIST will then have the value of the address of the start of the list, we can move the symbol LIST into R1 and move the starting address of the list into R1.

Thus, to insert this list at the end of the program (but before the .END statement, which indicates to the assembler that there is nothing more to assemble), we simply move down one line from the HALT instruction

    L

and begin inserting the list

    I....

All together the commands read

    LI;HERE IS THE LIST TO BE SUMMED
    LIST: 5
    6
    11.
    2
    13.
    :
    :
    etc.
    $$
    *EX$$

Then once back in the monitor, we can reassemble the program, yielding the following listing:

```
   1                         ; ADD 10 NUMBERS TOGETHER
   2 000000                  . ASECT
   3              001000      . =1000
   4 001000      012701       START:   MOV #LIST, R1      ; SET POINTER
   5 001004      012702                MOV #10., R2       ; SET COUNTER
   6 001010      005000                CLR R0             ; SET SUM TO 0
   7
   8 001012      062100       LOOP:    ADD (R1)+, R0      ; ADD EACH NUMBER INTO R0
   9 001014      005302                DEC R2             ; DECREMENT COUNTER
  10 001016      001775                BEQ LOOP
  11 001020      000000                HALT               ; STOP WITH SUM IN R0
  12
  13                          ; HERE IS THE LIST TO BE SUMMED
  14 001022      000005       LIST:    5
  15 001024      000006                6
  16 001026      000013                11.
  17 001030      000002                2
  18 001032      000015                13.
  19 001034      000042                34.
  20 001036      000020                16.
  21 001040      000007                7
  22 001042      000011                9.
  23 001044      000003                3
  24
  25              001000      . END START
. MAIN.   MACRO V03.01   PAGE 1-1
SYMBOL TABLE

LIST       001022              LOOP       001012              START   001000

  ABS.     001046         000
           000000         001
ERRORS DETECTED:   0
```

OBJ Files and the Linker

Now, since the program assembles without errors, we are ready to try to execute it. The files produced by MACRO in addition to the list is a file named ADD10.OBJ. This file is an *object* file intended to be linked with other files into a single executable program. In these simple examples, no other program files are needed for this program to execute. Nonetheless, we must still pass the file through the linker program to convert it from .OBJ format to that of an executable program, which is simply a picture of those numbers to be executed in memory. To do this, we simply type

LINK ADD10

This command invokes the program LINK which takes the file ADD10.OBJ and converts it to the program ADD10.SAV, which can be run directly from the monitor. When the linking is done, we type

    RUN ADD10

to load and execute the file ADD10.SAV. You can also combine the three commands MACRO, LINK, and RUN by just typing

    EXE ADD10

which does all three tasks.

First Execution of ADD10

When we type the RUN command on this program, we will find that the computer comes to a stop immediately, at the HALT instruction. On PDP-11's with a data register display, this shows the contents of R0 when the computer has halted, which in this case should be the sum of the 10 numbers. In this first execution, we see that the Data Register contains a 5, which is not the sum of these numbers.

We then study the program to see the cause of the error. We immediately note that the first number in the list of 10 is a 5 and that nothing has been added to it. This suggests that the program passed through the loop only once and that we should look into the looping logic. Here we discover the instruction BEQ LOOP, which will never branch back to loop since R2 will not be 0 the first time through. Recommenting the program to show what has happened, we might write

```
        DEC  R2        ;SUBTRACT 1 FROM R2
        BEQ  HALT      ;GO BACK TO LOOP IF R2 HAS BECOME 0
        HALT           ;ELSE HALT IF R2 IS NOT 0
```

The code we really meant was

```
        DEC  R2        ;SUBTRACT 1 FROM R2
        BNE  LOOP      ;GO BACK TO LOOP IF R2 IS NOT YET 0
        HALT           ;HALT IF R2 HAS BECOME 0
```

or, more compactly, we could have written

```
        SOB  R2, LOOP  ;DECREMENT R2, GO BACK TO LOOP IF NOT 0
        HALT           ;ELSE HALT IF R2 IS NOW 0
```

The simplest change we can make to correct this program is to replace
the two lines DEC R2, BEQ LOOP with the preceding code. We do this
by reentering TECO

    EDIT ADD10.MAC

and then searching for the line containing DEC

    SDEC$

printing out the line

    0LT$$

deleting the 2 lines

    2K

and then inserting the correct line

    ISOB R2, LOOP ;GO BACK IF R2 NOT 0
    $$

All together, this command string would be

    *SDEC$0LT$$

                    DEC R2   ;SUBTRACT 1 FROM R2

    *2KISOB   R2, LOOP   ;GO BACK IF R2 NOT 0
    $$
    EX$$

Reassembling ADD10

We can try to reassemble ADD10 without making a new listing and see
what the result is. This is done with the instruction

    EXE ADD10

which will assemble, link and run the program. The program will halt
with the value 152_8 in the Data Register, which is 72_{10}. This is the final
change in the program. The final listing is shown on the next page.

```
     1                          ; ADD 10 NUMBERS TOGETHER
     2  000000          . ASECT
     3                  001000   . =1000
     4  001000  012701  START:    MOV  #LIST, R1      ; SET POINTER
     5  001004  012702            MOV  #10., R2       ; SET COUNTER
     6  001010  005000            CLR  R0             ; SET SUM TO 0
     7
     8  001012  062100  LOOP:     ADD  (R1)+, R0      ; ADD EACH NUMBER INTO R0
     9  001014  077202            SOB  R2, LOOP       ; GO BACK IF R2 IS NOT 0
    10  001016  000000            HALT
    11
    12                          ; HERE IS THE LIST TO BE SUMMED
    13  001020  000005  LIST:    5
    14  001022  000006  6
    15  001024  000013  11.
    16  001026  000002  2
    17  001030  000015  13.
    18  001032  000042  34.
    19  001034  000020  16.
    20  001036  000007  7
    21  001040  000011  9.
    22  001042  000003  3
    23
    24              001000   . END START
MAIN.     MACRO V03.01    PAGE 1-1
SYMBOL TABLE

LIST     001020          LOOP     001012          START     001000

  ABS.   001044      000
         000000      001
ERRORS DETECTED:    0
```

Error Messages Generated by MACRO

The MACRO assembler generates single-letter error messages, which are printed to the left of the line numbers in the assembled listings. These error messages are listed below:

A   Addressing error. The addressing range is too far for a Branch or SOB instruction. This also occurs if the operand is not a symbol that can be addressed.

B   Boundary error. Attempt to assemble an instruction at an odd address. MACRO adds 1 to the location counter and continues.

D   Reference is being made to a symbol that has been multiply defined.

E   No .END directive found before the end of the file.

I   Illegal character detected. The listing will contain a "?" at the point where the character was detected.

L   Line buffer overflow. An input line has more than 132 characters. Only the first 132 are read.

M   Multiple definition of a label. A label is defined with an equals sign or colon in more than one place.

N   Number error. A number containing an 8 or 9 does not end in
       a decimal point.
O   Op code error. An instruction or assembler directive appears
       in an incorrect context.
P   Phase error. The value of a label differs from the first pass to
       the second pass.
Q   Questionable syntax. This can be caused by an incomplete in-
       struction, or by a Line Feed not following a Carriage Return.
R   Register symbol error. The program tries to refer to a register
       in an illegal way.
T   Truncation error. A number generated requires more than 16
       bits, or more than 8 bits if it is a .BYTE directive.
U   Undefined symbol.
Z   Warning. This instruction may not be compatible with all models
       of PDP-11's.

Assembler Directives

Commands that tell the MACRO assembler to perform certain operations
are called assembler *directives*. They differ from instructions to be assembled in that they generate no code, but instead tell the assembler *how*
to generate code. Some of the more common directives are as follows.

.ASCII   Convert the string that follows into ASCII characters,
                one per byte until a matching delimiter character is
                found. This directive has the form

                   .ASCII xANY CHARACTERSx

                where "x" is any character not used in the message.
.ASCIZ   Converts the string that follows into ASCII charac-
                ters, just as the .ASCII directive does, and adds a
                zero byte to the end of the string. Note that both
                directives may end on an odd byte and the .EVEN
                directive may then be required.
.ASECT   Tells the assembler that the code that follows is to
                be assembled as *absolute* code, having addresses
                starting at the location specified in the next ".=n"
                directive.
.BLKW n   Reserve *n* words of storage at this position.

.BLKB n	Reserve *n* bytes of storage here.
.BYTE	This directive tells the assembler that the numbers that follow are to be placed in adjacent bytes instead of being assigned one word each. May end up on an odd byte. This has the form

.BYTE 15, 12, 13

.CSECT	Tells the assembler to assemble the code that follows in relocatable format, so that it can be linked to run at whatever address is calculated by the linker.
.END	This is a signal to the assembler that the end of the source code has been reached, and the assembler is to scan no farther.
.EVEN	If the current location counter is odd, this directive tells the assembler to make it even, so that an instruction can be placed in the next word.
.FLT2	Converts the numbers that follow to floating point representation and places it in two successive words.
.GLOBL	The labels that follow this declaration are *global* labels and can thus be referred to between .OBJ modules. Such a global label may be defined either in this module or in another module.
.LIST, .NLIST	These directives turn on and off the listing of that section of the program.
.MACRO	Tells the assembler that the code that follows is part of a series of instructions to be inserted whenever that Macro name is encountered. This is described in Chapter 11.
.MCALL	Tells the assembler that the Macro names that follow are to be found in the special file SYSMAC.SML, the system Macro library. Most of these have to do with calls to the RT-11 monitor.
.PAGE	Skip to the top of the next page.
.TITLE	Print this title at the top of all successive pages. Also passed to the linker program as a module name.
.WORD	This tells MACRO to place each of the numbers that follow in a single word. This is the default operation for any number or string of numbers on a line separated by commas and is thus usually unnecessary

unless you wish to emphasize what you are doing.
For example,

.WORD 1234, 5312, 55372

has exactly the same effect as

1234, 5321, 55372

or as

1234
5321
55372

Exercises

7.1   What effect would the following TECO commands have on a file?
      JFSCHA$0LT$$
      Is this a reasonable command? What do you think would be better?
7.2   What characters would be put in memory by running the program
      assembled containing the following directives:
      .ASCII "123"
      .ASCIZ ATHIS IS A;MESSAGEA
7.3   Write, edit and assemble a program to add and subtract alternate
      numbers in memory until a zero sum is found. The program should
      then stop. What other constraint should you put on the program?
      Link and execute the program. It should be self-starting.

CHAPTER EIGHT

LOGICAL AND SHIFT INSTRUCTIONS AND MULTIPLY–DIVIDE INSTRUCTIONS

In this chapter we discuss several major instruction classes that are used in the PDP-11. The first of these, the logical instructions, allows the manipulation of particular bits of words for comparison. These bit manipulations are generally used when the programmer wishes to remember particular things in individual bits of a word. For example, one byte could be used to represent the spin states in one basis function in a nmr spectrum calculation. In this case we could let 0 represent spin alpha and 1 represent spin beta. Thus we could write $\beta\alpha\alpha\beta\alpha\beta\alpha\alpha$ as 10010100. The major logical operations that the PDP-11 performs are the inclusive OR, the exclusive OR, and a form of the AND. These are performed with the following instructions:

		Name		Description
BIT	b3arar	Bit test	BITB	src and dst→cond codes
BIC	b4arar	Bit clear	BICB	com(src) and dst→dst
BIS	b5arar	Bit set	BISB	src or dst→dst
XOR	074rar	Exclusive OR		reg xor dst→dst

The BIT or bit test instruction performs a logical AND between the source and destination operands and uses this result only to affect the condition codes. The BIT instruction is used mostly to mask a given operand with another operand to determine whether certain bits are set. For example, if we wished to determine whether bit 6 of register R3 was set, we could perform the operation

```
BIT R3, #100      ;TEST BIT 6 OF R3
BNE SET           ;GO TO "SET" IF RESULT IS NONZERO
```

This works as follows:

Suppose that R3 contains	1	011	100	111	010	111
If we AND this with 100,	0	000	000	001	000	000
we get a result which will	0	000	000	001	000	000
be 1 or 0, depending on						
whether or not bit 6 was set.						

This result affects the condition codes as follows:

$$N \quad Z \quad V \quad C$$
$$0 \quad 0 \quad 0 \quad 0$$

Thus the result is nonzero, as reflected in the Z-bit, and the test in the BNE instruction will cause a branch because the result is indeed nonzero. Note that neither operand is changed by this instruction; it affects only the condition codes.

The BIC or bit clear instruction also performs a logical AND but affects the destination as well as the condition codes. However, the BIC performs the AND between the one's complement of the source and the destination, leaving the result in the destination. Since the bits in the destination that were 0's in the source are set and the bits are cleared that were 1's in the source, the instruction is referred to as bit clear. Let us consider BIC #3, R1.

Suppose that R1	1	011	100	111	010	111	
contains and	0	000	000	000	000	011	Only bits 1 and 0
our source is							are set.

The processor first takes the one's complement of the source and then ANDs it with the destination, effectively doing the following:

R1:	1	011	100	111	010	111	
Complement of source:	1	111	111	111	111	100	All bits are
The AND result is:	1	011	100	111	010	100	set but 1 and 0

This instruction is the easiest way to clear particular bits of a word or register equal to 0 without affecting other bits.

The BIS or bit set instruction can be used to *set* bits in the destination operand, just as the BIC cleared bits in the destination operand. It simply performs the inclusive OR between the source and the destination. Recall that the inclusive OR sets bits in the result of a bit is set in *either* operand. Suppose that we wish to set bits 2, 3, and 8 of a particular word without affecting the remaining bits. Some of them may be set, but we wish to be sure that all three are set. We simply OR the destination with a word containing 2, 3, and 8 as 1's and the rest as 0's:

```
BIS #414, R5    ;SET BITS 2, 3 AND 8 of R5
```

Suppose that R5 contains	1	011	100	011	010	111	Bits 3 and 8 are 0's, bit 2 is 1
If we OR that with 414,	0	000	000	100	001	100	Bits 2, 3, and 8 are 1's
the result is	1	011	100	111	011	111	Bits 2, 3, and 8 are now all 1's

Finally, the more recent PDP-11's provide the exclusive OR function. This function can be used to find out when two bits are different, since this will set them to 1, whereas two bits the same will set the result to 0. The XOR instruction is always between a *register* and a destination, so you must occasionally move some number into a register before performing the XOR operation. The destination, of course, can be any word addressed by any of the addressing modes. We could perform the exclusive OR between register R4 and the word TEMP somewhere in memory by the instruction

```
XOR R4, TEMP
      ⋮
TEMP:  124631
```

If R4 contains 005301	0	000	101	011	000	001
and TEMP 124631,	1	010	100	110	011	001
the result is	1	010	001	101	011	000 or 121530

Condition Code Operations

There is a set of operations to set and clear the condition codes (cc's). They are used principally in conjunction with the rotate instructions that follow but occasionally in synthesizing results from unusual numerical representations. They are

CLC	000241	Clear C	SEC	000261	Set C
CLV	000242	Clear V	SEV	000262	Set V
CLZ	000244	Clear Z	SEZ	000264	Set Z
CLN	000250	Clear N	SEN	000270	Set N
CCC	000257	Clear all cc's	SCC	000277	Set all cc's
NOP	000240	No operation			

This instruction group has the op code 000240, and the bits are cleared if bit 4 is 0 and set if bit 4 is 1. The bits operated on have the same relative positions as in the processor status word (Figure 8.1). For example, we could set N and V by the code 000271.

Shift Instructions

The PDP-11 shift instructions are of two kinds: arithmetic shifts and rotates. A *rotate* shifts bits through the operand word or register to the right or left in an end-around fashion, including the C-bit (Figure 8.2).

The two rotate instructions are ROR and ROL. They operate either on words or on bytes:

ROR	b060ar	Rotate Right	RORB
ROL	b061ar	Rotate Left	ROLB

One shift to the left or right occurs with each instruction and *includes the C-bit*. The state of the C-bit must therefore be carefully controlled in rotate instructions. It can be used as an additional flag for remembering information, but it cannot be disregarded.

Rotates are of use principally when we wish to move bits near the high end of a word into the low end efficiently, since it takes fewer instructions

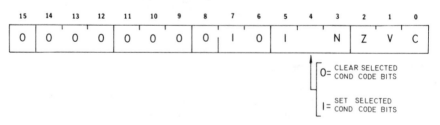

Figure 8.1 Bit assignments of condition code instructions.

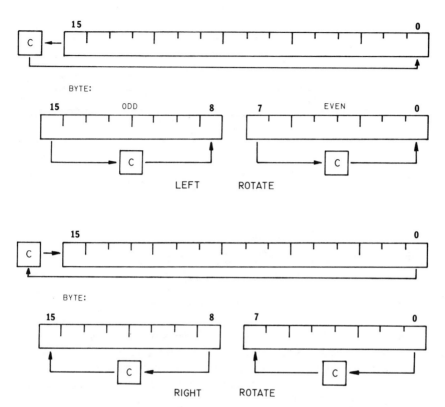

Figure 8.2 Illustration of left and right rotate instructions for word and byte addressing.

to rotate around to the left than to shift to the right. Rotates can also be used whenever the final word is to be examined bit by bit but left unchanged after the shifting is done. Byte rotates move the C-bit along with a byte.

The *arithmetic* shift instructions (Figure 8.3) have a somewhat different function: they are used principally to multiply and divide numbers by powers of 2. Since each bit in the binary computer word represents a power of 2, shifting to the left effects a multiplication by 2 and shifting to the right a division by 2. For example, if we shift 5 to the left,

$$101$$
we get 12_8 1   010

or 2×5. Similarly, if we shift 240 to the right,

$$010 \quad 100 \quad 000$$
we get 120   001   010   000   or $240/2 = 120$

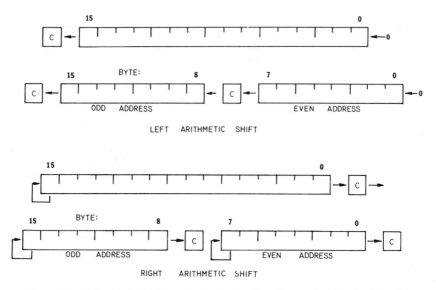

Figure 8.3 Left and right arithmetic shift instructions for word and byte addressing.

We must be careful about right shifts, however, because if the right shift were to include a signed number, the sign bit would be shifted to the right, leaving a 0 behind, so that the number would no longer be signed. For this reason right arithmetic shifts are defined to propagate the sign bit from the left. In other words, as we shift to the right, the sign bit remains in the leftmost position but is copied to the right with each shift. Bits running off the right end are lost in a right arithmetic shift, and bits falling off the left end are lost in a left arithmetic shift. The last bit shifted to the right or left in an arithmetic shift is left in the C-bit.

Byte arithmetic shifts are also possible. In right byte arithmetic shifts, bit 7 is propagated. The instructions are as follows:

ASR   b062ar   Arithmetic shift right   ASRB
ASL   b063ar   Arithmetic shift left   ASLB

Suppose that we wished to divide the number 171 by 2. We could do this by

MOV #171, R1
ASR   R1

This would accomplish the following:

171   001 111 001
ASR   000 111 100   (1 lost).   The result is 74_8.

Now let us suppose that we wanted to divide − 171 by 4. This is 177607. If we performed our shifts without the sign bit propagation, let us see what would happen. We would start with the two's complement of 171:

177607   1   111   111   110   000   111

and, after performing an ASR, if the sign did not propagate we would have

77703   0   111   111   111   000   011   (1 lost)

and after a second ASR we would have

37741   0   011   111   111   100   001   (1 lost)

The result is 37741, a *positive* number. To divide by 4 correctly, we must maintain the sign by propagation as shown here:

1	111	111	110	000	111	177607 or − 171
		ASR				
1	111	111	111	000	011	177703 or − 75
		ASR				
1	111	111	111	100	001	177741 or − 37

Note that the ASR and ASL instructions can shift a word addressed in any addressing mode.

The Arithmetic Shift and Arithmetic Shift Combined

Significant numbers of shifts in the PDP-11 can be very time-consuming since they must be written either in a line one after another or in a loop with a counter. Either method is quite slow and inefficient. For this reason, most recent PDP-11's have a single and double-precision multiple-shift instruction. These shifts are called ASH and ASHC (for arithmetic shift and arithmetic shift combined) and, like the multiply–divide instructions, have bit assignments of the forms 072rar and 073rar but syntax of the form

ASH   source, reg

where the register is shifted the number of places specified in the lowest 6 bits of the source word. Although only a register can be shifted, the number of shifts can be contained in a word addressed in any of the eight modes. The least significant 6 bits of the source operand are treated as a two's complement number, and the resulting shift is to the right if this number is negative and to the left if it is positive. The most common usage of the shift instruction is

```
ASH #5, R1      ;SHIFT R1 5 PLACES LEFT
ASH #-3, R4     ;SHIFT R4 3 PLACES RIGHT- PROPAG. SIGN BIT
ASH VSHIFT, R0  ;SHIFT R0 THE # OF PLACES IN LOCN "VSHIFT"
```

Bear in mind that these are still arithmetic shifts. The right arithmetic shift causes the sign bit to be propagated to the right, dropping bits off the right end, and the left arithmetic shift causes bits to be dropped off the left end, filling from the right with 0's.

The instruction ASHC, or arithmetic shift combined, shifts two registers to the right or left. The C-bit is not included. As before, if the source is positive, the shifts are to the left; if the source is negative, the shifts are to the right. The two registers shifted are R and R OR 1. This means that R0 will be shifted with R1, R2 with R3, and R4 with R5. Thus you always specify the lower of the two registers, and the upper one is automatically included. However, should you specify an odd-numbered register, then R OR 1 is the same register, and the arithmetic shift becomes a rotate. This can be slightly confusing and should be avoided unless you know exactly what you wish to accomplish.

In summary, the two multiple-shift instructions are

ASH   072rar Shift "r" by amount in word addressed by low 6 bits of "ar"—left if positive, right if negative.

ASHC 073rar Shift "r + r OR 1" by amount in word addressed by low 6 bits of "ar"—left if positive, right if negative.

Multiply and Divide Instructions

The multiply and divide instructions all have the bit configuration OPR reg, source, but are written as "OPR source, register" in assembly language. The multiply and divide instructions are

MUL 070rar Multiply R × source → R + R OR 1

DIV 071rar Divide R × R OR 1/source → / + R OR 1

The multiply instruction multiplies the contents of the register R by the contents of the source operand, addressed in any of the eight addressing modes, and puts the product in the register R OR 1 and, if it is large enough, the high-order bits in register R. Multiplying two 16-bit signed numbers together can result in a number that needs 32 bits to be represented. In this case the low-order bits are in R OR 1 and the high-order bits in R. If the register r specified in the instruction is odd, then only the low-order bits are stored; any overflow to the higher word is lost. This is reflected in the C-bit, which is 1 if two words are needed. The assembler syntax for the multiply instruction is

MUL   source,   register

in the opposite order from the way in which the bits are stored. For example, if we wished to multiply the contents of R4 by 12, we would write

MUL   #12,   R4

and we would find the result in R4 and R5, with the low-order bits in R5 and the high-order bits in R4. The previous contents of R5 are destroyed.

Unlike some computers, the PDP-11 allows the multiplication of signed numbers without special manipulations. Multiplication of this sort is essential to rapid processing of integer laboratory data.

The divide instruction operates much as would be expected. The contents of R and R OR 1 are divided by the number in the source operand. The quotient is left in R, and the remainder in R OR 1. Thus the quotient must be representable in 16 bits. If it is not, the V-bit is set and the instruction is aborted. If a division by 0 is attempted, the C-bit is set and the instruction aborted. Again the assembler syntax is reversed; the instruction reads: DIV source, register. Thus, if we wished to divide 74 by 14, we would do the following:

MOV #74., R5   ;PUT # IN LOW REGISTER
CLR R4   ;BE SURE HIGH REGISTER IS ZERO
DIV #14., R4

Note that the contents of the upper register (R4) must be 0, since the lower register contains the only useful information (R5). After the instruction is executed, the quotient will be in R4 and the remainder in R5. In this case this will be:

R4(5)
R5(4)

since $74/14 = 5$ with a remainder of 4. Note that this is integer division and that $4/4 = 1/2 = 0$.

Now let us consider the calculation of $(a \times b)/c$, where a, b, and c are stored in memory. We must put one number in a register, multiply, and then divide. We will then store the results back in two memory locations reserved for the quotient and remainder.

```
;CALCULATE (A*B)/C

          MOV  A, R0        ;PUT FIRST NUM IN R0
          MUL  B, R0        ;MULT A TIMES B- RESULT IN R0-R1
          DIV  C, R0        ;DIVD R0-R1 BY C
          MOV  R0, QUOT     ;QUOTIENT STORED FROM R0
          MOV  R1, REMNDR   ;REMAINDER STORED FROM R1

A:        xxxx              ;A STORED HERE
B:        xxxx              ;B STORED HERE
C:        xxxx              ;C STORED HERE
QUOT:     0                 ;QUOTIENT WILL BE PUT HERE
REMNDR:   0                 ;AND REMAINDER PUT HERE
```

The SWAB Instruction

The SWAB instruction has the form

0003ar   SWAB dst   Swap the bytes in the word addressed by dst

It is generally used in character manipulation by byte but is sometimes useful in other numeric manipulations. Note that it is roughly equivalent to eight rotate instructions, although it does not involve the C-bit as they do.

Exercises

8.1   Write a program to clear bits 5–7 and set bits 14, 12, and 1.

8.2   Write a program to perform the exclusive OR between memory locations labeled C and D. The result should be stored in E.

8.3   Write a program to solve $y = mx + b$ for m, x, and b stored in memory. Assume that y will be only single precision.

8.4   The exclusive OR can be simulated by subtracting the AND from the inclusive OR. Write a program to calculate the exclusive OR of memory locations C and D without using the XOR instruction.

8.5   What do the instructions
      ASR   R2
      ASL   R2
      ensure about register R2? How else could you write this?

8.6   Under what conditions would the instructions
      ASL   R3
      ASR   R3
      produce a different value in R3 than was present before they were executed?

□□□□□□□□ CHAPTER NINE □

INPUT–OUTPUT AND SUBROUTINES

The most common input/output device for the PDP-11 is the terminal. Today, most printing terminals print at 30 to 120 characters per second, and some video terminals are eight times faster than that. We will use the term "terminal" to refer to any keyboard-printer or keyboard-display device that is used as the primary communication with the minicomputer.

Most terminals have a rather slow output rate. Even 9600 baud (about 1200 characters per second) is much slower than the instruction cycle time of the lowliest microcomputer. Obviously, if we were to start sending characters at speeds of 1–5 microseconds each, as the central processor is able to do, we would quickly outstrip the terminal's printing or display capabilities. Therefore, it is necessary to slow down the computer to the speed of the terminal by checking a flag or *done bit* associated with each such device. We define the done bit as 0 if the device is still printing or reading and 1 at all other times. It thus indicates that the device is ready to transfer information.

Many computers employ a special class of input/output (I/O) instructions to communicate with peripheral devices of all sorts: disks, diskettes, tape reader-punch, keyboard, printer, and laboratory peripherals. The PDP-11 computer employs, instead, a special set of addresses called *device addresses* or *bus addresses*, which contain the status bits regarding the device and the buffer into which the characters are to be printed, punched or moved. These two device addresses are called the device *Status Register* and *Data Buffer* register. They are addressed just as if they were actual memory locations, and most of the instructions that operate on memory locations can be used to operate on the device addresses. A table of these addresses is given in Appendix V.

Status Registers

The most common configuration for a device Status Register includes the following standard meanings for 4 bits. Not all are utilized in any specific device, and many more bits are used in more complex devices.

Bit 15	Error	Some data transfer error has occurred.
Bit 7	Done	The device is ready to transfer data.
Bit 6	Interrupt Enable	The device can cause an interrupt when done if set.
Bit 0	Enable	Setting this bit "turns on" some devices that are not normally enabled.

Data Buffer Registers

These registers are simply addresses to which we move the data to be printed, punched, or stored, or locations from which we must read the data that have been sent by the external device.

ASCII Code

Most slow data rate devices deal in the transfer of 8-bit character representations through the use of the American Standard Code for Information Interchange (ASCII), colloquially pronounced "askee." This code uses 7 information bits and 1 parity bit to represent all printing characters as well as some control characters. The PDP-11, however, makes no use of this parity feature in any of its hardware or software, and therefore, we will treat ASCII codes as always having the leftmost (Parity) bit (bit 7 of 0–7) equal to zero. This is termed "Zero Parity." Systems in which bit 7 is always on are termed "Mark Parity." Those in which Parity is on to make the total number of bits on an even number are termed "Even Parity" and the converse "Odd Parity."

The alphabet is represented in ASCII by adding 100_8 to the number of the character:

A	101
B	102
C	103
:	:
Y	131
Z	132

These are, of course, *octal* numbers, so the codes run from 101 to 132_8. The numbers are biased by 60_8.

0	60
1	61
2	62
:	:
7	67
8	70
9	71

Similarly, the punctuation characters take on the following ASCII values:

Space	40
!	41
"	42
#	43
etc.	

Two characters that might not intuitively seem as if they need to be represented are

Carriage Return	15
Line Feed	12

These are the most important in printing messages, since we never know the position of the carriage when we start a program, and they allow us to *initialize* the carriage to a new, clean, line by printing a Return and Line Feed before starting the message.

Reading and Printing Characters

The status words of the terminal reader and printer can be interrogated to see whether a character is ready to be read. Note that the done bit of the status word is always bit 7 and that it can easily be examined by the TSTB instruction.

If bit 7 is a 1, the TSTB instruction will set the N-bit of the condition codes. The terminal Status Register, as for a DECwriter or video terminal is shown in Figure 9.1.

To read a character, we simply keep checking the N-bit with the TSTB

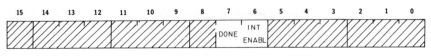

Figure 9.1 DECwriter Status word, TKS.

instruction, until it is set

```
TWAIT:   TSTB TKS       ;WAS A CHARACTER STRUCK?
         BPL TWAIT      ;NO, KEEP CHECKING
         MOV TKB, R0    ;READ CHARACTER INTO R0
```

When the done bit becomes 1, the TSTB instruction will set the N-bit, and the byte will no longer be positive. Then the BPL instruction will no longer be satisfied. To read the character, we simply move it from the data buffer to someplace else, in this case, R0. Any instruction that accesses the data buffer after the done bit is set clears done so it is best to copy the character somewhere else before testing or comparing it with anything else. The next character struck will replace the current contents of the data buffer, regardless of whether or not it has been read. In the case of the Teletype interface, no character can be read until the enable bit is set with an INC TKS instruction.

To print a character on the terminal, you simply move the ASCII character code from some word or register to the data buffer as soon as the done bit indicates that the printer is ready (not printing). This is most of the time unless a character has recently been printed. To print an "A" on the terminal, we test the done bit and then move the character as follows:

```
PWAIT:   TSTB TPS       ;IS THE PRINTER READY?
         BPL PWAIT      ;NO, KEEP CHECKING
         MOV #101, TKB  ;PRINT AN "A"
```

To write a simple program fragment to make the keyboard and printer act like a typewriter, we simply combine the preceding fragments:

```
START:   TSTB TKS       ;IS A CHARACTER READY?
         BPL START      ;NOT YET

PWAIT:   TSTB TPS       ;IS THE PRINTER READY?
         BPL PWAIT      ;NOT YET
         MOV TKB, TPB   ;YES, MOVE CHARACTER FROM READER TO PRINTER
         BR START       ;AND GET THE NEXT CHARACTER
```

Terminal Characteristics

The terminal consists of two separate sections: (1) the keyboard and (2) the printer or display. These are mechanically linked only when the terminal is in the "local" mode. However, in the "line" mode, typing a character at the keyboard will have *no effect* on the printer or screen unless a computer program is running to accept a character and send it to the printer. If the computer is not running such a program, striking the keyboard will have no particular effect.

The DECwriter consists simply of a keyboard and a printer. They have no connection at all, and a character struck on the keyboard will not be printed on the printer unless a program is running that tells the printer to do so. If the computer is not running such a program, striking the keyboard will not affect the printer in any way, unless the particular model has a Local switch, converting it into a typewriter.

Assembler Conventions

The assembler does not know any symbols for the device addresses of the teleprinter reader and punch. They must be defined to the assembler, using the equals sign. You may therefore name them by anything that you find convenient. The device addresses are as follows:

TKS	Reader status	177560	The same addresses are used
TKB	Reader buffer	177562	for either the Teletype or the
TPS	Printer status	177564	DECwriter.
TPB	Printer buffer	177566	

They can be defined to the assembler by

```
TKS = 177560
TKB = TKS + 2
TPS = TKB + 2
TPB = TPS + 2
```

The assembler has facilities for the conversion of ASCII characters into their octal equivalents. If we wish to print a single character, say B, on the printer, we could write

```
PWAIT:    TSTB TPS        ;READY?
          BPL PWAIT       ;NO
          MOV #'B, TPB    ;YES, PRINT B
```

The single quotation mark (or apostrophe) is used to convert a single character into its octal equivalent in a single 16-bit word. To manage the movement of this number into the proper register, we use the immediate mode, and the octal number for B, 102, is placed in the second word of the instruction.

Now let us write a program to print the characters H2O on the tele-printer. Such a program amounts to testing then moving, testing then moving, and testing then moving the characters H, then 2, and then O. We could write this program as follows:

```
PWAIT1:  TSTB TPS           ;PRINTER READY?
         BPL PWAIT1         ;NO
         MOV #'H, TPB       ;YES, PRINT H
PWAIT2:  TSTB TPS           ;PRINTER READY AGAIN?
         BPL PWAIT2         ;NO
         MOV #'2, TPB       ;YES, PRINT A 2
PWAIT3:  TSTB TPS           ;READY?
         BPL PWAIT3         ;NO
         MOV #'O, TPB       ;YES, PRINT O
```

This program gets the job done, but it is clearly quite inefficient. It would be terribly inconvenient if every time we needed to print out a message we had to write a separate print routine for each character. Obviously there must be a more efficient way, and it is called the *subroutine*.

Subroutines and the Stack

Intuitively the subroutine must be a section of code that can be called from many places in memory efficiently without tremendous overhead, so that when the code is completed it will be possible to return to the place from which the subroutine was called. The JSR (jump to subroutine) instruction provides a very efficient way of remembering the address from which a subroutine was called, so that a simple RTS (return from subroutine) instruction will allow the program to go back to the point in the main program where it left off.

The JSR instruction operates through the facility known as the *stack*. A stack is simply a set of memory locations that can be used for temporary storage of data such as return addresses, saved register contents, and other temporary constants or variables. The concept of the stack sounds arcane because of its terminology, but it is really quite simple. Let us assume that we choose some register to point to the items in the stack as we add and delete them. If we start with an empty stack at a high address and add items at lower addresses, we are said to be "pushing" items onto the stack. Later, when we wish to retrieve items from the stack or just delete them, we are said to be "popping" items from the stack. These pushing and popping operations are easily carried out with the benefit of the indirect autoincrement mode and the indirect predecrement mode.

Suppose that R6 is the register containing the address of the current item on the stack. If we wish to add an item to the stack, such as the number 547, we can do this by "pushing" it onto the stack, using the indirect predecrement mode:

MOV #547, −(R6)   ;PUSH 547 ONTO THE STACK

Later we might wish to retrieve the number that we stored here temporarily and put it into R0. We can do this by "popping" it off the stack, using the indirect autoincrement mode:

MOV (R6)+, R0   ;POP STACK ELEMENT OFF AND INTO R0

Now we usually do use the general register R6 for the Stack Pointer, but we generally refer to it, not as R6, but as SP to remind us that it is the Stack Pointer. Of course, since all register symbols must be defined in every program, we can call R6 anything we find convenient and easy to remember for that program. Furthermore, if we are going to use R6 or any other register as the Stack Pointer, we must *initialize* it to some value. Most commonly we start programs at address 1000 and allow the region from 400 to 776 to be the stack. If we believe that we may need a larger stack, we must start at a higher address, because if SP becomes less than 400, the processor takes special error action that will, in general, stop the program. In this error condition, the processor automatically loads a new PC (Program Counter) from location 4 and a new PS (Processor Status) from location 6. What these do is entirely up to the programmer or RT-11, but if they have not been allowed for, a crash will probably result.

The use of R6, rather than some other register, as the Stack Pointer is a conscious decision on the part of the programmer, because the hardware also uses the stack in some special cases. One of these is the JSR instruction.

The JSR Instruction

The JSR instruction is of the format

JSR reg, dest   JSR   004rar   Jump to subroutine at "ar"

where "reg" is any of the general registers, although not usually SP, and "dest" is the address of the subroutine, addressed in any of the addressing modes. When the JSR Rn, dest instruction is given, the hardware performs the following instructions automatically:

```
;HARDWARE PERFORMS THESE INSTRUCTIONS FOR A
;JSR Rn, dest
MOV Rn, −(SP)      ;PUSH Rn CONTENTS ONTO STACK
MOV PC, Rn         ;COPY PC INTO Rn
MOV dest, PC       ;JUMP TO DEST TO BEGIN EXECUTING
                   ;INSTRUCTIONS
```

In other words, what happens is that the contents of the register through which the subroutine is called are pushed onto the stack, the current PC is copied into that register, and the program jumps to the beginning of the subroutine.

After the subroutine is completed, the program can return to the main code from the subroutine by performing an RTS Rn instruction, where the register Rn is the same one through which the subroutine was called. The RTS Rn instruction does the following:

```
;HARDWARE PERFORMS THESE INSTRUCTIONS FOR AN
;RTS Rn
MOV Rn, PC         ;SET PC TO NEXT INSTRUCTION IN MAIN
                   ;PROGRAM
MOV (SP)+, Rn      ;PUT OLD CONTENTS BACK INTO Rn
```

The RTS instruction thus takes the address of the next instruction in the main program, puts it in the PC, and then puts the old contents of the register Rn back from the stack. This allows the subroutine to return to the main program.

Let us take a simple example of how this works by rewriting our short program fragment to print out the message "H2O." This program will print out the message using subroutines, and the register R0 will be used to contain the character that is to be typed each time the type subroutine is called. The program is

```
.=1000   ;SYSTEM WILL LOAD THIS PROGRAM AT 1000

1000    START:  MOV #., SP      ;INITIALIZE THE STACK POINTER
1004            MOV #'H, R0      ;PUT "H" IN R0
1010            JSR R5, TYPE     ;GO PRINT IT
1014            MOV #'2, R0      ;PUT "2" IN R0
1020            JSR R5, TYPE     ;PRINT THAT
1024            MOV #'O, R0      ;PUT "O" IN R0
1030            JSR R5, TYPE     ;AND PRINT THAT
1034            HALT             ;AND HALT

;TYPE SUBROUTINE BEGINS HERE
1036    TYPE:   TSTB TPS         ;IS THE PRINTER READY?
1042            BPL TYPE         ;NO, KEEP CHECKING
1044            MOV R0, TPB      ;YES, PRINT CHARACTER IN R0
1050            RTS R5           ;AND RETURN TO MAIN PROGRAM

.END START
```

When this program is executed, the following things happen: first, the Stack Pointer is initialized to 1000. Thus, when the first item is "pushed" onto the stack, it will be put into address 776. Then the octal code for the character H (110) is put in R0, and the instruction JSR R5, TYPE is executed at 1010. When both operands have been fetched and interpreted, the PC will point to address 1014, the *next* instruction after the JSR instruction. Then the hardware performs the following:

MOV R5, −(SP)   so that whatever was in R5 is saved in address 776,

MOV PC, R5   the current PC (= 1014) is put into R5, and

MOV 1036, PC   the address TYPE is put into the PC, starting the subroutine at 1036.

The subroutine proceeds as we have discussed earlier. The done flag of the printer is tested, and when the printer is no longer engaged in printing a character, it will set the Done flag to 1, causing the instruction MOV R0, TPB at 1044 to be executed. Then the RTS R5 instruction is encountered.

The RTS instruction performs the following operations.

MOV R5, PC   The contents of R5 (which are still 1014) are put in the PC, so that the next instruction to be executed will be the one after the address of the JSR instruction that called the subroutine.

MOV (SP)+, R5   The old contents of R5 are restored from the stack and the Stack Pointer is incremented back to 1000, thus "popping" the value off the stack.

The program then continues at 1014. In this case the instruction MOV #'2, R0 is executed, thus putting the ASCII code for 2 (62) into R0. Then the second JSR instruction is encountered. As before, the contents of R5 are saved on the stack, the address of the instruction following the JSR (1024) is put into R5, and the subroutine is begun at 1036. When the character 2 has been printed, the RTS R5 instruction puts R5 back into the PC, sets the PC to 1024, and puts the old contents of R5 back again, whatever they may be. The program proceeds again in line from the instruction after the JSR, moving the ASCII character O (117) into R0 and then calling the TYPE routine one more time. Again the contents of R5 are saved on the stack, the address of the instruction following the JSR (1034) is put into R5, and the TYPE routine at 1036 is executed. Again the address in R5 is put into the PC and R5 is restored. The last instruction executed, then, is that at 1034, causing the computer to halt.

In summary, we have seen that

1. The JSR instruction automatically uses the stack. The SP must therefore be initialized to a known value before a JSR can be used.
2. The JSR reg, dest instruction saves the contents of the register on the stack, puts the address following the JSR in the register, and begins executing instructions at the address "dest."
3. Exit from a subroutine is through a RTS Rn instruction, which puts the contents of the register into the PC and restores the register's old contents from the stack, thus allowing return to whatever the subroutine was called.

A Simpler JSR Form

The JSR Rn, dest instruction has a very general set of uses. The availability of a register that contains the address from which the subroutine is called can be very useful, as we will see later. However, it does prevent passing information directly in that register, and it also prevents use of the register for temporary storage during the subroutine. For this reason the register used in the JSR instruction can often be the PC, thus allowing free use of at least six registers. Let us suppose that we call a subroutine TYPE through the PC instead of one of the other registers. This will be in the following form:

```
        JSR PC, TYPE      ;CALL TYPE
          :
TYPE:   TSTB TPS          ;TYPE STARTS HERE
          :
        RTS PC            ;EXIT HERE
```

If we perform a JSR PC, TYPE instruction, this is what happens:

General Case	PC Case	
MOV Rn, −(SP)	MOV PC, −(SP)	;PC IS PUSHED ONTO ;THE STACK
MOV PC, Rn	MOV PC, PC	;NOTHING HAPPENS ;HERE
MOV dest, PC	MOV dest, PC	;DEST IS LOADED INTO ;THE PC

Then, when we are done with the subroutine, we perform the exit through the RTS PC instruction. What happens in this case is as follows:

General Case	PC Case	
MOV Rn, PC	MOV PC, PC	;NOTHING HAPPENS IN ;THIS CASE
MOV (SP)+, Rn	MOV (SP)+, PC	;RETURN ADDRESS ;GOES INTO PC HERE.

In the special case of the JSR PC, dest instruction, it turns out that the address following the JSR is stored on the *stack* rather than in a register, and that upon return the address is "popped" off the stack back into the PC for the subroutine to get back to the calling program. In both cases, a MOV PC, PC instruction is performed which has no net effect. Thus the JSR, PC, dest method allows free calling of subroutines and also permits free use of R0–R5.

The TYPE, READ, and CRLF Routines

The routines for reading from the keyboard and printing on the keyboard are so standard that we will give them here as JSR PC, routines. We will then refer to them throughout the rest of the text without specifically writing them out. After you have studied interrupts (Chapter 10) and RT-11 character handling (Chapter 12), you can substitute any method of character handling you wish in any program by simply rewriting these routines:

```
;SYMBOL DEFINITIONS FOR CHARACTER HANDLING ROUTINES
CR=15    ;CARRIAGE RETURN
LF=12    ;LINE FEED

;KEYBOARD READ ROUTINE

READ:   TSTB TKS        ;HAS KEYBOARD BEEN STRUCK?
        BPL READ        ;NOT YET, KEEP LOOKING
        MOV TKB, R0     ;YES, READ CHARACTER INTO R0
        BIC #-200, R0   ;CLEAR OUT PARITY BIT
        RTS PC          ;AND RETURN FROM SUBROUTINE

TYPE:   TSTB TPS        ;IS PRINTER READY?
        BPL TYPE        ;NOT YET
        MOV R0, TPB     ;YES, PRINT CHARACTER IN R0
        RTS PC          ;AND EXIT

CRLF:   MOV #CR, R0     ;PUT CR CHARACTER IN R0
        JSR PC, TYPE    ;AND TYPE IT
        MOV #LF,R0      ;THEN PUT LF IN R0
        JSR PC, TYPE    ;AND TYPE THAT
        RTS PC          ;THEN EXIT
```

Advanced Subroutine Concepts

Since the subroutine call always saves a return address on the stack or in a register, it is possible to write subroutines that call other subroutines to any level of complexity. This is illustrated most simply by the preceding CRLF routine. It is also possible to write subroutines that *call themselves* to any number of levels as long as there is sufficient stack space. This somewhat mind-boggling concept is called *recursion* and is of limited use in scientific programming, but of great value in the construction of assemblers and compilers.

It is sometimes necessary to utilize several registers when your subroutine is running, even though these registers may have been in use in the main program. When you believe that your subroutine may be using these same registers, you should save them on the stack, using entry calls and exit procedures such as those shown below.

```
SUBR:   MOV R0, -(SP)    ;SAVE R0-R5 AT START OF ROUTINE
        MOV R1, -(SP)
        MOV R2, -(SP)
        MOV R3, -(SP)
        MOV R4, -(SP)
        MOV R5, -(SP)
;main body of the subroutine goes here
;in which R0-R5 may be used for anything

;NOW BEGIN RESTORING THE REGISTERS
SBEXIT: MOV (SP)+, R5    ;NOTE THAT YOU MUST DO THIS
        MOV (SP)+, R4    ;IN THE OPPOSITE ORDER
        MOV (SP)+, R3
        MOV (SP)+, R2
        MOV (SP)+, R1
        MOV (SP)+, R0
        RTS Rn           ;AND EXIT THROUGH THE CALLING REGISTER
```

A more general case would be one in which we write a *subroutine* to save and restore the registers, since this register saving might go on in a number of places. This is tricky, however, since the calling of a subroutine itself pushes data into the stack. We can get around this by calling our register save routine through R0, which saves R0 on the stack, and then exiting through R0 with a JMP instruction:

```
;SUBROUTINE TO SAVE THE REGISTERS ON THE STACK
;CALLED WITH A JSR R0, REGSAV, PUSHING R0 ONTO THE STACK

REGSAV: MOV R1, -(SP)    ;SAVE R1-R5
        MOV R2, -(SP)
        MOV R3, -(SP)
        MOV R4, -(SP)
        MOV R5, -(SP)
        JMP @ R0         ;AND EXIT THROUGH R0
```

The restore routine is also called through R0, for symmetry, although it could be called through any register. We discard the pushed value of R0 from the stack at the outset and return through R0 to restore the old R0.

```
;REGISTER RESTORE ROUTINE
;CALLED BY JSR R0, REGRES

REGRES: TST (SP)+        ;POP OFF OLD VALUE OF R0
        MOV (SP)+, R5    ;RESTORE R5-R1
        MOV (SP)+, R4
        MOV (SP)+, R3
        MOV (SP+)+, R2
        MOV (SP)+, R1
        RTS R0           ;AND RESTORE R0 AND RETURN
```

Calling a Subroutine with Arguments

It is also possible to communicate arguments to the subroutine in ways other than through the general registers. For example, consider the task of printing text stored in successive bytes until a zero byte is found. This sort of subroutine might well be called with the address of the text in the word following the JSR, so that the text itself could be located elsewhere in memory.

Let us assume that we wish to print a message using the call

```
1200  JSR R5, UNPACK            ;CALL UNPACKING
                                ;SUBROUTINE
1204  MESG1                     ;ADDRESS OF MESSAGE1
      :
      :
4562  MESG1: .ASCIZ /HELLO/     ;MESSAGE PLACED HERE
                                ;WITH 0 BYTE ENDING
```

Now we can easily obtain the address and indeed the contents of the message bytes for the unpacking subroutine to use, because register R5 contains the address of the word following the call when the UNPACK subroutine is called. Such an unpacking routine would look like this:

```
UNPACK: MOV (R5)+, R1    ;GET ADDRESS OF MESSAGE INTO R1 AND INCREMENT
U1:     MOVB (R1)+, R0   ;GET EACH BYTE OF TEXT INTO R0
        BEQ UEXIT        ;QUIT IF THIS IS THE 0 BYTE
        JSR PC, TYPE     ;ELSE PRINT CHARACTER IN R0
        BR U1            ;GO BACK TILL 0 BYTE FOUND

UEXIT:  RTS R5           ;IF 0 BYTE, EXIT FROM SUBROUTINE

TYPE:   TSTB TPS         ;CHECK FOR PRINTER READY
        BPL TYPE         ;GO BACK IF NOT READY
        MOV R0, TPB      ;PRINT CHARACTER
        RTS PC           ;AND EXIT
```

In this UNPACK routine, register R5 starts out containing the address of the word after the JSR call, in this case 1204. Address 1204 contains 4562, the address of the message MESG1. It is referenced right away by the instruction MOV (R5)+, R1, which puts the address 4562 into R1. Register R5 is also incremented to point to the real return address, 1206. Now R1 contains the address of the message bytes, and they are accessed and printed one by one in the loop at U1. If the next byte is a 0, printing ceases, and control is transferred to UEXIT, where the RTS R5 instruction returns the program to address 1206, the second word after the JSR.

In this way, any number of arguments can be transferred into the subroutine from their positions in a list following the JSR instruction. It is up to the programmer, however, to make use of the correct number of arguments so that the register points to the true return address when the RTS is executed.

Exercises

9.1   Write a program to type out "THIS PROGRAM WORKS!" on the terminal. Programs generating each character separately in the immediate mode are not acceptable.

9.2   Write a program to type out a Christmas tree. Use the following design or one of your own.

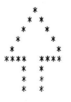

9.3   Write a program to accept characters from the teleprinter until a dollar sign is struck ($). Then the entire list should be printed out, after which a new list can be entered.

9.4   Modify Exercise 9.3 so that each Return automatically generates a Line Feed and so that striking an ampersand (&) causes the computer to halt.

9.5   What will be displayed in R0 when the following program finishes?

```
              .=1000
    START:    MOV #.,  SP
              CLR R0
              MOV START,  R0
              JSR PC,  DUMMY
              ADD  -(SP),  R0
              HALT
    DUMMY:    RTS  PC
```

The Switch Register

The console Switch Register is relatively unique in that it has no Status Register or done bit. The values of the switches can be read at any time by simply reading the contents of device address 177570. This could be done by

```
SR = 177570      ;DEFINE SYMBOL SR
:
        :
:
MOV SR,  R3     ;READ SWITCH REGISTER INTO R3
```

Printing Out Octal Numbers

Since the computer has the ability to print only one character at a time on the teleprinter, it should be obvious that a program of some complexity is required to print out an entire number. Since it is easiest to print out numbers in octal, we will start with this case. If we consider the way we read an octal number from the binary digits, we immediately see how a program can be written to print out the number. First we divide the binary number conceptually into groups of 3 bits from the right side. In a 16-bit number, this will produce five 3-bit groups and 1 bit left over on the left. Therefore, in reading the number out loud, we read out that leftmost bit first and then the groups of 3 bits, starting with bits 12–14. Thus, if we have the binary number

    1   011   101   100   000   101

we will want to print out a 1, then a 3, then a 5, and so forth.

To do this we will write a program to print out the single leftmost bit as a single digit and then print out five numbers from the five groups of 3 bits. We can do this most efficiently as follows: Use the rotate instruction to place successive bits into the lowest order bits of the register, clear out all others with the BIC instruction, and then print each digit by adding the ASCII bias of 60_8 and moving the resulting ASCII code into the printer buffer. We will write this printout routine, a subroutine called by JSR PC, OCTOUT, where the number to be printed is in R1 on entry to the subroutine. The flowchart is shown in Figure 9.2, and the subroutine follows.

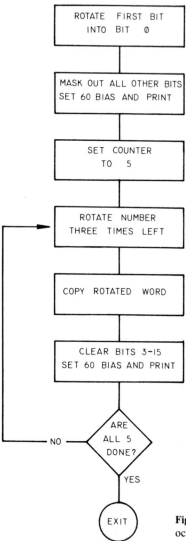

Figure 9.2 Flowchart of program to print out octal numbers.

```
;SUBROUTINE FOR PRINTING OUT AN OCTAL NUMBER
;ENTER WITH NUMBER TO BE PRINTED IN R1
;CALL BY JSR PC, OCTOUT

OCTOUT: CLR RO            ;CLEAR REGISTER WE WILL SHIFT INTO
        ASHC #1, RO       ;SHIFT BIT 15 INTO RO
        JSR PC, OPRINT    ;ADD 60 AND PRINT
        MOV #5, R2        ;SET COUNT FOR 5 MORE DIGITS

OLOOP:  ASHC #3, RO       ;SHIFT IN AN OCTAL DIGIT
        JSR PC, OPRINT    ;PRINT IT
        SOB R2, OLOOP     ;GO BACK UNTIL ALL 5 DONE
        RTS PC            ;THEN EXIT

;SUBROUTINE TO PRINT OCTAL DIGIT
OPRINT: BIC #177770, RO   ;CLEAR OUT ALL BUT DIGIT
        ADD #60, RO       ;ADDD ON ASCII NUMBER BIAS
        JSR PC, TYPE      ;PRINT IT
        RTS PC            ;AND EXIT
```

Decimal Printout

In a similar manner, it is also possible to print out numbers in the decimal radix by dividing by 10 each time instead of by 8. This is most easily done by using the hardware multiply and divide instructions MUL and DIV. To print out an integer as a decimal number, it is necessary to divide by the largest possible power of 10 that can be contained in a 16-bit word (10,000) and then print this quotient. The procedure is repeated for each smaller power of ten on the remainders. We will also see in Chapter 11 that this can be carried out using the floating point math package FPMP11.

Exercises

9.6 Write a program to read the Switch Register and halt displaying it in R0.

9.7 Write a program to print out *signed* octal numbers.

9.8 Write a program to read octal numbers in from the teleprinter keyboard and to store them in memory.

9.9 Write a program to print out signed decimal numbers.

9.10 Write a program to print out a number as either an octal or decimal value, depending on whether bit 15 of the Switch Register is up or down.

□□□□□□□□□□ CHAPTER TEN □

TRAPS AND INTERRUPTS

It is sometimes desirable to allow the computer to carry out a particular task, such as a long calculation or the display of data as a main program and then have various peripheral devices (e.g., the teleprinter and high-speed punch) serviced only when they are ready to transfer data. One way of accomplishing this if only one device is to be serviced is to simply poll its done bit periodically, for example, at the end of the largest loop in the program. However, this polling may not occur very often if the time to perform one pass through the loop is long; and if several devices are to be serviced, the polling may become rather complicated. For this reason, most computers have an *interrupt* facility that allows a device to interrupt the main program wherever it is running in memory whenever the device needs service.

One of the great advantages of the PDP-11 method of handling interrupts is the ease with which the programmer can cause and utilize interrupts. Many computers of earlier designs have one or only a few interrupt channels; and once an interrupt has been caused, the entire set of possible devices must be examined to find out which one caused the interrupt. In the PDP-11, however, this is never necessary, as each device can be programmed either to cause or not cause an interrupt. When the device is ready, if its interrupt capability is enabled, it automatically causes the processor to push the current PC *and* the current PS onto the stack and to take a new PC and PS from a pair of locations unique to that device. These locations are called interrupt *vector* locations and are as follows for the teleprinter:

60   Teleprinter keyboard vector address
62   New PS for keyboard interrupt

64	Printer interrupt vector address
66	New PS for printer interrupt
70	New PC for high-speed reader interrupt
72	New PS for high-speed reader interrupt
74	New PC for high-speed punch interrupt
76	New PS for high-speed punch interrupt

The vectors for all devices are listed in Appendix V.

Thus the programmer need only put the address of the device service routine in the first of the two vector locations and the new value of the PS in the second location for the interrupt service routine for that device to be defined. This interrupt will occur, and the service routine will be executed if three conditions are satisfied:

1. The done bit for the device in question must be a 1.
2. The interrupt enable bit for the device must be set.
3. The processor must be running at a priority *lower* than that of the interrupting device.

Recall that the priority of the processor is controlled by bits 5–7 of the PS register. If these bits are all 0's, devices having priorities from 1 to 7 can interrupt. If these bits are 100, devices having priorities from 5 to 7 can interrupt the processor. If these bits are all 1's, then no device can interrupt the processor, and this state is equivalent to turning the interrupt facility off.

Most PDP-11's have only five of the eight processor levels implemented; only the PDP-11/45 and higher numbers have the remaining levels available. The lower PDP-11's can distinguish levels 4, 5, 6, and 7 but treat levels 0, 1, 2, and 3 all as equivalent to level 0. This is not an undue limitation, however, because most of the common peripheral devices interrupt at levels above 3. Some of the bus request levels for devices are

Teleprinter reader	4
Teleprinter printer	4
High-speed tape reader	4
High-speed tape punch	4
Analog-to-digital converter	6
LPSKW clock	5
RF11 and RK11 disks	5
KW11 real time clock	6
LP11 line printer	4

Bus request levels are given for all devices in the *PDP-11 Peripherals Handbook,* and for many devices in Appendix V.

When the PDP-11 is started from a HALT, the INIT signal is sent through the processor, which, among other things, sets the PS to 0, thus setting the processor priority to 0 as well. When the processor is running at zero priority, any device whose interrupt is enabled can interrupt it. When such an interrupt occurs, however, we generally want to prevent further interrupts at that level until the device that has caused the interrupt has been serviced. We can do this by setting the value of the new PS equal to that interrupt level. For example, if we wished to set up an interrupt service routine for the teleprinter keyboard, we would set the new PS to 200, signifying interrupt level 4. Then neither the teleprinter keyboard nor any other device operating at level 4 could interrupt the processor further. Remember that to cause an interrupt, a device must have an interrupt level *greater* than that of the processor.

When an interrupt actually occurs, two things happen. First the current PS *and* the current PC are pushed onto the R6 stack, by the equivalent of

```
MOV PS, –(SP)
MOV PC, –(SP)
```

Then the new values of the PS and PC are loaded from the two interrupt vector locations. When the interrupt service routine is complete, we simply give the RTI (return from interrupt) instruction, which reverses this process, restoring the old PS and PC from the stack by an effective

```
MOV (SP)+, PC
MOV (SP)+, PS
```

This interrupt process should be differentiated from the subroutine call process, which pushes only the contents of one register onto the stack, whereas the interrupt puts two items onto the stack. Furthermore, return from a subroutine is always through a register, while return from an interrupt does not require reference to any register.

Latency

Clearly the processor cannot respond instantaneously to an interrupt. It can respond only at certain specified times so that no information is lost. Generally the processor responds to an interrupt at the end of the current instruction. For the PDP-11/05 and 11/10 this means a latency of 7 microseconds, maximum. In the PDP-11/35 and 11/40, the latency is 5.42

microseconds, maximum. If the 11/35 or 11/40 has floating point hardware instructions built in, the latency may increase to 25.50 microseconds if the floating point instruction is being executed and is near completion. The actual time for the stack manipulations and the transfer to the service routine is an additional 5.42 microseconds in the 11/35 and 11/40 and 7 microseconds in the 11/05 and 11/10.

Writing Programs for Interrupts

Again, we should emphasize that the value of the interrupt facility is greatest when several devices need service or when one device may need service erratically. In these cases the rules for writing an interrupt program are simple to follow:

1. Put the address of the service routine and the value of the new PS in the interrupt vector locations.
2. Set the processor PS to a priority lower than that of the device to cause the interrupt.
3. Set the device's interrupt enable bit, usually bit 6 of the Status Register.

The interrupt service routine, then, need not poll the flag of the device, since it must have been set high to have caused the interrupt. It must simply

1. Transfer the information needed to or from the device.
2. Test to see whether more information is to come through the device. For example, there may be only a few pieces of information expected, after which the interrupt is disabled. Take special action if done.
3. If more information is expected, perform an RTI instruction.

Let us take an example of programming the teleprinter reader to cause an interrupt and to print the character that has been read. We will not have the printer causing an interrupt as well, as this could be more complex.

First, we must set the values of the interrupt vector addresses. If we give the first location of the interrupt service routine a name such as RDRINT, we can simply put this label in the address and let the assembler decide where the service routine will ultimately be assembled. If we are using a paper tape system (without any magnetic tape or disks), we can simply put the values in the addresses by setting the location counter:

```
.=60
KBINT      ;ADDRESS OF KEYBOARD INTERRUPT
200        ;NEW PS DURING READER SERVICE ROUTINE
```

However, if our program is to be loaded (or may ever be loaded) by a disk operating system of some sort, we must MOV the values into these locations from within the main program above location 400, as the loaders in these operating systems prevent direct loading of any values below location 400. In this case, we simply write

```
MOV #KBINT, @#60       ;PUT SERVICE ROUTINE ADDRESS IN LOCN 60
MOV #200, @#62         ;PUT NEW PS IN 62
```

Then we must write the code to enable the interrupt, by setting bits 6 and 0 of the keyboard Status Register:

```
MOV #101, TKS          ;ENABLE KEYBOARD AND INTERRUPT
```

The interrupt service routine, then, simply obtains the character from the buffer and prints it as follows:

```
;KEYBOARD INTERRUPT SERVICE ROUTINE

KBINT:  TSTB TPS       ;IS PRINTER ALSO READY?
        BPL RDRINT     ;IT NEARLY ALWAYS WILL BE
        MOV TKB, TPB   ;YES, PRINT CHAR IN KEYBOARD BUFFER
        RTI            ;AND RETURN FROM INTERRUPT SERVICE
```

The interrupt is thus reenabled upon return from the service routine; and each time the keyboard is struck the done bit will go high, causing an interrupt and the printing of the character.

Disabling an Interrupt

It is sometimes desirable to turn on an interrupt for a specified time and to turn it off when some event occurs, as when a specified number of characters have been entered. To turn off the interrupt, we must disable the interrupt of the device in question. In some cases we may also wish to transfer control or to jump to some special routine when the interrupt is disabled. In this case we cannot perform an RTI instruction, but must jump somewhere after first removing the old PS and PC from the stack. We may either place the old PS back into the PS or set it some new value.

In the following example, we read 100 characters into a memory region that we will call a *buffer*, and then jump to an output routine. We also make use of the PDP-11 instruction WAIT, which simply causes the processor to stop executing new instructions and wait at that address for an interrupt. When an interrupt occurs, however, the processor pushes the address after the WAIT instruction onto the stack, so that, if no main program instructions at all are to be executed during the wait for interrupts, a two-instruction loop must be written:

```
W1:  WAIT      ;WAIT FOR INTERRUPT
     BR W1     ;GO BACK
```

This loop will be executed far fewer times than the equivalent pair of instructions:

```
W2:  NOP       ;NO OPERATION
     BR W2     ;GO BACK
```

because the processor does not continually execute the pair of them, but goes onto the BR W1 instruction only after the completion of each interrupt service routine. This WAIT instruction, in addition to having the pedagogic value of allowing us to write very short main programs when we are really concerned with illustrating the interrupt facility, also has the advantage of using much less of the central processor's time. This is important when we are expecting a very high rate of interrupts from one device or from a number of devices.

Here, then, our main program will just be a WAIT instruction, and our service routine a program to accept characters from the keyboard and place them in 100 successive bytes of memory. When all 100 have been filled, the service routine disables further interrupts, restores the proper PS, and instead of returning to the main program at W1 goes to a routine at BUFFIL, which takes some appropriate action.

Note that in this example the counter and pointer are addresses rather than registers. Although this is not strictly necessary for this example, it would be required if we were writing an interrupt service routine which would operate in conjunction with a main program that made use of the registers. The alternative would be that the registers to be used in the subroutine would have to be saved on the stack each time the subroutine was entered and restored before exit from the service routine. In this simple case, however, it would be foolish to waste so many instructions

```
;PROGRAM TO ACCEPT 100 CHARACTERS FROM THE KEYBOARD
;AND STORE THEM IN MEMORY

.ASECT
.=1000

START:  MOV #., SP               ;INITIALIZE THE SP
        MOV #100., COUNT         ;SET A MEMORY COUNTER TO 100
        MOV #BUFFER, POINT       ;SET PTR TO START OF BUFFER
        MOV #KBINT, @#60         ;SET KEYBD INTERRUPT VECTOR
        MOV #200, @#62           ;PRIORITY 4 IN PS
        MOV #100, TKS            ;ENABLE INTERRUPT

;HERE THE MAIN PROGRAM JUST WAITS FOR AN INTERRUPT
;A MORE COMPLEX PROGRAM WOULD PROBABLY HAVE A DISPLAY
;ROUTINE RUNNING HERE

W1:     WAIT;                    ;HANG UNTIL INTERRUPT
        BR W1                    ;GO BACK FOR MORE INTERRUPTS

;KEYBOARD INTERRUPT SERVICE ROUTINE
KBINT:  MOVB TKB, @ POINT        ;GET CHARACTER FROM KEYBD AND PUT IN MEMORY
        INC POINT                ;GO ON TO NEXT BYTE
        DEC COUNT                ;COUNT IT-- ALL DONE?
        BEQ KBDONE               ;YES, TAKE SPECIAL EXIT
        RTI                      ;NO, JUST RESTURN FROM THE INTERRUPT

;THIS CODE EXECUTED ONLY IF ALL 100 CHARACTERS
;HAVE BEEN ENTERED

KBDONE: TST (SP)+                ;POP INTERRUPT RETURN OFF STACK
        MOV (SP)+, PS            ;RESTORE PS FROM STACK
        CLR TKS                  ;DISABLE INTERRUPT
        BR BUFFIL                ;BUFFER FULL- TAKE ACTION

;CONSTANTS USED BY PROGRAM
POINT:  0           ;POINTER TO CHARACTER BUFFER
COUNT:  0           ;BYTE COUNTER
BUFFER: .BLKB 100.  ;BYTES STORED HERE
```

when there is no advantage to moving the pointer and counter into registers to use them.

Interrupt-Driven Tape Duplication

One of the more common uses for the interrupt structure is in applications where two or more devices may be allowed to cause interrupts. One of the most useful of these occurs when paper tapes are to be duplicated using the high-speed reader and punch. The high-speed reader will read characters at 300 characters per second, and the punch will punch them at 50 characters per second. Thus a program that simply reads a character and then punches it is perfectly adequate, but uses neither the reader nor the punch at its full speed. Although the small speed advantage to be gained by using the punch at a slightly higher rate is not really worth the trouble, most readers and punches suffer from much greater error rates when they are used in this one-character-at-a-time mode, where neither serves at its full capability. For this reason alone, an interrupt-driven reader punch routine is advantageous, since it will allow somewhat faster and much more error-free duplication.

The routine given in Figure 10.1 again has no main program, although it is perfectly possible to write a program that reads and punches and performs some other calculations simultaneously. Instead, the main program is again a WAIT instruction, and the subprograms are simply read

```
;TAPE DUPLICATION PROGRAM- ILLUSTRATES THE USE
;OF TWO INTERRUPTS FOR DEVICES RUNNING AT DIFFERENT RATES

;DEFINE SYMBOLS OF TELEPRINTER AND READ-PUNCH
HSRS=177550
HSRB=HSRS+2
HSPS=HSRB+2
HSPN=HSPS+2
TKS=HSPB+2
TKB=TKS+2
TPS=TKB+2
TPB=TPS+2

;DEFINE USEFUL CONSTANTS
CR=15    ;CARRIAGE RETURN
LF=12    ;LINE FEED

INTBIT=100       ;INTERRUPT ENABLE BIT
BR=200           ;BUS REQUEST PRIORITY 4
BUFLEN=512.      ;TAPE BUFFER LENGTH

.ASECT
.=1000

START:  MOV #., SP       ;INITIALIZE STACK POINTER
        MOV #RSERV, @#70        ;READER SERVICE ROUTINE
        MOV #BR, @#72    ;AND PRIORITY
        MOV #PSERV, @#74       ;PUNCH SERVICE ROUTINE
        MOV #BR, @#74    ;AND PRIORITY
        CLR TFLAG        ;CLEAR OUT OF TAPE FLAG
        JSR PC,CRLF      ;INITIALIZE CARRIAGE

;NOW WAIT FOR A RETURN AT THE KEYBOARD
T1:     TSTB TKS         ;WAS KBD STRUCK
        BPL T1           ;NOT YET
        MOV TKB, R0      ;READ IT
        BIC #-200, R0    ;STRIP OUT PARITY BITS, IF ANY
        CMP R0, #CR      ;WAS IT A RETURN?
        BNE T1           ;NO, KEEP WAITING
        JSR PC, CRLF     ;YES, BEGIN DUPLICATION
;SET UP INITIAL VALUES OF READER AND PUNCH BUFFER POINTERS
        MOV #PBUF, R0    ;R0 IS READER ROUTINE POINTER
        MOV R0, R1       ;R1 IS THE PUNCH BUFFER POINTER
        MOV #INTBIT+1, HSRS      ;ENABLE READER AND INTERRUPT
        MOV #INTBIT, HSPS        ;AND PUNCH INTERRUPT

W1:     WAIT             ;AND WAIT HERE FOR INTERRUPTS
        BR W1
```

Figure 10.1 Listing of tape duplication program. The logic of this program depends on the reader interrupting more frequently than the punch.

```
;READER INTERRUPT SERVICE ROUTINE
RSERV:  TST HSRS            ;ANY TAPE IN READER?
        BMI TAPOUT          ;NO, DUPLICATION DONE
        MOVB HSRB, (RO)+         ;YES, STORE CHAR IN BUFFER
        CMP RO, #PBUF+BUFLEN  ;IS THE BUFFER FULL?
        BGE RDROFF          ;YES, SHUT OFF READER FOR A WHILE
        INC HSRS            ;OTHERWISE TURN READER BACK ON
RDROFF: RTI                 ;AND RETURN TO WAIT FOR MORE INTERRUPTS


;OUT OF TAPE CONDITION
TAPOUT: CLR HSRS            ;DISABLE FURTHER READER INTERRUPTS
        INC TFLAG           ;SET OUT OF TAPE FLAG
        RTI                 ;AND RETURN, STILL ALLOWING PUNCH INTERRUPTS


;PUNCH SERVICE ROUTINE
PSERV:  MOVB (R1)+, HSPB        ;PUNCH ONE CHARACTER
        TST TFLAG           ;OUT OF TAPE?
        BNE LAST1           ;IF OUT, THIS IS LAST TIME
        CMP R1, #PBUF+BUFLEN  ;END OF PUNCH BUFFER?
        BEQ RDRON           ;YES, TURN READER BACK ON
        RTI                 ;ELSE KEEP PUNCHING


;TURN READER BACK ON
RDRON:  MOV #PBUF, RO       ;RESET READER TEXT BUFFER POINTER
        MOV RO, R1          ;AND PUNCH POINTER
        INC HSRS            ;RE-ENABLE READER
        RTI                 ;AND ALLOW NEW INTERRUPTS


;PUNCHING LAST BUFFER
;COMPARE RO WITH R1 FOR LAST CHAR TO BE PUNCHED
LAST1:  CMP R1, RO          ;HAS PUNCH CAUGHT UP WITH READER?
        BGT RESTART         ;QUIT WHEN R1>RO
        RTI                 ;OTHERWISE KEEP PUNCHING


RESTAR: CLR HSPS            ;DISABLE FURTHER PUNCH INTERRUPTS
        TST (SP)+           ;POP OLD RETURN ADDRESS OF STACK
        MOV (SP)+, PS       ;RESTORE OLD PS
        BR START            ;AND RESTART


TFLAG:  0          ;OUT OF TAPE FLAG
PBUF:   .BLKB BUFLEN        ;START OF READER-PUNCH CHARACTER BUFFER

        .END START
```

or punch, depending on whether the reader or punch is ready. One important concept of this program, however, is the dual buffer pointer scheme. In this program the punch and reader pointer start out pointing to the start of the text buffer; and as soon as a character has been put into the buffer by the reader, the punch interrupt is enabled. The reader goes on until it has filled the character buffer to its maximum (100 characters), and then the reader interrupt is disabled until the punch ''catches up.'' At that point, the reader is reenabled, and the buffer pointers for both the reader and punch are set to the top of the buffer once again. The only special case occurs when the tape runs out. In this case, the reader is disabled, and the punch is stopped when it finishes the final buffer, which may contain any number of characters from 0 to 100.

Traps

The PDP-11 has a trap facility that operates just like an interrupt in that the current PS and PC are pushed onto the stack and a new PS and PC taken from two memory locations. However, traps occur either from software instructions or from specific hardware conditions.

Software Traps.   The four software trap instructions are really ways of calling specific routines from anywhere in memory in a single instruction. They also allow return to the main program with a single-word RTI (return from interrupt) instruction. The software trap instructions are as follows:

104000	EMT	Emulate trap	—vector addresses 30 and 32
104400	TRAP	Trap	—vector addresses 34 and 36
000003	BPT	Breakpoint trap	—vector addresses 14 and 16
000004	IOT	Input–output trap	—vector addresses 20 and 22

The EMT instruction is used by the manufacturer's disk and tape monitor system software, and the IOT instruction in a similar fashion for both disk and paper tape software. The BPT instruction is used in the octal debugging programs ODT-11 and ODT-11X. None of these instructions should be employed by the user in programs that may run in a disk monitor system. The TRAP instruction is reserved for the user and is used only for the floating point package FPMP11 discussed in the next chapter.

Hardware Traps.   Under certain hardware conditions, the PDP-11 processor also causes traps through specific vector locations. The more common of these are shown in the following list.

Locations	Description
4 and 6	Bus error trap. Caused by addressing a word or instruction at an odd address, illegal or nonexistent memory or device, or Stack Pointer decremented below 400_8.
10 and 12	Attempt to execute an illegal, reserved, or unimplemented instruction.
14 and 16	Trap after each executed instruction when the T-bit is set in PS.
24 and 26	Power fail trap. Whenever ac power drops below 95 volts or outside 47–63 hertz, this trap occurs and 2 milliseconds is allowed for the saving of registers before a halt will occur.

The T-Bit of the PS and the RTT Instruction

All models of the PDP-11 have a bit in the Processor Status word called the *T-bit* (bit 4). This is the trace trap bit; and when it is set, the processor will automatically execute a TRAP instruction through vectors 14 and 16 after the completion of each instruction. Since the new PS in location 16 presumably does not have the T-bit set, no new traps will occur while the trap is being processed. This feature is ideal for writing debugging programs to detect which instructions are executed by reporting the address of each instruction as it is executed.

Return from the trace trap may be through the RTI instruction, but in this case the RTI instruction itself will cause a new trace trap. This can be avoided in all PDP-11's except the 11/05 and 11/10 by using the RTT (return from trap) instruction, which inhibits the trace trap until the *next* instruction is executed. The trace trap feature has been well used in a DEC-supplied debugging program called ODT-11X and is described in detail in Chapter 13.

The RESET Instruction

When the PDP-11 is started using the Load Add, Start switches, one of the start-up conditions is the resetting of a large number of device flags and registers to a power-up state by means of an internal pulse, called the INIT pulse. This pulse does the following:

1. Sets the interrupt priority to 0.
2. Sets each external device to its most common condition.

Occasionally, it is necessary to completely reset the machine while a program is running so that these start-up conditions again hold. This can be accomplished using the RESET instruction. The exact effect of INIT and RESET on each bit of each device status word is described in the *PDP-11 Peripherals Handbook*. For example, INIT effects the Teletype registers as follows:

1. Clears all bits of TKS.
2. Clears bits 6 and 2 and sets bit 7 of TPS.

Exercises

10.1 Write a program to increment a memory location continuously starting from 0 until a key on the teleprinter is struck. The current value of this register should then be displayed in R0 as the computer halts. Be sure to use interrupts.

10.2 Modify Exercise 10.1 to print out the value of the counting register on the teleprinter and to continue counting. Be sure that both the keyboard and the printer utilize interrupts.

10.3 Suppose that you have a frequently called subroutine in location 7006, and you wish to call it as efficiently as possible in a one-word instruction. If no registers are available to store its address in, what other technique could you use?

10.4 Some programs start by putting 0's in locations 6, 12, 16, 22, 26, 32, and 36 and in locations 62, 66, 72 and 76, and then putting these numbers in locations 4, 10, 14, 20, 24, 30, and 34 and in locations 60, 64, 70, and 74. What purpose does this serve?

CHAPTER ELEVEN

THE FLOATING POINT MATH PACKAGE FPMP11

The 16-bit computer word will allow representation of unsigned numbers from 0 to $2^{16} - 1$ or 0 to 65535. If we assign 1 bit to the sign, we can represent numbers from -32768 to 32767. While these large numbers are represented to four or five significant figures, smaller numbers like 12 or -5 are represented to only one or two significant figures.

Furthermore, this integer mode does not allow any method for the representation of fractions. If we were to simply assign some bit position in the computer word to the right of which fractions would be represented and to the left of which integers would be represented, we would markedly diminish the size of the integers that could be represented.

This problem is solved in scientific notation. We simply adjust numbers of any size to lie between 1 and 9.9999 . . . and adjust the power of 10 accordingly. Thus 5236 becomes 5.236×10^3, and -0.000642 becomes -6.42×10^{-4}. There is no reason why a similar approach cannot be adapted for representing numbers in computers, and this, in fact, is what has been done.

The format chosen for the PDP-11 is one in which two words are used for each floating point number and double-precision floating point numbers are represented in four words.[1] In the following discussion, only the single-precision (two-word) mode will be emphasized.

The PDP-11 floating point format arranges numbers as shown in Figure 11.1. The sign bit contains the sign of the overall number, and the exponent the power of 2 to which the mantissa must be raised to reach the proper

126

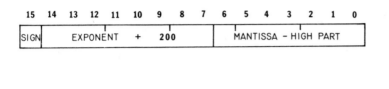

Figure 11.1   Floating point number format.

magnitude *plus* 200_8. Thus the exponent in 8 bits ranges from 0 to 377, representing 2^{-128} to 2^{127}.

The mantissa is a number normalized to lie between 0.5 and 1.0. This is illustrated below for the number 5.

$$
\begin{aligned}
5 &= 101 \\
  &= 10.1 \quad \times 2^1 = 2.5 \quad \times 2 \\
  &= 1.01 \quad \times 2^2 = 1.25 \quad \times 4 \\
  &= 0.101 \times 2^3 = 0.625 \times 8
\end{aligned}
$$

Thus the exponent in this case would be 011 (3), and the mantissa 0.101. The sign bit of the floating point number is 1 for negative numbers and 0 for positive numbers. However, the mantissa is stored, not in two's complement form, but as an absolute value. Furthermore, since the mantissa must always be between 0.5 and 1.0, the bit immediately to the right of the binary point will always be a 1, since $0.5_{10} = 2^{-1} = 0.100$. Since only 23 bits are allocated for the mantissa, this bit, which is always 1, is not part of the floating point representation. Instead it is merely understood, used in intermediate calculations but shifted out in storing the final floating point number.

For example, the floating point representation of 5 is shown in Figure 11.2. Similarly, -1.75 would be represented as shown in Figure 11.3.

$$
\begin{aligned}
-0.875 \times 2^1 = \; &-1 \times 2^{-1} \quad 0.5 \\
&-1 \times 2^{-2} \quad 0.25 \;\times 2^1 = 0.111 \times 2^1 \\
&-1 \times 2^{-3} \quad \underline{0.125} \\
&\phantom{-1 \times 2^{-3} \quad} 0.875
\end{aligned}
$$

Floating point numbers are handled in some minicomputers completely by software. No hardware floating point instructions are used in these cases. The software packages developed, then, are designed to free the user completely from the necessity of understanding how they work, and they usually perform simple and some transcendental operations. Floating

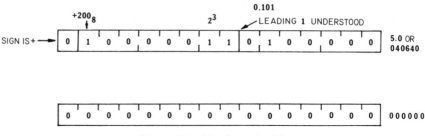

Figure 11.2   Floating point 5.0.

point calculations performed by software simulation are necessarily much slower than equivalent integer operations.

The floating point math package for the PDP-11 (FPMP11) is designed to support the floating point hardware instructions of the PDP-11/70, but not the simpler set of the PDP-11/34. The FPMP11 can easily be modified, however, to use these instructions to conserve space and improve speed. It provides routines for the single-precision functions listed in Table 11.1. Most are available as double-precision modes as well, but are not included in the version available to students, since for most calculations they are unnecessary.

Most of these functions can be called in a single instruction, including possible addressing modes, through an executive routine called the *trap handler*, which is part of FPMP11.

The TRAP instruction is one of a special class in the PDP-11 that pushes the current PS and PC onto the stack and takes its new PS and PC from two special memory locations, in this case locations 34 and 36. It is the responsibility of the user of FPMP11 to initialize these locations. The TRAP instruction has octal code 104400, where bits 0–7 are not decoded by the processor and can thus be used to transmit information to the routine which is trapped to. This feature is used by the trap handler routine, which decodes the TRAP instructions for FPMP11. It treats the

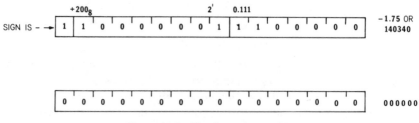

Figure 11.3   Floating point −1.75.

Table 11.1   Arithmetic Operations Performed by the Floating Point Math Package

Operation	Function	Name	Trap Handler Value	Common Name
Opr + FLAC→FLAC	add	$ADR	12	FPADD
ln(FLAC)→FLAC	ln	ALOG	53	FLOG
log(FLAC)→FLAC	log	ALOG10	54	FLOG10
tan⁻¹(FLAC)→FLAC	arctan	ATAN	42	ATAN
FLAC-Opr→cond codes	compare	$CMR	17	FCMP
cos(FLAC)→FLAC	cosine	COS	37	FCOS
FLAC/opr→FLAC	divide	$DVR	25	FPDIV
exp(FLAC)→FLAC	exponential	EXP	51	FEXP
float(SP)→top 2 words of stack	float	$IR	···	···
fix(2 words of stack)→stack	fix	$RI	···	···
Opr x FLAC→FLAC	multiply	$MLR	21	FPMULT
sin(FLAC)→FLAC	sine	SIN	36	FSIN
FLAC-opr→FLAC	subtract	$SBR	13	FPSUB
sqrt(FLAC)→FLAC	square root	SQRT	46	FSQRT
tanh(FLAC)→FLAC	hyperbolic tangent	TANH	50	TANH
Opr→FLAC	load	$LDR	71	FPGET
FLAC→Opr	store	$STR	73	FPSTR
TTY→FLAC	floating input	FLIP[a]	···	FLIP
FLAC→TTY	floating output	FLOP[a]	···	FLOP

[a] Added to the standard package.

instruction as having the format shown in Figure 11.4. Bits 0–5 are used as routine numbers, and bits 6 and 7 as address mode bits, where these modes are specially defined by software and are not to be confused with the hardware modes of the PDP-11. They are as follows:

00	Stack mode	The operand is on top of the stack.
01	@ R0 mode	The address of the operand is in R0.
10	Immediate mode	The operand follows the instruction.
11	Relative mode	The address of the operand minus the PC follows the instruction.

The destination of all floating point operations is the FLAC, or Floating Accumulator, a pair of words within FPMP11 that act as a floating point register.

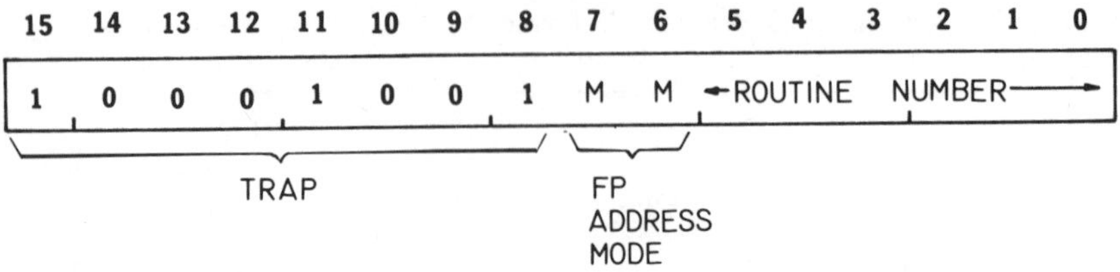

15	14	13	12	11	10	9	8	7	6	5	4	3	2	1	0
1	0	0	0	1	0	0	1	M	M	←ROUTINE		NUMBER			→

TRAP   FP
ADDRESS
MODE

Figure 11.4   Format of trap handler instructions.

Initialization

The TRAP instruction, when summed with the numbers for the floating point routine and addressing mode, allows calls of floating point routines. This instruction automatically causes the processor to push the current PC and PS onto the stack and take the new PC and PS from locations 34 and 36. Location 34, then, must contain the address of the trap handler, and location 36 the processor priority while FPMP11 is executing. Since FPMP11 is not reentrant, it is safest to prevent interrupts during FPMP11 calculations by setting the PS to 340, or BR7. Further, since these are stack operations, it is *crucial* that we initialize the stack pointer before we do anything else. These initialization operations should be at the beginning of each program and take the form

```
               .ASECT
               .=1000
      START:   MOV #., SP      ;SET THE STACK POINTER
               MOV #TRAPH, @#34 ;SET THE TRAPH HANDLER
               MOV #340, @#36   ;AND BR7
```

Defining the FPMP11 Codes to MACRO

The TRAP instruction is recognized by MACRO, but neither the floating point routines nor the special addressing modes are known to it. Thus the instruction to perform the floating point addition of a number to the FLAC in the immediate mode would be coded as

TRAP + 200 + 12   ;ADD IMMEDIATE

where

104400	TRAP
200	Immediate mode
12	Add routine
104612	

A more efficient way to do this is to define the floating point routines you will need and suitable addressing mode symbols:

```
FPADD    =        TRAP+12
FPDIV    =        TRAP+25
FPMULT   =        TRAP+21
         ;ETC.

;DEFINE ADDRESSING MODES

;STACK MODE IS 0 AND NEED NOT BE DEFINED
ARM      =        100      ;@R0 MODE
IMM      =        200      ;IMMEDIATE MODE
RELM     =        300      ;RELATIVE MODE
```

These definitions are often collected into a single definition file, which is either combined with the program file with TECO or assembled with it using MACRO. At Tufts, this file is SY:DEFNS.MAC. Be sure not to use FADD, FSUB, FMUL, and FDIV, as these are floating point *hardware* instructions.

MACRO also performs conversion into two-word or four-word floating point, using the directives .FLT2 and .FLT4. The format of the directive, when the location is labeled is

KHUND: .FLT2 100   ;FLOATING 100

A list of floating point numbers may occupy the same line:

.FLT2 1, 10, 100, 1000, 10000

Detailed Descriptions of the Addressing Modes

Mode 0: Stack Mode.   In this mode, the operand is assumed to have been pushed onto the top of the stack. It is popped off by the operation and combined with the FLAC as directed. Pushing such an operand can be accomplished with MOV instructions or with some FPMP11 subroutines to be discussed later. The destination of all floating point operations is the FLAC, a FPMP11 pseudoregister. Thus, if we wished to add 3.14 into the FLAC, the following code could accomplish this. Note that we do not even have to define the stack mode by a label, since both address mode bits of the TRAP instruction would be 0 anyway.

```
FPADD=TRAP+12    ;DEFINE CODE

.ASECT
.=1000
START:  MOV #., SP          ;INIT STACK POINTER
        MOV K314+2, -(SP)        ;PUSH LOW ORDER WORD FIRS
        MOV K314, -(SP) ;PUSH HIGH ORDER WORD
        FPADD                ;ADD TOP OF STACK TO FLAC
        :
        :

K314:   .FLT2 3.14       ;WILL OCCUPY 2 WORDS
```

After a stack mode call, the Stack Pointer will be advanced by two words, popping that value off the stack. This mode is most useful when a number of values can be pushed onto the stack and popped off, one per operation. Before exit to the calling program, the trap handler automatically sets the condition codes to reflect the current state of the FLAC, except after a floating compare:

After Any Arithmetic Operation						After Compare				
FLAC	N	Z	V	C: Condition Codes			N	Z	V	C
< 0	1	0	0	0		FLAC < OPR	1	0	0	0
= 0	0	1	0	0		FLAC = OPR	0	1	0	0
> 0	0	0	0	0		FLAC > OPR	0	0	0	0

Numbers are always pushed onto the stack high-order word last, so that the sign can be examined by TST(SP) and similar instructions.

Mode 100: @R0 Mode.   In this mode the register R0 contains the absolute address of the operand. The contents of R0 are not modified by the FP (floating point) operation. This mode is most useful when an address has been looked up in a table or calculated. Then R0 points to the first word of the two-word FP number, and this word is, by convention, that containing the sign of the number, the exponent, and the high-order part of the mantissa. In the following example the address of THOUSN is put in R0 and multiplied by the contents of the FLAC. The result is left in the FLAC.

```
        FPMULT=TRAP+21   ;DEFINE CODE
        ARM=100          ;AND ADDRESSING MODE

        MOV #THOUSN, R0  ;MOVE ADDRESS TO R0
        FPMULT +ARM      ;MULT IN @R0 MODE
            :
            :
THOUSN:  .FLT2 1000      ;FLOATING PT 1000
```

More common, however, would be the call to some subroutine that calculates or looks up an address, returning this address in R0:

```
    FPMULT= TRAP+21 ;DEFINE CODE
    ARM=100         ;AND MODE

    JSR PC, ADRCLC  ;CALCULATE ADDRESS OF VALUE- LEAVE IN R0
    FPMULT+ARM      ;MULTIPLY FLAC BY IT
```

Mode 200: Immediate Mode.   In the immediate mode the actual operand is in the two words following the FP instruction. The program continues following the constant. This mode is most useful for a constant that is used only once in a calculation. The following example divides the FLAC by 6.28.

```
FPDIV=TRAP+25
IMM=200
FPDIV+IMM       ;DIVIDE FLAC BY FP FOLLOWING NUMBER
.FLT2 6.28      ;DIVISOR IS HERE
```

Mode 300: Relative Mode.   In this mode the address following the FP TRAP instruction contains the address of the FP constant minus the current PC. This is, in other words, the *relative* address of the datum and may be positive or negative, depending on whether the location is before or after the current location. For example, to subtract 900.37 from the FLAC, we could write

```
FPSUB=TRAP+13
RELM=300

FPSUB+RELM      ;SUBTRACT 900.37 FROM FLAC
KNINEH-.        ;ADDRESS-CURRENT LOCATION
    :
    :
KNINEH: .FLT2 900.37
```

The notation LABEL–. is used to tell the assembler to subtract the current location from the address value of the label. This addressing mode is the most generally useful, particularly when the desired constants and variables occur several times. This allows you to refer to an address *relative* to the PC to write position-independent code.

Loading and Storing the FLAC

Since the destination of all FP arithmetic operations is the FLAC, it must be loaded with the first value and be able to store the final value somewhere in memory. This can be accomplished using the load and store routines in FPMP11, which we will refer to as FPGET and FPSTR.

```
FPGET = TRAP + 71
FPSTR = TRAP + 73
```

We will assume that these, like the arithmetic routine and addressing modes are defined in the definitions file attached to the program and need not be repeated with each example in this chapter.

All four addressing modes can be used with the load and store routines:

```
FPGET+RELM          ;LOAD FP CONSTANT INTO FLAC
CONST-.             ;RELATIVE ADDRESS OF CONSTANT

FPGET+IMM           ;GET IMMEDIATE CONSTANT
.FLT2 xxx           ;STORED HERE

FTSTR               ;PUSH FLAC ONTO STACK

FPGET+ARM           ;GET FP NUMBER PTD TO BY R0
```

We are now in a position to write a floating point program. This example calculates $y = 5x + 3z$.

```
        .=1000
START:  MOV #.,SP        ;INITIALIZE THE SP FIRST EVERY TIME
        MOV #TRAPH, @#34 ;INIT TRAP HANDLER
        MOV#340, @#36
        FPGET+RELM      ;LOAD X INTO FLAC
        X-.             ;RELATIVE ADDRESS
        FPMULT+IMM      ;MULTIPLY BY 5
        .FLT2 5
        FPSTR           ;PUSH ONTO STACK TEMPORARILY

        FPGET+RELM      ;GET Z
        Z-.
        FPMULT+IMM      ;MULTIPLY BY 3
        .FLT2 3

        FPADD           ;STACK MODE ADDITION OF 5X TO 3Z
        FPSTR+RELM      ;STORE IN Y
        Y-.
        ;etc.
```

Similarly, we could calculate the quadratic formula, where a, b, and c are already stored in memory. Note that the function FSQRT operates on the FLAC and leaves the result in the FLAC.

```
;CALCULATE Y=(-B + SQRT(B^2 -4AC))/2A

FGET+RELM           ;GET C
C-.
FPMULT+RELM         ;AC
A-.
ADD #400, FLAC      ;MULTIPLY BY 4 BY ADDING 2 TO EXPONENT
FPSTR               ;PUSH 4AC ONTO STACK
FPGET+RELM          ;GET B
B-.
FTSTR               ;PUSH A COPY OF B ONTO THE STACK
FPMULT              ;MULT B ON STACK BY B IN FLAC = B^2
FPSUB               ;SUBTRACT 4AC
FSQRT               ;SQRT(B^2-4AC)
FPSTR               ;PUSH DISCRIMINANT
FPGET+RELM          ;GET B
B-.
ADD #100000,FLAC            ;NEGATE
FPADD               ;-B+SQRT(B^2-4AC)
FPDIV+RELM          ;DIVIDE BY 2
A-.
SUB #200, FLAC      ;DIVIDE BY 2
FPSTR+RELM          ;STORE IN Y
Y-.
```

Since the exponent represents a *power of 2*, the fastest way to multiply or divide by a power of 2 is by adding or subtracting that number from the exponent bits of the number, usually contained in the FLAC. This can be done only if the FLAC is nonzero. A zero FLAC is all 0's, and changing the exponent will make it nonzero. Furthermore, negation can be accomplished by adding 100000 to the FLAC, complementing the sign bit, again only if the FLAC is nonzero.

Floating Input and Output

The DEC-supplied version of FPMP11 does not do any input–output (I/O) to the terminal. It converts ASCII strings stored in memory to FP numbers on the stack and FP numbers to ASCII strings. On input, the ASCII strings are assumed to be error free, and on output, the exact format must be specified each time, as well as the actual I/O routines. Although this method allows tremendous versatility, it is extremely complex for the average user who must write his or her own I/O routines, as well as do all formulating and error checking. Complete discussions of the routines for performing these conversions are given in the FPMP11 manual. They are all called in the JSR PC, routine mode and appear in the following list.

$ECO	Single precision to ASCII E-format
$FCO	Single precision to ASCII F-format
$GCO	Single precision to ASCII G-format
$ICI	ASCII to integer
$ICO	Integer to ASCII
$OCI	ASCII to octal
$OCO	Octal to ASCII
$RCI	ASCII to single precision, ASCII in any format

FLIP and FLOP

To simplify use of the floating point I/O, routines for ASCII input and output for the Teletype or DECwriter have been written, providing simple floating point I/O capabilities. The floating point input routine is called FLIP, and the output routine FLOP. They are called by JSR PC, FLIP (or FLOP) and operate on the FLAC. Listings of these routines are given in Figure 11.5. They can be conveniently assembled as part of the FPMP11 package or linked with it using the LINK program.

```
;FPMP11 I/O MODULE

;CONSTANTS
CR=15                   ;CARRIAGE RETURN
RUBOUT=177
BACKSL='\
PLUS='+
MINUS='-
DOT='.
BUFLEN=30.              ;NUMBER OF BYTES IN I/O BUFFER
JSW=44                  ;JOB STATUS WORD
BIT12=10000             ;BIT TO SET FOR SPECIAL I/O

;GLOBAL VALUES USED BY FPPIO
.GLOBL $RCI, $GCO, FLAC

;GLOBAL VALUES CONTAINED IN FPPIO
.GLOBL FLIP, FLOP, VFLAG, TCHAR, WIDTH$, DEC$

;RT-11 SYSTEM CALLS
.MCALL .PRINT .TTYIN, .TTYOUT


;FLOATING POINT INPUT ROUTINE
;TAKES THE FIRST LEGAL NUMBER IN THE RT-11 BUFFER
;AND CONVERTS IT TO A FLOATING POINT NUMBER IN THE FLAC
;EXITS AT THE FIRST ILLEGAL CHARACTER, WHICH IS LEFT IN TCHAR
;LOCATION VFLAG INDICATES WHETHER  VALID DATA WAS ENTERED
;BEFORE A TERMINATING CHARACTER OCCURRED.

FLIP:   JSR R0, REGSAV  ;SAVE THE REGISTERS
        MOV @#JSW, -(SP)         ;PUSH CURRENT JSW
        BIS #BIT12, @#JSW       ;SET SPECIAL I/O BIT
        JSR PC, INIT    ;SET UP POINTERS TO BUFFER
FLIP10: CLR VFLAG       ;CLEAR VALID DATA FLAG
        CLR TCHAR       ;TERMINATING CHARACTER

FLIP20: MOVB #'0, (R2)+ ;SET WHOLE BUFFER TO ASCII ZEROES
        SOB R3, FLIP20
        JSR PC, INIT    ;RESET POINTERS
        CLR R1          ;COUNT NUMBER OF CHARACTERS

FLIP30: JSR PC, READ    ;READ A CHARACTER INTO R0
        CMP R0, #RUBOUT ;DELETE ALL IF THIS WAS A RUBOUT
        BEQ RUBBIT
        CMP R0, #'0     ;WAS IT > ASCII 0?
        BGE FLIP40      ;YES

;LESS THAN ASCII 0, CHECK SPECIAL CHARACTERS
        CMP R0, #PLUS
        BEQ FLIP50      ;+ IS OK
        CMP R0, #MINUS  ;- IS OK
        BEQ FLIP50
        CMP R0, #DOT
        BEQ FLIP50      ;. IS OK
        BR FLIPXT       ;ELSE EXIT FROM FLIP

;R0 > ASCII 0, IS IT < ASCII 9?
FLIP40: CMP R0, #'9
        BGT FLIP70      ;YES, ONLY 'E IS ALLOWED
```

Figure 11.5   Listing of the FLIP and FLOP routines.

```
; ONLY 0-9 AND +,-, AND . GET HERE
FLIP50: INC VFLAG          ; SET VALID DATA FLAG
        MOVB R0, (R2)+      ; PUT CHARACTER INTO BUFFER
        INC R1             ; COUNT CHARACTERS
        SOB R3, FLIP30     ; GO BACK TILL BUFFER FULL

; TOO MANY CHARACTERS, ISSUE RUBOUT AND START AGAIN
RUBBIT: MOV #BACKSL, R0 ; PRINT BACKSLASH
        JSR PC, TYPE
        BR FLIP10

; EXIT FROM INPUT AND CONVERT STRNG IF ANY
FLIPXT: CMP R0, #CR        ; WAS THIS AN CR?
        BNE FLIPX5         ; NO
        JSR PC, READ       ; YES, READ THE LF, TOO
FLIPX5: TST R1             ; ANY CHARACTERS ENTERED?
        BNE FLIP60         ; YES
        MOV #BUFLEN, R1 ; NO, SET COUNTER

; SET UP STACK FOR CALL TO $RCI
FLIP60: MOV R0, TCHAR      ; SAVE TERMINATING CHARACTER
        MOV #ABUF, -(SP)            ; PUSH BUFFER ADDRESS
        MOV R1, -(SP)      ; PUSH TOTAL FIELD WIDTH
        CLR -(SP)          ; PUSH D
        CLR -(SP)          ; PUSH P SCALE FACTRO
        JSR PC, $RCI       ; CONVERT ASCI TO # ON STACK
        MOV (SP)+, FLAC ; POP INTO FLAC
        MOV (SP)+, FLAC +2
        MOV (SP)+, @#JSW           ; RESTORE PREVIOUS JSW
        JSR R0, REGRES     ; RESTORE THE REGISTERS
        RTS PC             ; AND EXIT

; CHECK FOR "E"
FLIP70: CMP R0, #'E        ; IS IT E?
        BEQ FLIP50         ; YES
        BR FLIPXT          ; NO


; ROUTINE TO INITIALIZE POINTER AND COUNTER
INIT:   MOV #ABUF, R2    ; POINTER
        MOV #BUFLEN, R3 ; COUNTER
        RTS PC
```

Figure 11.5 (Continued)

FLIP will accept a maximum of 30 input characters. A single Rubout deletes the entire number and prints a back slash (\). It is called by JSR PC, FLIP, after which characters are accepted from the teleprinter until an illegal character is typed. After exit occurs from FLIP, the converted value is found in the FLAC. If any legal value has been entered, VFLAG is set to the number of characters entered; if no legal value has been entered, VFLAG will be 0. The character that caused FLIP to exit is stored in TCHAR. The previous contents of the FLAC are destroyed.

FLOP prints out the current contents of the FLAC in FORTRAN F format. The values of w and d are set to F10.3 (10-character total field width, three places to the right of the decimal) but can be changed by altering WIDTH\$ and DEC\$. These constants remain set to their new values until changed again. The FLAC is not changed by calling FLOP.

```
; INPUT BUFFER
ABUF:    .BLKB BUFLEN
VFLAG:   0        ; VALID DATA FLAG
TCHAR:   0        ; TERMINATING CHARACTER
WIDTH$:  10.      ; WIDTH FO FIELD
DEC$:    3        ; NUMBER OF DECIMAL PLACES

; FLOATING POINT OUTPUT ROUTINE
; PRINTS OUT THE CONTENTS OF THE FLAC
; IN AN F W.D FORMAT WHICH IS DEFINED BY THE
; CONTENTS OF WIDTH$ AND DEC$.
; IF THE NUMBER IS TOO LARGE FOR THIS FORMAT
; THE OUTPUT ROUTINE REVERTS TO SCIENTIFIC NOTATION

FLOP:    JSR R0, REGSAV  ; SAVE REGISTERS
         JSR PC, INIT    ; SET POINTER AND COUNTER
; PUT 200'S IN BUFFER
FLOP10:  MOVB #200, (R2)+         ; AS OUTPUT TERMINATOR
         SOB R3, FLOP10
; SET UP FOR $GCO
         MOV #ABUF, -(SP)         ; PUSH BUFFER ADDRESS
         MOV WIDTH$, -(SP)        ; PUSH W
         MOV DEC$, -(SP)          ; PUSH D
         CLR -(SP)                ; PUSH P-SCALE FACTOR
         MOV FLAC+2, -(SP)        ; PUSH FLAC
         MOV FLAC, -(SP)
         JSR PC, $GCO    ; CONVERT
         .PRINT #ABUF    ; PRINT BUFFER ON TERMINAL
         JSR R0, REGRES           ; RESTORE REGISTERS
         RTS PC

; CHARACTER INPUT ROUTINE
READ:    .TTYIN ; READ CHARACTER INTO R0
         BIC #-200, R0   ; REMOVE PARITY AND SEGUE TO OUTPUT

; TERMINAL OUTPUT ROUTINE
TYPE:    .TTYOUT
         RTS PC

; REGISTER SAVING ROUTINE
REGSAV:  MOV R1, -(SP)   ; SAVE R1-R5
         MOV R2, -(SP)
         MOV R3, -(SP)
         MOV R4, -(SP)
         JMP @ R0

; REGISTER RESTORING ROUTINE
REGRES:  TST (SP)+       ; DUMP OLD R0
         MOV (SP)+, R4
         MOV (SP)+, R3
         MOV (SP)+, R2
         MOV (SP)+, R1
         RTS R0          ; RESTORE R0 AND EXIT

    END
```

Figure 11.5 (Continued)

Both FLIP and FLOP utilize terminal input and output routines that are shown using calls to RT-11 functions .TTINR and .PRINT. These are discussed in Chapter 12. If you prefer, you can write your own keyboard input and output routines.

For example, to print out the contents of the FLAC according to F10.3, we simply call

  JSR PC, FLOP

but to print out according to any other format, we must change it by

```
MOV #8., WIDTH$   ;SET OVERALL WIDTH
MOV #2, DEC$      ;AND NUMBER DECIMAL PLACES
JSR PC, FLOP      ;PRINT THE NUMBER
```

To read in two numbers and multiply them together, we store the first one on the stack. Then we print the product out using FLOP.

```
JSR PC, FLIP     ;GET 1ST NUMBER
JSR PC, CRLF     ;PRINT CR-LF
FPSTR            ;STORE VALUE ON STACK
JSR PC, FLIP     ;GET SECOND NUMBER
JSR PC,CRLF      ;TYPE A CRLF
FPMULT           ;MULTIPLY STACK BY FLAC
JSR PC, FLOP     ;AND PRINT OUT THE RESULT
```

Calling Routines in Polish Mode

Polish mode is named after the Polish mathematician Lukasiewicz, who developed it. The nomenclature arose because chauvinistic European mathematicians couldn't be troubled to learn to pronounce his name. The method amounts to representing a string of arithmetic calculations by listing the numbers first and then the operations to be performed, so that $(A+B) \times C - D$ becomes

  AB+ C× D−

The terms are always evaluated so that the current operation (e.g., multiplication) operates on the last two numbers. Polish notation is commonly used in some pocket calculators, such as the HP series, in which (5 + 3) × 4 − 2 could be evaluated by

−2	ENTER
4	ENTER
3	ENTER
5	+
	×
	+

The FPMP11 package also allows Polish calls to many of its routines in which all of the operands are first pushed onto the stack. The result of each Polish mode operation is the shortening of the stack by one item, with the result on *top of the stack, not* in the FLAC. We take up Polish mode here because it is one convenient way to convert between floating and fixed point numbers. Polish mode can be extremely fast because the high overhead of the trap handler is avoided.

The Polish mode is entered by a call of the form

```
JSR R4, $POLSH     ;ENTER POLISH MODE
$ADR               ;ADD 2 ITEMS TOGETHER
$MLR               ;MULTIPLY BY 3RD
.+2                ;EXIT
```

It *must* be called through R4. The words in the addresses following the entry to the Polish mode are the actual addresses of the FPMP11 routines. It is not generally necessary to know their absolute values, as these can be inserted by the linker. To exit from Polish mode, the last entry in the list should be the address of the next instruction. This is specified by entering ". + 2." The Polish mode interpreter simply performs a JMP to the address pointed to by R4. When that address is the next in line, it jumps there and effectively exits from the Polish mode. The routines that can be called in Polish mode by FPMP11 in single precision are as follows:

$ADR	Add
$CMR	Compare
$DVR	Divide
$IR	Integer to real (float)
$MLR	Multiply
$RI	Real to integer (fix)
$SBR	Subtract

In all cases, the result is on the top of the stack.

Caution: The Polish mode does not save any registers, and whatever registers the called routines use are destroyed.

Floating Numbers. To convert an integer to a floating quantity, the followng code can be used:

```
;FLOAT NUMBER IN R0
         MOV R0, -(SP)      ;PUSH NUMBER ONTO STACK
         JSR R4, $POLSH     ;ENTER POLISH MODE- MUST BE THRU R4
         $IR                ;CONVERT INTEGR TO REAL
         .+2                ;ADDRESS OF EXIT= NEXT INSTRUCTION
         FPGET              ;POP REAL NUMBER OFF STACK
```

The preceding example pushes the integer value onto the stack and calls the Polish interpreter, which then jumps to the $IR pointed to by R4. When $IR is complete, it jumps to the next address, which is the address of the next instruction in line, the FPGET, and exits from the Polish interpreter.

Fixing Numbers.   Occasionally, it is desirable to convert a floating number to integer form. This is done exactly as shown in the preceding example except that the FP number is pushed onto the stack and the integer result popped off upon exit from Polish mode. In the following example, the number is obtained using FLIP:

```
ENTER:   JSR PC, FLIP      ;GET A NUMBER FROM THE TERMINAL
         TST VFLAG         ;SEE IF LEGAL VALUE
         BEQ ENTER         ;IF NOT, GET ANOTHER ONE
         FPSTR             ;PUSH FLAC ONTO STACK
         JSR R4, $POLSH    ;ENTER POLISH MODE
         $RI               ;CALL FIXING ROUTINE
         .+2               ;EXIT FROM POLISH MODE
         MOV (SP)+, R0     ;GET NUMBER FROM STACK
```

There are two other types of calls to FPMP11 routines that are less frequently used, but are described in the *FPMP11 Manual* (pp. 3-10–3-14). These are the JSR R5 mode and the JSR, PC mode.

Fixing Numbers with Fractional Parts

It is sometimes desirable to convert a real number to a fixed point representation having a fractional part that can be used in certain fixed point calculations. The following fixing routine converts a number in the FLAC from floating point to fixed point format, leaving the integer part in R0 and the fractional part in R1. These registers, of course, must be saved if they contain useful information when FIX is called.

```
;FIXING SUBROUTINE
;RETURNS INTEGER PART IN R0
;AND FRACTIONAL PART IN R1

FIX:      MOV FLAC, R0      ;GET FLAC IN R0-R1
          BIC #-200, R0     ;CLEAR OUT EXPONENT AND SIGN BITS
          BIS #200, R0      ;RESTORE "HIDDEN" BIT
          MOV FLAC+2, R1    ;GET LOW ORDER PART
          MOV FLAC, R2      ;GET EXPONENT PART
          BIC #100177, R2   ;CLEAR OUT SIGN AND MANTISSA BITS
          ASH #-7, R2       ;SHIFT TO LOW ORDER BITS
          ASHC #7, R0       ;SHIFT TO PUT A BINARY POINT AT LEFT
          SUB #217, R2      ;REMOVE 200 BIAS AND ALLOW FOR SHIFTS
          TST FLAC          ;NOW CHECK SIGN
          BGE FIX10         ;ALREADY POSITIVE
;NEGATE BY COMPLEMENTING AND INCREMENTING
          COM R0            ;COMPLEMENT HIGH ORDER PART
          COM R1            ;AND LOW ORDER PART
          INC R1            ;THIS WILL SET C-BIT
          ADC R0            ;ADD IN CARRY IF ANY

;NOW SHIFT TO FIX IT
FIX10:    TST R2            ;IGNORE IF TOO BIG TO FIX
          BGE FIXEXT
FIX20,    ASHC #-1, R0      ;SHIFT BOTH WORDS RIGHT
          SOB R2, FIX20     ;GO BACK UNTIL R2 IS 0
FIXEXT:   RTS PC            ;EXIT- R2 WILL BE NON-ZERO IF UNFIXABLE
```

An analogous routine can then be written to float such fractional numbers. One of the most common representations of fractional numbers is that with the binary point between bits 14 and 15. Register R1 can then be converted to this format by a simple right shift.

Use of GLOBL Values

The fact that certain addresses in the FPMP11 package must be referred to by their absolute values has already been mentioned. Trap handler calls automatically find the right address, no matter where a routine is loaded, but the address of the trap handler itself must be put in location 34 by the calling program. Further, the addresses of FLIP, FLOP, DEC$, WIDTH$,VFLAG, $IR and $RI have been referred to by a label, and of course, that label must eventually have an absolute address value.

This problem could be solved by assembling the entire FPMP11 package with each program that uses it, but this would take a fair amount of assembly time as well as disk or diskette space, because FPMP11 is fairly long and complex. Furthermore, FPMP11 contains so many labels that the likelihood of duplicating one by accident is rather high.

Another possible method would be to consult a listing or symbol table of FPMP11 and determine the absolute addresses of these symbols. This has the disadvantage that a lot of six-digit numbers must be accurately transferred to your own program and that FPMP11 must then always reside at the same spot in memory.

The most desirable way to accomplish this connection with FPMP11 is to let some computer program do it for you. The most likely method is to have a file that contains binary program information, including the addresses of a few commonly used symbols, and then allow these symbols to be passed to other programs so that references to them can be calculated.

Such a file is called an *object file*, and it contains PDP-11 binary code in relocatable form, having no absolute addresses but with the relative position of certain symbols specified in an accompanying symbol table. When this file is combined with one or more other object files to produce a loadable program file, the absolute addresses of these symbols are calculated and inserted throughout the modules. Then the absolute binary file is produced in a form that can be RUN in the usual fashion.

The object modules are produced by the MACRO assembler including a table of any symbols that are declared to be .GLOBL and are to be passed to the linker. Then the files are linked together into a loadable program file by a linking program called LINK. The output of the linker program is a file in absolute binary format that has a .SAV extension as usual.

Thus, if we wished to call FLIP, FLOP and refer to TRAPH and $RI, we would declare in our program that these symbols were to be resolved later by the linker using the .GLOBL declaration

.GLOBL FLIP, FLOP, TRAPH, $RI

Then, when these symbols are encountered by MACRO, no error messages will be generated even though their values are not yet known. Instead, a flag is set in the object module indicating that these are .GLOBL addresses and that a value must be inserted there by the linker.

The absolute address of the entire program module may be decided at assembly time or at load time. If it is decided at assembly time, the MACRO directive .ASECT is used, followed by the absolute load address. If the modules are to be located at linking time, the MACRO directive .CSECT is used. Then the programs can be linked together using LINK.

The following examples indicate how the floating point package FPMP11, the input and output modules FPPIO, and your main program can be linked together.

Instructions for Assembling and Linking Programs Using FPMP11

1. Create your file using TECO. Any symbols such as TRAPH, FLIP, and FLOP that you wish to reference to FPMP11 must be declared .GLOBL. It is usual to start your program with

    ```
                    .ASECT
                    . = 1000
    START:          MOV #., SP
                    MOV #TRAPH, @#34
                    MOV #340, @#36
    ```

    You can also start your program with a .CSECT and let the LINKer locate it for you, but the listing will then start at 0 instead of 1000.
2. Assemble your program using MACRO. You must either include the file DEFNS.MAC in your source program or list it as part of the assembly instructions. Either type

    ```
    .R MACRO
    ```

    ```
    *MYFILE,TT: = SY:DEFNS,DK:MYFILE
    ```

or

    ```
    .MACRO MYFILE/LIST:TT:/NOSHOW:BEX
    ```

    after you have combined your file with the DEFNS file.

3. To produce an executable file, link your file with FPPIO and FPMP11. This is most easily done by

    LINK MYFILE,FPPIO,FPMP11

4. Finally, to execute your file, type

    RUN MYFILE

Calling Routines Directly Through R5

A number of the routines in the transcendental functions and certain other floating point utility routines can be called directly by a JSR R5, routine call. In these cases the argument, if floating point, must be in R0 and R1, and, if integer, in R0 when the call is made. The results are returned in R0 if integer or R0 and R1 if floating. Double-precision routines operate on four-word numbers in R0–R3. The routines are described in the following list.

Name	Function
ALOG	ln (argument)
ALOG10	log (argument)
AINT	sign (arg) times largest integer $\le$ | arg |
ATAN	arctangent (argument)
ATAN2	arc (arg1/arg2)
DBLE	converts single precision in R0–R1 to double precision in R0–R3
DLOG	double-precision ln (argument)
DLOG10	double-precision log (argument)
DCOS	double-precision cosine (argument)
DSIN	double-precision sine (argument)
DSQRT	double-precision square root (argument)
DATAN	double-precision arctan (argument)
DATAN2	double-precision arctan (arg1/arg2)
DEXP	double-precision exp (argument)
EXP	exp (argument)
FLOAT	floats R0 into R0–R1
IFIX	fixes R0–R1 into R0
IDINT	double-precision INT
INT	integer part of largest integer $\le$ | arg |
SIN	sine (argument)
COS	cos (argument)
SNGL	rounds double precision in R0–R3 to single precision in R0–R1
TANH	hyperbolic tangent (argument)

Buffered Input and Output Routines

As mentioned earlier, FPMP11 does no output itself. The functions FLIP and FLOP have been added for convenience by the author. There are times, however, when the ability to read or write characters into memory buffers and perform the conversion internally is desirable. The input and output routines of FPMP11 do just that. The input routines take a string of ASCII characters from a predefined memory region and convert them to a floating point or other form of number on top of the stack. The output routines convert a number on top of the stack to a string of ASCII characters in memory. Since the entire FPMP11 package was lifted from a FORTRAN operating system, it is not surprising that the features and limitations of the I/O routines are those defined by FORTRAN.

Input Routines

$DCI	Double-precision ASCII to FP
$ICI	ASCII to integer
$OCI	ASCII to octal
$RCI	ASCII to FP

Output Routines

$FCO	F—format output
$GCO	G—format output
$ECO	E—format output
$DCO	D—format double-precision output
$ICO	I—format integer output
$OCO	O—format octal output

All the input and output buffer routines are called by pushing various calling parameters onto the stack, and then calling the routines by a JSR PC, routine call. For input routines the result is on top of the stack, and for output routines in ASCII in a string of bytes in the memory buffer that was specified by the calling parameters. For example, to call the ASCII to FP routine $RCI, you must push the address of the start of the ASCII memory buffer area, the length of the ASCII field in bytes, the width of the decimal field or the position of the assumed decimal point, and a dummy argument called the P-scale, which is overridden by whatever is found and is therefore usually 0. This is done as follows for a 10-

byte field:

```
MOV #ASCBUF, -(SP)    ;PUSH START OF ASCII FIELD
MOV #10., -(SP)       ;PUSH WIDTH = 10 BYTES
MOV #3, -(SP)         ;NUMBER OF DECIMAL PLACES—
                      ;D IS OVERRIDDEN IF A DECIMAL
                      ;POINT IS FOUND
CLR -(SP)             ;PUSH P-SCALE = 0
JSR PC, $RCI          ;CALL REAL INPUT CONVERSION
                      ;ROUTINE
. . .                 ;RETURN WITH FP NUMBER ON
                      ;TOP OF STACK
```

Upon output, both the d-scale and the P-scale have meaning. The d-scale is the number of decimal places, and the P-scale is defined as follows. For D-, E-, and G-format conversions, the P-scale is the number of places to the left of the decimal, and the exponent is adjusted accordingly, For F-conversions, the P-scale is the power of 10 by which the number is multiplied before output. For F-format conversion, the P-scale, therefore, should always be 0. On input, the P-scale is meaningless.

The severe limitations of the format conversion routines derive from the card-oriented input FORTRAN specifications. Blanks are always treated as 0's, no matter how silly this might seem; and if a six-character field is specified and only the first character is nonblank, as in 6bbbbb, the result is treated as 600000 rather than 6. This calls for careful buffer manipulation before conversion.

Input Routines	Push	Then Call	Result in
$DCI	Address of ASCII field Length in bytes (w) Number of decimal places (d) P-scale (0)	JSR PC, $DCI	Stack
$ICI	Address of ASCII field Length of field in bytes	JSR PC, $ICI	Stack
$OCI	Address of ASCII field Length of field in bytes	JSR PC, $OCI	Stack
$RCI	Address of ASCII field Length of field in bytes Number of decimal places P-scale (0)	JSR PC, $RCI	Stack

Output Routines		Push	Then Call	Result in
$DCO, $ECO	Address of ASCII field			
$FCO, $GCO	Length of field in bytes			
	Number of decimal places			
	P-scale (0)	JSR PC, $DCO	ASCII	
	Number to be converted	$ECO	field	
		$FCO		
		$GCO		
$ICO, $OCO	Address of ASCII field	JSR, PC, $ICO	ASCII	
	Length of ASCII field in bytes		field	
	Integer to be converted	$OCO		

The only difference between $DCI and $RCI is that in the former a four-word rather than a two-word FP number ends up on top of the stack. Similarly, $DCO differs from the others in that a four-word FP number must be pushed onto the stack before conversion. All the input routines save and restore the general registers. The output routines destroy R0–R3 but save R4 and R5. Errors detected by the input routines leave a 0 on the stack for the FP or other number and return with the C-bit set.

Error Handling in FPMP11

The floating point package, as initially configured, has no special facilities for the handling of errors. Instead, all error conditions, such as overflow of a number or division by 0, automatically cause the processor to halt with the error code displayed in R0. These codes contain an error number in bits 15–8 (the upper byte) and an error class number in the lower byte, bits 7–0. The values of these error codes in decimal and octal are shown in Table 11.2. Error classes 0–4 automatically cause a halt. Errors of class 5 are normally ignored, but the FPMP11 package can be reassembled with a switch called CLASS5 set = 1 to cause halts on class 5 errors as well.

Naturally, it is often more desirable for the programmer to maintain control of these error conditions than to let the processor halt. This can be done by moving the address of an error routine the program into the global location $ERVEC as follows:

```
.GLOBL ERVEC      ;DEFINED IN FPMP11

MOV #ERROR, $ERVEC      ;INSERT YOUR ERROR ROUTINE ADDRESS
    :
    :
ERROR: MOV R1, -(SP)      ;SAVE ANY NEEDED REGISTERS AND HANDLE THE ERROR
```

Table 11.2   FPMP11 Error Codes

Class, No.	R0 Display	Issued by	Description
0, 0	0	TRAPH	Illegal trap instruction
3, 1	403	$ADD	Exponent overflow
3, 2	1003	$ADR	Exponent overflow
3, 3	1403	$DVD	Division by 0
3, 4	2003	$DVD	Exponent overflow
3, 5	2403	$DVI	Division by 0
3, 6	3003	$DVR	Exponent overflow
3, 8	4003	$DVR	Division by 0
3, 10	5003	$MLD	Exponent overflow
3, 11	5403	$NEG	Exponent overflow
3, 12	6003	$MLR	Exponent overflow
3, 14	7003	$MLI	Product larger than 16 bits
3, 22	13003	$RI	Real too big to fix
3, 23	13403	$DR	Exponent overflow
4, 2	1004	DEXP	Argument greater than 87
4, 3	1404	DLOG	Argument less than or equal to 0
4, 4	2004	DSQRT	Argument less than 0
4, 5	2404	EXP	Argument greater than 87
4, 10	5004	ALOG	Argument less than or equal to 0
4, 11	4404	SQRT	Argument less than 0
4, 12	6004	SNGL	Exponent overflow
5, 1	2403	$ADD	Exponent underflow (warning)
5, 2	1005	$ADR	Exponent underflow (warning)
5, 3	1405	$DVR	Exponent underflow (warning)
5, 4	2005	DEXP	Argument less than -88.7 (warning)
5, 5	2405	EXP	Argument less than -88.7 (warning)
5, 6	3005	$MLD	Exponent underflow (warning)
5, 7	3405	$MLR	Exponent underflow (warning)
5, 8	4005	$DVD	Exponent underflow (warning)

The error handling routine can be written to go on without losing control of register contents or the stack by saving any registers other than R0. Register R0 is saved in the FPMP11 $ERRA routine before it jumps to your error handler. Then, after testing for error type, you can continue where appropriate by restoring all registers other than R0 and executing an RTS PC instruction to go on from there. The only error from which it is impossible to continue is error 0,0, a trap handler error.

PDP-11/34 Floating Point Hardware

This section describes the floating point instructions of the PDP-11/34 and 11/40. The PDP-11/45 and higher-numbered machines have a much more sophisticated floating point processor; and since these minicomputers are used less often in the laboratory, the reader is referred to the PDP-11/45 manual for further details.

The PDP-11/34 allows manipulation of floating point numbers on the stack or on any new floating stack defined by any of the general registers. The instructions are as follows:

07500r	FADD reg	Adds two FP items on stack—result in second set of addresses.
07501r	FSUB reg	Subtracts first item on stack from second item—result in second item's addresses.
07502r	FMUL reg	Multiplies two FP items on stack together—result in second set of addresses.
07503r	FDIV reg	Divides first FP words *into* second FP words on stack.

Symbolically we can represent the FADD operation as

FADD (r)&(r + 2) + (r + 4)&(r + 6)   Put in (r + 4)&(r + 6)

where we use the symbols (r)&(r + 2) to represent the two words of a floating point number in the stack at the addresses pointed to by register "r" and in the following word. For example, if we wished to perform a floating addition between two floating point numbers stored in memory, we would first push the two numbers onto the stack, low-order word first, and then execute a FADD instruction:

```
;ADD TOGETHER NUMBERS F3.5 AND FTEMP

          MOV FTEMP+2, -(SP)      ;BEGIN PUSHING NUMBERS
          MOV FTEMP, -(SP)
          MOV F3.5+2, -(SP)       ;PUSH 2ND NUMBER
          MOV F3.5, -(SP)
          FADD SP            ;FLOATING ADDITION ON STACK
          MOV (SP)+, FTEMP        ;RESULT ON TOP OF STACK
          MOV (SP)+, FTEMP+2      ;2ND WORD

FTEMP:    .FLT2 xxxx
F3.5:     .FLT2 3.5
```

Note that although any register can be used for a stack, there are advantages to using SP so that FPMP11 can also easily communicate with the results. It then becomes identical with FPMP11 stack mode. Note especially that in subtraction you push the minuend onto the stack first. The number to be subtracted (the subtrahend) is pushed second. Likewise, during division, the dividend is pushed first, followed by the divisor. In all cases one of the two double-word operands is popped off the stack by the floating point instruction leaving the register pointing to the result in the locations formerly occupied by the first numbers pushed.

PDP-11/34 Floating Point Hardware Errors

If a floating point instruction results in an error such as overflow, underflow, or divide by 0, a trap occurs through location 244, and the condition codes have the following meanings:

	V	N	C	Z
Overflow	1	0	0	0
Underflow	1	1	0	0
Divide by 0	1	1	1	0

Modifying FPMP11 to Use the PDP-11/34 Hardware Instructions

It is possible to modify the FPMP11 source package to use the PDP-11/34 floating point instructions, but this cannot be done by a simple assembly switch, as no such provisions were made by the authors of the package. Instead, the add, multiply, and divide instructions must be inserted at the beginning of the routines, where the .IFDF FPU conditional assembly parameters are shown. The .IFDF FPU for $ADR04, $MLR05, and $DVR08 must be interchanged with the .IFNDF FPU assembly instructions. This causes the software simulation section to be ignored. Then the instructions for addition, multiplication, and division are changed from the 11/45 to the 11/34 codes. For the $ADR04 module, these instructions are as follows:

11/45 Instructions		11/34 Instructions
.WORD 170001	;SETF	FADD SP
.WORD 172426	;LDF	
.WORD 172026	;ADDF	
.WORD 174046	;STF	
JMP @ (R4)+		JMP @ (R4)+

Similarly, for the multiplication and division modules, the several 11/45 instructions are removed and replaced with FMUL SP and FDIV SP. Replacement of these instructions will not perceptibly speed up the FPMP11 because of the high trap handler overhead, but it will significantly decrease the space needed. The only way to speed up the actual calculation times is to use the instructions directly in your code.

Errors in FPMP11

Early versions of FPMP11 have the wrong constant in $STR01 and $STD01, the single- and double-precision store routines. The instruction preceding labels LP$46 and LP$47 should be MOV #15, R2. This is shown correctly in the source code of the printed manual but incorrectly in the assembled octal numbers as a 13. Early source tapes had MOV #13, R2 instead in both places.

Some copies of trap handler source are mispunched, so that the text shown as line 298 of the source code in the printed listing reads as .IFNDF CND$14 instead of .IFNDF CND$44.

Exercises

11.1   Write a program to calculate $y = mx + b$ for m and x entered at the Teletype and b stored in memory. The program should print out "M = " and "X = " before obtaining the values, and print out the value of y for the result.

11.2   Write a program to solve the quadratic formula $y = ax^2 + bx + c$ that makes use of Polish mode.

11.3   Write a program to convert from Fahrenheit to centigrade if the entered number is terminated with an "F," and from centigrade to Fahrenheit if the entered number is terminated with a "C."

11.4   Write a subroutine INOUT that can be used to print out and allow the modification of floating point constants in your program. The call should be

```
JSR R5, INOUT
address of constant
return here
```

When INOUT is called, it should print out the current value of the number whose address follows the call and then allow entry on the same line of a new value or just a Return. The number should remain unchanged if no new value is entered. If a valid number is entered, the value of that constant should be changed in memory.

□□□□□□□□CHAPTER TWELVE□

ADVANCED PROGRAMMING CONCEPTS AND TECHNIQUES

Position-Independent Code

One specific advantage of the PDP-11 instruction set over those of many other computers is that many of its addressing modes utilize PC-relative addressing. Thus in many cases entire blocks of code can be relocated in memory without any changes and run exactly as they stand without reassembly. To achieve this goal, however, it is necessary to remember which addressing modes are allowed and which are not. This is easily figured out by simply remembering how each mode works. In brief, the instructions that cannot be used are those which refer to an absolute address in a relative way so that, although a relative addressing mode is used, an absolute location, such as one of the trap or interrupt vectors, is actually referenced.

Consider the instruction

MOV #200, @#60

This clearly says to place the number 200 in location 60. But will it be assembled as the same code regardless of its position in memory? Let us

153

see. The addressing mode #200 is PC mode 2 and corresponds to

```
MOV (PC)+, xxx
    200
```

The value 200 is always in the location following the instruction, and thus the program will execute this instruction in the same way regardless of its location. The addressing mode @#60 is PC mode 3 and is again a relative mode equivalent to writing

```
MOV (PC)+, @(PC)+
  200
   60
```

No changes need be made to use this, regardless of its location. However, let us consider the instruction

```
KBVECT = 60
MOV #200, KBVECT
```

which has the same effect as the preceding one but is not assembled in the same way. Here, PC mode 6 is called for, requiring the code

```
MOV (PC)+, X(PC)
   200
   X
```

The value of the index constant X *varies with the position of the code in memory!* Thus PC mode 6, the PC-index mode, and PC mode 7, the indirect PC-index mode, must be avoided in writing position-independent code. This is more of a problem in coding input–output routines than in coding these vector loading instructions, however, since we have made it conventional to write such things as the teleprinter output routine as

```
TYPE:    TSTB TPS
         BPL TYPE
         MOV R0, TPB
         RTS PC
```

Note that these are PC mode 6 relative mode instructions, and the index constants would indeed vary with the distance between the instruction and the bus address of TPS and TPB. This can be avoided by simply rewriting the instructions as

```
TYPE:    TSTB @#TPS
         BPL TYPE
         MOV R0, @#TPB
         RTS PC
```

Note that there is no change in the definitions of the symbols TPS and
TPB. In both cases, these symbols represent the addresses of the Status
and Buffer Registers, 177564 and 177566.

Reentrant Routines

Sometimes it is necessary, particularly in time-sharing systems, to write
a subroutine that will be used by two separate program segments which
have no cognizance of each other. This might occur when different interrupt routines both call the same subroutine. In such a case it makes
sense to conserve memory space by sharing as many subroutines as
possible, even though there will be no control over such things as register
contents or intermediate storage registers. Instead, all such intermediate
data are stored on the stack. This sometimes means that the user must
write programs to pluck data out of the middle of the stack, as in the
preceding register saving example, and must carefully clean up the stack
when done.

Let us consider just the simple teleprinter output routine given earlier
as an example, although we must realize that this routine is usually handled by interrupts. In our case we will use it as a typical method of passing
data to a reentrant routine. The calling routine will be

```
MOV xxx, -(SP)     ;PUSH CHARACTER TO BE TYPED
                   ;ONTO STACK
JSR PC, TYPE       ;AND CALL TYPE ROUTINE
```

We can then proceed without worry, for no matter how many interrupts
occur between the two instructions, the stack will have returned to the
same state when the subroutine TYPE is finally reached. Then we must
pluck the datum from inside the stack and print it:

```
TYPE:   TSTB @#TPS       ;WAIT FOR TTY READY
        BPL TYPE
        MOV 2(SP), @#TPB ;PLUCK AND PRINT
        MOV (SP)+, (SP)  ;MOVE RETURN POINTER AND CLEAN UP STACK
        RTS PC           ;AND EXIT WITH CORRECT RETURN POINTER
```

Note that we pluck the character from the second stack location and then
copy the return pointer to this same position, popping the original return
pointer at the same time to prevent the stack from filling up.

Incrementing and Decrementing

Finally, let us consider the most efficient methods of incrementing and
decrementing registers. Of course, if we are accessing a particular data

word in a list only once, the most efficient way to go on to the next one is to use mode 2, or the indirect autoincrement mode:

CLR (R0)+

However, if the address is accessed several times, it is not always possible to increment the pointer after any given access to the word. In this case, the register must be incremented separately later.

Two less efficient methods of incrementing are

INC R0   or   ADD #2, R0
INC R0

These require two locations to increment the register by 2. The most efficient way to do this is by making a dummy test of the register:

TST (R0)+

which, while it affects the condition codes, adds 2 to the register in one instruction. Similarly, we can add 4 to the same register:

CMP (R0)+, (R0)+

2 to two different registers:

CMP (R0)+, (R1)+

or increment one register while decrementing another:

CMP (R0)+, −(R1)

all in a single-word instruction. Note that the values in the registers must be existing addresses for these instructions to work without causing a bus error trap.

Use of the Stack for Recursive Subroutine Calls

One of the principal reasons for selecting the stack architecture for a computer system is the ability to use the stack for recursive calls to subroutines. In brief, this means calls to subroutines that may call themselves or call other subroutines that call the first one. In the scientific laboratory this is of limited utility, although the stack storage mode can still be well utilized. One common example of a subroutine that does call itself, however, is a subroutine which converts data to another form in reverse order to that in which they will be printed out or transmitted to other routines. Such a subroutine is one to convert positive numbers to decimal for printing on the printer. The following subroutine DECOUT has these characteristics.

```
;CALLING ROUTINE
START:  MOV #.,  SP
        MOV NUM, RO      ;LOAD RO WITH NUMBER TO BE PRINTED OUT
        JSR PC, DECOUT   ;PRINT IT
S1:     HALT             ;AND HALT

;DECIMAL OUTPUT ROUTINE
DECOUT: MOV RO,  R1      ;COPY NUMBER INTO LOW ORDER WORD FOR DIVISION
        CLR RO           ;CLEAR HIGH ORDER WORD
        DIV #10., RO     ;DIVIDE INPUT NUMBER BY 10
        MOV R1,  -(SP)   ;PUSH REMAINDER ONTO STACK
        TST RO           ;IS QUOTIENT ZERO?
        BEQ DEC2         ;YES, START PRINTOUT OUT NUMBER
        JSR PC, DECOUT   ;ELSE RECALL THE SUBROUTINE TO FINISH THE CONVERSION

;WHEN ALL CALLS ARE DONE, RETURNS START HERE
DEC2:   MOV (SP)+, RO    ;POP LAST REMINDER OFF STACK
        ADD #60, RO      ;ADD ASCII BIAS
        .TTYOUT          ;TYPE THIS DIGIT
        RTS PC           ;AND RETURN TO DO NEXT DIGIT OR EXIT
;NUMBER TO BE PRINTED
NUM:    4096.
```

In this routine the input number is successively divided by 10 with the remainder saved on the stack each time. Then the routine is again called, pushing the return address DEC2 onto the stack each time. When R0–R1 has been divided down to 0, a branch to DEC2 occurs, where the last remainder will be popped off the stack. If the decimal number 4096 is to be converted, the stack will look like this when DEC2 is first reached.

460	4
462	DEC2
464	0
466	DEC2
470	9
472	DEC2
474	6
476	S1

Then the last remainder is popped off, biased, and printed. When the RTS PC instruction is encountered, the Stack Pointer contains 462, pointing to the start of the routine DEC2 again. The RTS PC instruction then pops this address off the stack into the PC, returning the program to address DEC2. At DEC2 the second digit of the output is popped from the stack, biased, and printed, and DEC2 is again called by the RTS PC instruction. Finally, after the "6" is printed, the last RTS PC instruction will put the contents of location 476 into the PC, returning the program to location S1, where it halts.

This simple example shows the great power of recursive calling of subroutines when this can be included in the program. Compare the complexity of this program with the answer to Exercise 9.9. We are grateful to Professor George Meyfarth of Tufts University for pointing out this simple example to us.

Command Files

One of the great powers of RT-11 is the ability to create .COM *command files* that can be used as part of RT-11 command strings. For example, it quickly becomes tedious to type

.MACRO FILE/LIST:TT:/SHOW:TTM/NOSHOW:BEX

but, using the command file construction, you can represent the part of this string after the filename by a commercial sign (@) followed by the name of a command file. Further, the MACRO command itself can be represented by the shortened command "M." Thus, if you create a file using TECO having the name MX.COM and the contents

/LIST:TT:/SHOW:TTM/NOSHOW:BEX

you can type the shortened command

M FILE @MX

MACRO Commands in the Assembler

It is sometimes useful to be able to generate repetitive sections of code within a program for various purposes. These might be simple routines or templates for tables or file entries. Such sections of code can be defined with .MACRO calls and referred to by the name you give them. The recurring code sequence is defined by

.MACRO name dumarg1, dumarg2,..., dumargn
statements
.ENDM

and the code is invoked by

NAME arg1, arg2,...argn

For example, suppose you wished to define some simple code to take the absolute value of an integer in memory. You could define the Macro to the assembler by

```
.MACRO SWAP A, B
       MOV B, -(SP)      ;PUSH B ONTO STACK
       MOV A, B          ;COPY A INTO B
       MOV (SP)+, A      ;COPY OLD B INTO A

.ENDM
```

Then you can generate the code for this swapping function anywhere you want by just writing

SWAP J, K   ;SWAP J AND K

The assembler treats the arguments you pass in the .MACRO calls as character strings and the dummy arguments in the .MACRO definitions as slots for such strings. It combines them and generates the actual in-line code for each call to the Macro.

This MACRO feature is considerably more elaborate than that outlined here, and includes the ability to nest Macros, generate variable expressions and generate local labels. These features are discussed in the *RT-11 Advanced Programmer's Guide*.

RT-11 System Addresses

Locations 0 and 40–57 are used by RT-11 to indicate the status of the job and monitor. These words have the following meanings:

0   System Restart address. If RT-11 is intact after a halt, you can restart at location 0 without re-bootstrapping.

40   Starting address of program.

42   Initial value of SP for program. Defaults to 1000 or top of .ASECT, whichever is larger.

44   Job Status Word (JSW). The bits of the JSW are set and examined by RT-11. They have the following meanings:

Bit	Meaning
15	USR Swap bit. Set if USR need not be swapped.
14	Disables conversion of lower case in EDIT.
13	Reenter bit. If set, program may be restarted from terminal
12	Special TT input mode. See .TTYIN below.
11–10	Used only in Foreground/Background monitor.
9	Overlay bit. Set if program uses Linker Overlay structure
8	CHAIN bit. If set, chaining takes place. See *RT-11 Guide*.
7	Set if you want to halt on an I/O error
6	Inhibit TTWAIT Bit. Foreground/Background only.
5–0	Unused.

46   Load address for USR.

50   Highest address used by program.

52   Byte 52 contains the EMT error code after an RT-11 Programmed Request.

53   User program error status.

54   Address of start of RT-11 monitor

56   Character used for filling for terminals requiring extra delays during long carriage returns. (Even byte only.)

57   Number of fill characters to be generated if required.

Swapping the USR

RT-11 is a minimum length monitor that must call in an overlay called the User Service Routine (USR). This swapping is done so that memory is saved on disk if the region selected is part of your program. You tell RT-11 where your program and its data arrays end by the contents of location 50. It is set to the top of your program by the Linker, but you can change it using the .SETTOP request described later. You can also specify an alternate location (instead of high memory) for the USR using address 46. If it is nonzero, it specifies such an alternate address.

Generally, you will not need to specify an alternate address. You may need to use .SETTOP to prevent the USR from overlaying some acquired data you want to work on.

Calling RT-11 Functions Through MACRO

The MACRO assembler does not in itself recognize special commands or characters for calling RT-11 system functions. However, the assembler does have an associated System Macro Library, called SYSMAC.SML, that does contain .MACROs for calling RT-11 functions. In general, these functions preserve all registers except R0, make their calls through a trap handler that traps the EMT instruction (through locations 30 and 32) and indicate completion through the clearing of the C-bit. Errors in these functions are indicated by the C-bit being set and the contents of *byte 52*.

To study the complete set of RT-11 calls, refer to the *RT-11 Advanced Programmer's Guide*. Some of the basic RT-11 calls are shown in Table 12.1.

Table 12.1 Common RT-11 Programmed Requests

.CLOSE	Close a specified file.
.CSIGEN	Call Command String Interpreter.
.CSISPC	Call Command String Interpreter but do not open files.
.ENTER	Create a new file for output.
.EXIT	Exit to RT-11.
.LOOKUP	Open an existing file.
.PRINT	Print a string on the terminal.
.READW	Read data from disk(ette) into memory.
.SETTOP	Set the top address your program will use.
.TTYIN	Read character from keyboard into R0.
.TTINR	Check the keyboard input.
.TTYOUT	Print character on terminal.
.WRITW	Write data from memory onto disk(ette).

Telling MACRO to Use RT-11 Requests

The MACRO directive .MCALL looks up the directives that follow in the file SYSMAC.SML and puts them in MACROs symbol table. Thus to use .EXIT and .TTYOUT, for example, you should put the statement

.MCALL .EXIT, .TTYOUT

at the beginning of your program.

Exiting to RT-11 from Your program

The RT-11 call .EXIT generates the code

EMT 350

which causes an exit to RT-11. Any I/O is completed first, and all I/O channels are released.

Terminal Input

The calls

.TTYIN and .TTINR

are used to get characters from the terminal. This allows simplified interrupt-controlled handling of characters and leaves RT-11 in control, allowing use of CTRL/C and in some cases of the CTRL/U and Delete keys.

The call

.TTYIN

waits for a character to be available and returns it in R0. The call

.TTINR which expands to EMT 340

returns with Carry set if there is no character and with Carry clear and the character in R0 if there is one. Thus the call

.TTYIN expands to   EMT 340
                      BCS . -2

Under the usual input mode, the characters are only available to your program after a Return has been struck. Then they are taken as requested from the RT-11 line buffer. If you wish the characters to be available immediately, without waiting for a Return, you must set bit 12 of the Job Status Word (Location 44) before calling the .TTINR or .TTYIN calls.

```
BIS #BIT12, @#JSW      ;SET SPECIAL INPUT MODE
.TTINR                 ;CHARACTER READY?
BCS LOOP               ;NOT YET
```

This bit stays set until you exit from your program. In this special mode, you cannot use the Delete (Rubout) or CTRL/U keys for editing your input line.

Character Output

The call .TTYOUT prints the character in R0 on the terminal. If you have a string of characters to print in a message, you can print them using the .PRINT call, where the argument following the call must be the address of the string.

```
.PRINT #MESG
    :
    :
```

       MESG: .ASCIZ /THIS IS THE MESSAGE/

If the message is terminated with a zero byte, .PRINT will print a CRLF at the end of the message. If the message is terminated with a 200 byte,

```
.PRINT #MESG2
    :
    :
```

       MESG2:   .ASCII /MESSAGE WITHOUT CR/
                 .BYTE 200
                 .EVEN

the text will be printed without the Return.

Arguments to RT-11 Requests

All the symbols used in RT-11 programmed requests are sources for MOV instructions and thus must be presented so that the correct value will be MOVed. For example, in the .PRINT request, the address of the message is moved into R0. Thus

    .PRINT #MESG1   expands to   MOV #MESG1, R0
                                         EMT 351

Note that the directive .PRINT MESG1 would generate the expansion

    ;WRONG EXAMPLE OF PROGRAM REQUEST
    .PRINT MESG1   expands to   MOV MESG1, R0
                                         EMT 351
MESG1: ...

and this would put the *contents* of the first word of the message in R0, rather than the address of the message.

Handling Files from Your Programs

If your program needs to read or write files that you have acquired as data or calculated, you must use the calls provided by RT-11 to access these files. If a file already exists, you must give the command .LOOKUP to find is position on the disk and the .READW command to read it into memory.

    If the file is to be created by your program, you must use the .ENTER command to reserve a space for it on disk and the .WRITW command to write it onto the disk.

    In either case, you must eventually tell RT-11 that you are done with the file by giving a .CLOSE command.

    The .ENTER and .LOOKUP commands are called with R0 pointing to a data area containing several constants that are needed by the monitor. These constants include the filename and device, an RT-11 channel number to be used later by the read, write and close commands. These constants are moved into this data area by code generated by the .LOOKUP and .ENTER Macro calls found in the file SYSMAC.SML.

    RT-11 files are composed of 256-word (512 byte) logical *blocks* regardless of their physical arrangement on a particular device. The .READW and .WRITW commands read a minimum of one block at a time. Thus you must always read at least 256 words from a file at a time.

The .LOOKUP Call

The RT-11 .LOOKUP call has the form

    .LOOKUP area, channel, fileblk

where area is a three-word block used by the call to store the constants
          and pointers,
        channel is an arbitrary channel number between 0 and 377,
        fileblk is a four-word block containing the device name, file-
          name and extension in a packed format called .RAD50, which
          packs 3 characters per word.

You don't need to put anything in particular in the *area*. The constants
are put there by the expansion of the .LOOKUP Macro. You can select
any channel number you want in the range of 0–377 as long as it has not
already been used within your program. These channel numbers simply
provide a way for other read and write calls to refer to this particular file
succinctly.

The *fileblk* must be a block that actually contains the filename. For
example, if you want to look up the file DX1:ETPYR.DAT, that file block
would contain

```
.RAD50 "DX1"
.RAD50 "ETP"
.RAD50 "YR"
.RAD50 "DAT"
```

A complete .LOOKUP call for this file would then be

```
.LOOKUP #LAREA, #2, #LFILE
        :
        :
        :
LAREA:  .BLKW 3 ;AREA WHERE CONSTANTS
                ;WILL BE STORED
LFILE:  .RAD50 "DX1"
        .RAD50 "ETP"
        .RAD50 "DAT"
```

This .LOOKUP call is expanded by the .LOOKUP Macro to

```
MOV #LAREA, R0      ;ADDRESS OF AREA IN R0
MOV #2, (R0)        ;CHANNEL NUMBER IN LOW BYTE
MOVB #1, 1(R0)      ;CALL TYPE IN HIGH BYTE
MOV #LFILE, 2(R0)   ;ADDRESS OF THE FILENAME BLOCK
EMT 375
```

where the LAREA block then contains

```
.BYTE 2,1   ;TYPE OF CALL AND CHANNEL NUMBER
LFILE       ;ADDRESS OF FILENAME BLOCK
```

The third word is only used by magnetic tape and cassette handlers and will not concern us here.

Before you can perform a .LOOKUP on a device, it is necessary that the device handler for the device containing the file be in memory. If the file is on the same type of disk or diskette RT-11 is running from, the appropriate device handler will always be present. Otherwise you must give the .FETCH call that loads it first. This call has the form

.FETCH address, devicename

where "address" is the address where the handler can be safely loaded (usually 256 words are sufficient) and "devicename" is a pointer to the .RAD50 device name.

If the first word of the filename in a .LOOKUP call is 0, the lookup will assign the channel so that the block numbers are absolute without regard to any files already present and allow you to write any 256-word block to any block on the disk. This is usually very dangerous!

On return from a .LOOKUP, R0 contains the number of 256-word blocks in the file just looked up. R0 will be zero if the filename had a 0 first word as shown earlier.

The .RAD50 Format

The RAD50 packing scheme allows you to pack 3 characters in 16 bits according to the rules

Character	RAD50 Code	ASCII Code
space	0	40
A–Z	1–32	101–132
$	33	44
.	34	56
undefined	35	—
0–9	36–47	60–71

Then the 3 characters are packed according to the equation

$$RAD50 = [(C1 \times 50) + C2] \times 50 + C3$$

Errors in RT-11 Calls

All RT-11 calls return any error conditions in location byte 52. The number of each possible error is given with the command. If an error occurred, the C-bit will be set, and you should examine byte 52 to find out which error occurred. The errors for .LOOKUP are

0　　Channel already open.
1　　File not found.

Reading a File into Memory

The .READW request has the form

.READW area, channel, buffer, wordcount, blknum

where　area is a five-word block,
　　　　channel is the channel number requested in .LOOKUP,
　　　　buffer is the address where the data is to be read,
　　　　wordcount is the number of words to be read by the call,
　　　　blknum is the number of the block to be read starting with the
　　　　　beginning of the file as block 0.

A typical call would be

```
.LOOKUP #LAREA, #2, #LFILE        ;AS BEFORE
MOV R0, WRDCNT                    ;SAVE THE COUNT
BCS LOKERR                        ;LOOKUP ERROR
.READW #RAREA, #2, #DATA,WRDCNT, #0      ;READ IN THE FILE
BCS RDERR                         ;READ ERROR HANDLER
```

Upon return, R0 will contain the number of words actually read. Possible errors are

0　　Attempt to read past end of file
1　　Hardware read error
2　　Channel not open

The .ENTER Request

The .ENTER request is analogous to the .LOOKUP request except that it is used when you want to create a file on disk. You can choose the size of the file or open the largest one available. As before, you must select an unused channel number to go with your request. The .ENTER request has the form

.ENTER area, channel, fileblk, length, count

where   area is a four-word argument block,
        channel is a channel number between 0–377,
        fileblk is a four-word filename block in .RAD50,
        length is 0 to open half the largest empty, −1 to open the
          largest empty, and m to open a file of *m* 256-word blocks,
        count is an optional argument used only with mag tape and
          cassettes.

The file created is not permanent until a .CLOSE is given, and the
final size is not determined until that time. On return, R0 contains the
number of blocks actually allocated. The error codes are

0   Channel already in use
1   No space ≥ *m* found or directory full

The .WRITW Request

You can write on disk or diskette with the .WRITW request, which is
entirely analogous to .READW. You can write to a file whether it is
existing and opened with .LOOKUP or new and created with .ENTER.
The form of the call is

.WRITW area, channel, buffer, wrdcnt, blknum

where   area is a five-word argument block,
        channel is a channel number,
        buffer is the address in memory where the transfer is to begin,
        wrdcnt is the number of words to be transferred,
        blknum is the number of the block to be written.

Upon return, R0 contains the number of words actually transferred.
The possible errors are

0   Attempt to write past end of file
1   Hardware error
2   Channel not open

The Command String Interpreter

Thus far, we have seen how to manipulate files whose names are specified
in .RAD50 blocks defined within the program. The Command String Interpreter (CSI) allows these file specs to be entered at the keyboard, in

the form

output1, output2, output3 = input1, input2,..., input6

The format of the CSI call in the general mode is .CSIGEN devspc, defext, command

where   devspc is the address of the memory region above the program where device handlers can be loaded. No memory used if all files come from the system device. Always put some address there to satisfy the Macro expansion, however.
defext is the address of a four-word block containing the default extensions for the input and output files:

```
DEFEXT: .RAD50 'FID'    ;DEFAULT EXTN FOR ALL INPUT FILES
        .RAD50 'SPC'    ;DEFAULT EXTN FOR OUTPUT FILE 1
        .RAD50 'DAT'    ;DEFAULT EXTN FOR OUTPUT FILE 2
        0               ;DEFAULT FOR OUTPUT 3 OR 0 IF NONE
```

command is the address of a buffer containing the command, or #0 if the command is to come from the keyboard.

.CSIGEN opens all the files and assigns channel numbers. The three possible output files are assigned channels 0–2, and the six possible input files channels 3–10. Possible errors are

0   Illegal command syntax
1   Illegal device
2   Unused
3   Directory full
4   Input file not found

Calling the CSI in the Special Mode

The .CSISPC call interprets the filenames and converts them to RAD50, but does not do any lookups or enters. Instead, it leaves the filenames in a 39-word file block, consisting of four words each for the six possible input files and five words each for the three possible output files.
The format of the .CSISPC call is

.CSISPC outspc, defext, comand

where   outspc is the address of the 39-word block where the file names are placed. The output file blocks contains the four filename words and a fifth word containing a size request, if one was entered in brackets after that filename. The output files are listed first in the block, followed by the four-word input blocks, starting at byte 36.

Reserving Memory with .SETTOP

The .SETTOP request has the form

.SETTOP addr

where addr is the address of the upper limit of the program and any data. On return, R0 and word 50 contain the highest word assigned to your program. You can also change address 50 directly and have the same effect. The only difference is that RT-11 checks to make sure that the address you wish to enter is legal.

Closing a File

The .CLOSE request closes a file on a particular channel. It has the form

.CLOSE channel

where *channel* is the channel to be closed. This makes all .ENTERed files permanent in the directory and stores their final size.

Writing Modules That Can Be Called by FORTRAN

One easy way to handle calculation segments in a program is to perform them in FORTRAN and reserve the assembly routines for the actual display and data acquisition. Any subroutine that is not mentioned in the FORTRAN source code will automatically be treated as a .GLOBL and can be linked with the LINKer program. The MACRO subroutines are called in the JSR PC, calling mode, with R5 pointing to a block of arguments as follows:

R5 →   number of arguments
        address of arg1
        address of arg2
        address of arg3
            etc.

You may use all of the registers without restriction. FORTRAN FUNC-TIONS return integer values in R0 and floating point values in R0–R1.

CHAPTER THIRTEEN

PROGRAM STRUCTURE AND DOCUMENTATION

Once you have written a program for some laboratory or general computing purpose, it is important to realize that it must be well documented or it will never be used. It must, in addition, be easy to use if anyone, including you, is ever going to use it. In this chapter we discuss how to structure a program and how to describe its use.

Program Structure

All programs should start out with a large block of comments describing what they are intended to do and how this is accomplished. In no case should these comments *replace* the actual how-to-use-it documentation, but they should explain how the program works in some detail.

Although you might think that comment frequency diminishes with programming experience, exactly the reverse is true: Experienced programmers often start out their programs with a two- or three-page "letter to Mom" describing the program. Then they need only mark each section with smaller sets of comments germane to that program section.

Programs also should contain a section where all symbolic constants that are not part of the standard MACRO definitions are defined. These could include

```
CR = 15          ;ASCII FOR CARRIAGE RETURN
TAB = 11         ;ASCII TAB CHARACTER
BUFSIZE =256.    ;SIZE OF DISK BUFFERS

;etc.
```

Following this should be any .GLOBL symbols to be passed between program modules and MACROs:

.GLOBL FLIP, FLOP, FLAC, TRAPH
.MCALL, .EXIT, .TTYINR, .TTOUTR

Then the program code itself should start with the assembly mode (.ASECT OR .CSECT) and starting address, if any. The first section of the program should be the *initialization* section, where all constants and hardware registers are set to their starting values, and the carriage itself is initialized.

```
            .ASECT
            .=1000   ;STARTING ADDDRESS

    START:  MOV +., SP     ;INITIALIZE THE STACK POINTER
            MOV #TRAPH, @#34 ;INITIALIZE THE TRAP HANDLER
            MOV #340, @#36  ;PREVENT INTERRUPTS
            JSR PC, CRLF    ;AND THE CARRIAGE
```

Then, the program should print out some sort of "prompting" character to tell the user that it is ready for commands. Since RT-11 uses a period (.) for a prompt and most system programs an asterisk (*), we suggest that you use some other prompting character such as a greater than sign (>) or a number sign (#) to prompt within your program. If at all possible, you should preserve the RT-11 interrupt system so that CTRL/C can be used to exit from or abort your program.

The Command Interpreter

Most programs will have a command decoder that takes one-letter or multiple-letter commands and that has some sort of vectoring scheme to recognize a command and jump to the proper routine. A simple prompt and command decoder could simply test for single letters of the alphabet. A slightly more complicated one might put two characters into two bytes and compare them with a table of two byte commands. These command schemes are discussed in the following sections.

The Single-Character Command Interpreter

This simple routine idles until a character is struck. It then compares it directly with one of a few command characters and branches directly to the routine specified. This simple method is suited to programs having only a few commands, especially when all of them are near the command interpreter so that they can be reached with a BR instruction.

Since the program usually has nothing to do while waiting for a command, it is common to insert a display routine in laboratory systems displaying the data that has been acquired or is about to be processed. We will just refer to this routine by name for now and show what it contains in Chapter 15.

```
;SINGLE CHARACTER COMMAND  INTEPRETER

         BIS #BIT12, @#JSW ;TTY SPECIAL INPUT MODE
PROMPT:  JSR PC,CRLF       ;INITIALIZE THE CARRIAGE
         MOV #'>, R0        ;AND PRINT A PROMPTING '>'
         .TTYOUT           ;RT-11 PRINT FROM R0

DLOOP:   JSR PC,DISPLAY    ;REFRESH THE DISPLAY ONCE
         .TTYINR           ;AND SEE IF A CHARACTER HAS BEEN STRUCK
         BCS DLOOP         ;NOT YET, CONTINUE DISPLAYING

;CHARACTER HAS BEEN STRUCK AND IS IN R0
;COMPARE IT WITH LEGAL COMMANDS
         CMP R0, #'C       ;WAS THIS A C?
         BEQ CONTR         ;YES CONTRACT THE DISPLAY
         CMP R0, #'X       ;WAS IT AN X?
         BEQ XPAND         ;YES, EXPAND THE DISPLAY
         CMP R0, #'P       ;WAS IT A P?
         BEQ PLOT          ;YES, BEGIN PLOTTING

;IF NOT ANY OF ABOVE COMMANDS, PRINT A '?'
         MOV #'?, R0
         .TTYOUT
         BR PROMPT         ;AND WAIT FOR A NEW COMMAND CHARACTER
```

The Two-Character Command Interpreter

This command structure is suitable for programs having more commands, which may be at some distance from the interpreter. The routine uses a *command table* consisting of the two-character command in the two bytes of one word followed by the address of the routine in a second word. The table is terminated by a pair of 0's, which allows the scanner to exit on finding a 0 without knowing how many entries there will be in the table.

The command is received as two characters. The program idles in a display routine until the first character is struck and then puts the first character into the lower byte of a variable COMD. Then it idles until a second character is struck and puts this character into the upper byte of the COMD word. Then, this word containing two characters is compared with the entries in the command table and if a match is found, the interpreter jumps to the address contained in the next higher word in the table.

```
;TWO CHARACTER COMMAND INTERPRETER

          BIS #BIT12, @#JSW ;TTY SPECIAL INPUT MODE
PROMPT: JSR PC,CRLF      ;INITIALIZE THE CARRIAGE
          MOV #'>, R0      ;PRINT A PROMPTING CHARACTER
          .TTYOUT
DLOOP:  JSR PC, DISPLA ;DISPLAY THE DATA
          .TTYINR         ;WAS A CHARACTER STRUCK?
          BCS DLOOP       ;NO, KEEP DISPLAYING AND CHECKING
          MOV R0, COMD    ;PUT COMMAND CHARACTER IN LOWER BYTE
DL1:    JSR PC, DISPLA ;DISPLAY AND WAIT FOR 2ND CHARACTER
          .TTYINR         ;CHARACTER STRUCK?
          BCS DL1         ;NOT YET
          MOVB R0, COMD+1 ;PUT 2ND CHAR IN HIGH BYTE

;COMMAND ASSEMBLED, NOW SCAN THE COMMAND TABLE
          MOV #COMTAB, R0 ;SET POINTER TO TOP OF TABLE

CMLOOP: CMP COMD, (R0)+ ;DOES IT MATCH THIS COMMAND?
          BEQ MATCH       ;YES, EXECUTE THAT ROUTINE
          TST (R0)+        ;ELSE ADVANCE R0 AND KEEP LOOPING
          BEQ ENDCOM      ;IF 0 THAT WAS THE END OF THE TABLE
          BR CMLOOP       ;ELSE KEEP LOOKING

;COMMAND MATCH FOUND, JSR TO THAT ROUTINE
MATCH:  JSR PC, (R0)
          BR PROMPT       ;WHEN DONE RETURN HERE - RESTART INTERPRETER

;NO MATCH FOUND, TYPE '?'
ENDCOM: MOV #'?, R0      ;PUT ?-MARK IN R0
          .TTYOUT         ;PRINT IT
          BR PROMPT

COMD:    0               ;COMMAND ACCUMULATED HERE

;COMMAND TABLE HAS THE FOLLOWING FORM
COMTAB: 'PL          ;PL MEANS PLOT
          PLOT         ;ADDRESS OF PLOT ROUTINE
          'GO          ;GO MEANS START DATA ACQUISITION
          GO           ;ADDRESS OF ACQUISITION ROUTINE
          'FT          ;FT MEANS DO FOURIER TRANSFORM
          FOURIER ;ADDRESS OF FFT ROUTINE
          0,0          ;TERMINATORS TO TABLE
```

Structure of Individual Routines

Each routine in the program should be reasonably compact and readable without jumping around over several pages of code. It should call subroutines for common functions to shorten and structure the overall appearance of the main routines.

The concept of *structured programming* is one in which programs are written in a *top-down* modular form. This means that the entire command routine may be only five or six lines long; each line is a call to a subroutine. The subroutines can then be written and tested separately without disturbing the structure of the command itself. This also makes it possible for more commands to use the same routines.

The structured programming methodology also is one in which a lot of "jumping around" is avoided by laying out the subprograms in a logical, linear, readable fashion. In addition, it is sometimes desirable to further indent sections of code that are inner loops to show that they are subsidiary to the outer loops and the main program code.

Any routine in the program that has a long execution time, such as a data acquisition routine, should be interruptable from within the program without causing exit to the monitor and loss of the constants associated with the current processing and acquisition functions.

Program Gullibility

A "gullible" program believes whatever it is told. A more sensible program is one that checks carefully for the consistency and logic of what it has been told to do. The most common example of such gullibility lies in the *zero case*. For example, you might tell your program to plot an array when no array has yet been defined, resulting in a plot of zero points. A gullible plot routine may treat zero points as a counter which needs to be decremented 65,536 times before it again reaches 0, leading to troublesome problems that may not be recoverable.

Other cases might be unreasonable commands or data. We have already seen how a command interpreter checks for a legal command and prints an error character if there is no match. More sophisticated routines might check for the physical reasonableness of the parameters associated with the data. For example, continuing to add base to an aqueous solution having a very high pH will not usually cause anything more to happen, and lowering a temperature controller below 0° in a water sample may have disastrous results for the glass sample tube and sensing devices.

Documentation of Programs

Once you have written and tested your program, it must still be *documented*. Even programs that you intend only for your own use should be documented, since you can easily forget in a few months exactly how the program works. It is also an important logical exercise to document a program, since you will often find logical holes in how it operates in attempting to describe it. In fact, it can easily be shown that you might well write the documentation *first* and *then* write the program since this documentation will be a detailed series of specifications for the program's operation.

For novice or sometime programmers, however, it is still better to document the program after it is written than not at all. You should also resist the temptation to write a "self-documenting" program that prints out reams of messages describing how to use it, since the new user usually will not know how to start it or how to get at these messages. Further, reading output while running a program is hardly as satisfactory as coming to the lab prepared to the use the program knowledgeably.

Outline of a Program Document

The following list provides a brief outline of the major points to be covered in documenting a program. Many of these points might require whole chapters in complex programs, but the typical novice program will require only a page or two to document.

1. Simple program title. This gives the reader a name to hang on the program (if it works) or hang the program by if it doesn't.
2. Date or version number and author's name and address.
3. A one- or two-sentence abstract of the program's purpose.
4. Hardware requirements for running this program: disks, displays, analog-to-digital converters, and so on.
5. Software requirements: any related programs or monitors.
6. Instructions for loading and running the program.
7. Command conventions: how to enter commands. What to do if you make a mistake.
8. List of commands and subcommands with brief one-line descriptions of each.
9. Detailed explanations of each command's purpose.
10. Examples of use.
11. Exiting from the program and leaving the system.

Documentation for the SCSQRT Program

The SCSQRT Program by Ada Lovelace 2/7/79.   The SCSQRT program is used to find the sine, cosine and square root of numbers entered at the terminal. Illegal values cause a question mark to be printed.

The SCSQRT program is stored on diskette JWC287 and thus requires a computer having a dual diskette drive, running the RT-11 monitor system. A video or hard copy terminal is also required. The program contains and utilizes the FPMP11 floating point math package.

Loading and Running the Program. To use the program, place the diskette JWC287 in drive 1. RT-11 must be running. If typing a Return does not produce a "." prompting character, bootstrap the system in the usual way. When RT-11 starts and prints its prompt character, run the program by typing

RUN SCSQRT

The program will start and print a ">" prompt character.

Using SCSQRT. To use the program, type a number after the prompt, and terminate it with one of the following command characters:

C   Take the cosine
Q   Take the square root
S   Take the sine

SCSQRT will perform the specified function and print out the result preceded by an equal sign. All values are legal for sine and cosine and are entered in radians. The program will type a "?" if you try to take the square root of a negative number. After the command is executed, the program will print a Return and a new ">".

Exiting. To exit from the program, type a CTRL/C. This will return you to the RT-11 monitor. Take out the diskette and store it in its jacket.

```
; PROGRAM SCSQRT

;        BY ADA LOVELACE, 2/7/89

; THIS PROGRAM TAKES THE SINE, COSINE OR SQUARE ROOT
; OF A NUMBER ENTERED AT THE KEYBOARD
; THE COMMANDS ARE
;        C-       TAKE COSINE
;        Q-       TAKE SQUARE ROOT
;        S-       TAKE SINE

; ALL OTHER CHARACTERS CAUSE A ? TO BE PRINTED
; THE PROGRAM REQUIRES FPPIO AND FPMP11 TO WORK PROPERLY

; FPMP11 GLOBALS
. GLOBL FLIP, FLOP, TCHAR, TRAPH, FLAC

; CONSTANTS
CR=15
LF=12
; FPMP11 DEFINITIONS
FSQRT=TRAP+46
FPSIN=TRAP+36
FPCOS=TRAP+37

; RT-11 MACROS
. MCALL . TTYOUT

. ASECT
. =1000
START:  MOV #. , SP      ; INITIALIZE THE STACK
        MOV #TRAPH, @#34
        MOV #340, @#36   ; AND THE TRAP HANDLER
```

```
;GET A NUMBER FOLLOWED BY A COMMAND CHARACTER
LOOP:    JSR PC, CRLF      ;ALWAYS START ON A NEW LINE
         MOV #'>, R0       ;PRINT PROMPTER
         .TTYOUT
         JSR PC, FLIP      ;GET NUMBER INTO FLAC
         CMP TCHAR, #'S    ;WAS THER TERMINATOR AN S?
         BEQ SINE          ;YES, TAKE THE SINE
         CMP TCHAR, #'C    ;WAS IT A C?
         BEQ COS           ;YES, TAKE THE COSINE
         CMP TCHAR, #'Q    ;WAS IT A Q?
         BEQ SQRT          ;YES, TAKE A SQUARE ROOT

;ERROR IF NONE OF THE ABOVE
;PRINT A QUESTION MARK AND GET A NEW COMMAND
ERR:     MOV #'?, R0       ;PUT ?-MARK IN R0
         .TTYOUT
         BR LOOP           ;AND GO BACK FOR NEXT COMMAND

;TAKE THE SINE OF THE NUMBER IN THE FLAC
SINE:    FPSIN
         BR OUTPUT         ;AND PRINT THE RESULT

;TAKE THE COSE OF THE NUMBER IN THE FLAC
COS:     FPCOS
         BR OUTPUT         ;AND PRINT IT


;TAKE THE SQRT OF ANY + NUMBER IN THE FLAC
SQRT:    TST FLAC          ;IS IT +?
         BMI ERR           ;NO, PRINT A ?-MARK
         FSQRT             ;ELSE TAKE THE SQUARE ROOT

;OUTPUT ROUTINE, PRINTS AN = AND THE RESULT
OUTPUT:MOV #'=, R0         ;PRINT =-SIGN
         .TTYOUT
         JSR PC, FLOP      ;PRINT CONTENTS OF FLAC
         BR LOOP           ;AND GO BACK FOR NEW COMMANDS

;CARRIAGE RETURN LINE FEED ROUTINE
CRLF:    MOV #CR, R0       ;PRINT CR
         .TTYOUT
         MOV #LF, R0       ;PRINT LF
         .TTYOUT
         RTS PC  ;AND EXIT

    .END START
```

CHAPTER FOURTEEN

DEBUGGING
USING ODT-11

As soon as your programming becomes at all complex, it becomes apparent that programs do not work perfectly the first time. There is nothing embarrassing in this, and it should be recognized that almost no nontrivial program is without "bugs." In fact, one of the "laws of programming" states that a sufficient condition for triviality is that the program has no bugs.

Because of this "natural law," fairly complex programs will generally have to be "debugged" thoroughly as part of the testing procedure before they can be used for their intended purposes. This usually means that you must have a method of assuring yourself that a particular program is indeed producing the right answers. Most commonly this requires that a suitable hand-calculated case be available for comparison with the results of your program. This case should, in fact, be calculated using the same method or "algorithm" so that comparison of intermediate results is also possible. With the advent of hand calculators, this requirement is no longer as onerous as it once was.

Nonetheless, once a program has run and has not produced the expected answer, the question arises of how to debug it. This usually means stepping slowly through the program, or at least through the offending sections of it, in such a way as to be able to check on the intermediate results. It would be possible to simply place the computer in single-instruction mode by placing the Halt switch down and depressing Continue for each instruction to be executed, but in the PDP-11 this allows the programmer to examine only the address from which each instruction is being executed. This display of data in R0 is not active in the single-step mode. Furthermore, the examination of specific registers and memory locations can be extremely tedious and, with the misuse of the Load, Add, and Start switches, can easily lead to severe programmer frustration and error. Finally, if the program bugs lie far within the program, the amount of single stepping necessary far exceeds the patience of most human beings.

One possibility for debugging such programs is the insertion of a HALT instruction into the program code or binary instructions at a position where an error is suspected. However, this method is again prone to error, not only because of the complexity of examining and changing various locations, but also because the instruction must then be reinserted in place of the HALT and a new HALT inserted somewhere without disturbing any other registers. This can be a bit tricky.

For these reasons, the manufacturer has designed a series of programs called ODT for the examining, changing and debugging of programs. They are different in length and complexity, with the original "paper tape" version of ODT occupying about 2100 bytes. The longer and much more elaborate version of ODT supplied with RT-11 as an .OBJ file has a number of advantages, but since it occupies substantially more memory, we will not discuss it here. A summary of its commands are given at the end of this chapter.

A slight change in ODT-11 for calling subroutines directly from ODT, which increases its debugging versatility considerably, is described later in this chapter.

ODT has two main uses: (1) the examination and change of memory locations and (2) the insertion of program breakpoints to simulate halts in the program in a more efficient fashion. When ODT is RUN or started from the Switch Register, it prints an asterisk to indicate that it is ready for commands. Thus the program to be debugged must be in memory before you start ODT, and you must be sure that the memory regions of ODT and the program do not conflict. At Tufts, we have modified RT-11 to leave the last 2048 words for utility programs.

134000	RT-11 bootstrapper
135000	ODT
137500	Paper tape binary loader
137744	Paper tape bootstrap loader

Thus, at any time, you can load a program into memory and start RT-11 by

```
.GET MYPROG
.STA 135000
*
```

The commands for ODT are as follows:

n/   Print out the contents of location *n* and allow modification.
      The new value must be terminated with one of the following
      characters:
              Return: enter new value in memory.
              Line Feed: enter new value in memory and open location
                        n + 2.
                  mO: calculate the word and byte offsets between
                        location *n* and address *m*. Neither is entered
                        in memory unless retyped.
              ←or _: take contents of opened location, add contents
                        of PC to it, and open that location.
/   Open last examined location. Works only if no P command has
      been given since last location was examined.
nB   Set a breakpoint at location *n*.
B   Remove the breakpoint.
nG   Begin executing instructions at address *n*.
P   Proceed from the last breakpoint.
nP   Proceed from the last breakpoint, and pass through the break-
      point *n* times before allowing a new breakpoint to occur.
$n   (*n* = 0–7) Open the contents of registers R0–R7.
$B   Find address of current breakpoint.
$M   Set Mask value.
nW   Search for words that, when XORed with *n* and ANDed with
      the Mask are 0.
nE   Search for words that reference effective address *n*.
$S   Examine and change saved PS.
$P   Examine and change ODT's priority.

RT-11 ODT expects to find a semicolon as part of each command to
separate parts of the commands from other parts. Therefore, to maintain
compatibility, ODT-11 ignores semicolons any place in the command. All
numbers entered in ODT are octal and are truncated to 16 bits from the
left. Excess digits, therefore, are ignored.

Examining and Changing Locations

To examine any memory location, simply type the address of the location
or device, followed by a slash. The address must be even in ODT, whereas
RT-11 ODT allows odd (byte) addresses.

500/12706

ODT prints out the underlined portion, followed by a space, allowing you
the opportunity to enter a new number or type a Return or Line Feed.
A Return simply closes the location without modification. A Line Feed
closes the location and opens the next sequential location (N + 2):

    500/12706 (LF)
    000502/000500

The process can be repeated as many times as desired. If a new number
is entered before the Return or Line Feed, it replaces the contents of that
memory location:

    000502/000500   600 (Return)
    *

In this example the contents of location 502 are changed from 500 to 600.

If a back arrow or underline character is typed (← or _), the contents
of the memory location are added to the address of that location to form
a new address, which is opened for examination and modification. On
some terminals the underline character appears, and on others the back
arrow. The two have the same ASCII value in different versions of the
ASCII code. This feature can be used in calculating and checking offsets.

If a slash is typed by itself in response to ODT's asterisk, the last-
examined location is reopened for modification. This applies only while
ODT is running. If a command is given to start a program and that program
returns to ODT through a breakpoint, the address of the last-opened
location is lost, and it must be reentered.

Caution: If ODT-11 is used to examine a nonexistent memory location
or device address, a trap through location 4 will occur. This can lead to
a crash.

Breakpoints

It often happens that the only way to find out where a program is mis-
behaving is to step through it carefully. This can be done very easily using
ODT. Typing the command nnnnB places a *breakpoint* in a program at
location *nnnn*. This breakpoint is actually a BPT (breakpoint trap) instruction, which causes a trap through location 14 and a transfer of control
to ODT. ODT remembers the contents of address *nnnn* and either restores
them or simulates them when debugging of the program continues. Note
that the address of the breakpoint is the first instruction in sequence *not*
to be executed. When the processor encounters the breakpoint, it jumps
immediately to ODT. Breakpoints cannot be placed in any memory lo-

cation that is changed by the program or in any but the first word of a multiple-word instruction.

In ODT-11 only one breakpoint may exist at a time. The principal difference in RT-11 ODT is its ability to handle eight breakpoints simultaneously.

When a breakpoint occurs, ODT-11 types out the address of the breakpoint and awaits any of its legal commands. Usually the programmer examines some memory locations or some registers, perhaps changing some, and occasionally altering some instructions, and then allows the program to continue by typing the P (proceed) command. Often the breakpoint is moved by the programmer before typing the P command so that the program will continue from the old breakpoint to the new one when the P command is given. Remember that the P command always continues from the breakpoint in force at the time ODT-11 was entered. Thus you can change the breakpoint, and then change it again if you wish, and the P command will still work correctly.

Examining Registers.   The $n command will allow the examination and change of the saved internal registers. Whenever a breakpoint occurs, the contents of R0–R7 and the PS are saved by ODT, so that its operation will not destroy them and they can be restored when a P command is given. It is sometimes useful to examine and change these registers by typing the $n command. For example, to examine R3, we type

$3/000123

and ODT-11 prints out the saved contents of R3. They can be changed in the usual fashion by typing a new number before the Return, and the remaining registers can be examined sequentially using the Line Feed command. Similarly, the saved PS can be examined and changed with the $S command.

Examples of the Use of the Breakpoint Feature.   Let us suppose that we think we have written a program to add together 10 numbers and halt with the sum displayed in R0. The following program purports to do this. However, we find upon running it that it never stops. We must find out why.

```
      ;PROGRAM TO ADD TEN NUMBERS TOGETHER
500 012701          MOV #1000, R1  ;SET POINTER
502 001000
504 012702 LOOP:  MOV #12, R2     ;SET COUNTER
506 000012
510 005000          CLR R0          ;CLEAR SUM
```

512 062100	ADD (R1)+, R0	;ADD EACH NUMBER
		;INTO SUM
514 005302	DEC R2	;DONE?
516 001372	BNE LOOP	;NO, GO BACK
520 000000	HALT	;YES, STOP AND
		;DISPLAY RESULT IN R0

Having loaded this program and established that it does not work, we reload it without starting it, so that in case it has modified itself we will have a fresh copy to work with. Then we load and start ODT. As a start let us put a breakpoint at address 504, just to make sure that the program starts correctly:

*504;B   Put a breakpoint at 504.
*500;G   Start the program at 500.
B;000504 A breakpoint occurs at 504, and the program jumps to ODT.

Now we might examine some things to see whether anything has yet changed.

500/12701 Location 500 is examined and found to be correct.
$1/001000 Register R1 is examined and found to contain 1000 as expected.

Now let us move the breakpoint on to location 514 and examine the running sum.

*514;B   Put a breakpoint at 514.
*;P   Proceed from the last breakpoint until a new one is found.
B;000514 Breakpoint is encountered at 514.

Now we should see how the sum, pointer, and counter look. The sum is in R0, the pointer in R1, and the counter in R2, so we just examine these three registers sequentially:

$0/000005 (LF)   The sum is now 5. A new register will be opened if a Line Feed is struck.
xxxxxx/001002 (LF)   The pointer now points to 1002. The number typed out, represented by the x's, depends on where ODT is loaded. It is the internal address where that register is saved.
xxxxxx+2/000012 (Return) The counter is now 12 as expected, since it has not yet been decremented. Remember that the instruction in the loca-

tion containing the breakpoint is not executed.

At this point all things still look fine, so let us go on to see what happens after two more summations have been performed. This can be accomplished by typing the command ";P," waiting for a breakpoint to occur, and then typing ";P" again, or it can be accomplished by typing "2;P." The 2;P command means "Continue executing the program until a second breakpoint is found." This is accomplished as follows:

*2;P	Proceed until a second breakpoint is found. In this case location 514 will be passed through once, and when it is encountered a second time will cause a break to occur.
B;000514	A break occurs at 514. We again examine the registers.
$0/000123 (LF)	The sum is now 123.
xxxxxx/001006 (LF)	The pointer is now 1006.
xxxxxx + 2/000012 (Return)	The counter is 12.

As we study the preceding register list, we discover that the sum is reasonable and the pointer is correct. However, the counter has not been decremented at all. This means either that the DEC R2 is not being encountered or that somehow R2 is continually being reset. Looking back at the code, we discover that the instruction BNE LOOP branches back to location 504, where R2 is set to 12 in each pass through the loop. This can easily be corrected, however, by changing the branch instruction accordingly. This is discussed in the following section.

Changing Offsets

The PDP-11 processor utilizes two types of offsets. The branch instructions contain a word offset, which is the number of *words*, positive or negative, over which the jump is to take place. Index mode instructions utilize a byte offset, which is the number of *bytes* that must be added to a register to obtain the final effective address. Both of these types of offsets can be calculated by ODT, which makes tedious binary or octal subtraction and shifting unnecessary. To use ODT to calculate an offset, simply type the address of the location that is to contain the offset, followed by a slash. When its contents are printed out, type the address to which you wish to determine the offset, followed by the letter "O." The word and byte offsets will be printed out.

For example, to change the offset of the BNE LOOP instruction, we examine location 516 and calculate the offset to location 512:

516/001372 512O 177772 177775

The byte (177772 or -6) and word (177775 or -3) offsets are printed out when the O command is given. The base of the instruction BNE is 1000, and the negative 8-bit offset must be added to it to produce the new branch instruction. This new instruction is 1375, and, if typed on the same line, it will automatically replace the current contents of 516, which is still open:

516/001372 512O 177772 177775 1375 (Return)

We have corrected a major error in the program, and now we can test the program by restarting it at 500 to see whether it runs correctly. First, we remove the breakpoint, and then we restart:

*;B   Remove the breakpoint
*500;G   and restart the program at 500.

The program now halts with the correct sum in R0.

Byte Offsets

The difference between the current location + 2 and the entered address in bytes was also printed out by the preceding O command. It can be used in changing PC-relative and other index mode instructions. Let us suppose that we wish to change the instruction MOV A, B in the following example to MOV A, C:

```
. = 2400
2400   016767   MOV A, B
2402   000054
2404   000054
   ⋮
2460   xxxxxx   A:  xxxxxx
2462   yyyyyy   B:  yyyyyy
2464   zzzzzz   C:  zzzzzz
```

It is worth noting in passing that the offsets to A and B are both 54 because the distances from 2404 to 2460 and from 2404 to 2462 are both 54. Recall that the PC always points one address beyond the current word of the instruction. If we want to change MOV A, B to MOV A, C, we can do this by changing the second index constant of the instruction, in location 2404. Therefore, we examine 2404, type 2464O, and look at the results:

   *2404/000054 2464O 000056 000027 56 (Return)

Here we find that the byte offset is 56 and the word offset is 27. We enter the 56 followed by a Return, and the instruction is changed. If we now wish to test this, we can reopen 2404 and, using the back-arrow command, see what address it refers to:

   *2404/000056 ← 002464/zzzzzz

Spurious Breakpoints

If a program has been debugged with ODT and the breakpoints have not been removed, starting the program will cause new breakpoints when the BPT (000003) instruction is encountered in the program. This is a trivial problem, as typing a "B" by itself will remove the breakpoint so that the program can proceed unmodified. However, if a new copy of ODT has been loaded where the previous breakpoints have not been removed, as might be the case when a partial crash occurs, these breakpoints are not "known" by the new copy of ODT and will cause a BE (bad entry) error message to be printed when ODT is entered. Similarly, if the current copy of ODT is started at its starting address from the Switch Register, the record of all breakpoints is destroyed and BE messages can occur.

Searches for Various Values in Memory

ODT can search for locations that equal a given number or address a given word, using the W and E commands. The n;W command searches between specified limits for all words that, after being XORed with n and ANDed with the Mask are equal to 0. The Mask and limits are specified by typing the $M command and entering or changing the Mask. The search limits are contained in the two locations that follow the Mask and are accessed by typing two Line Feeds. For example, suppose we wish to find all locations between 500 and 526 that end in 7. If we XOR any word xxxxx7 with n equal to 000007, we will always produce xxxxx0, which when ANDed with a Mask equal to 000007 will produce 000000. We enter these values as follows:

   $M/000000   7 (LF)   Sets the search Mask to 7 and goes on
                                  to change the search limits

x̲x̲x̲x̲x̲x̲/̲0̲0̲0̲0̲0̲0̲	500 (LF)	Sets lower search limit to 500
x̲x̲x̲x̲x̲x̲/̲0̲0̲0̲0̲0̲0̲	526 (Return)	Sets upper search limit to 526
*̲7̲;̲W̲		Searches for all words between 500
000520/012767		and 526 that have bits 0–2 equal to 7
0̲0̲0̲5̲2̲4̲/̲0̲0̲0̲2̲0̲7̲		(111). They are printed out on the tele-
		printer.

The underlying principle of the search Mask and search word n is that the exclusive OR returns 0's if the bits are the same and 1's if the bits are different. Thus an exact match between the search word n and the word in memory being examined will return all zeros in the bits of interest. These bits are then Masked out with the Mask so that the entire word will be zero when a match is found. These words are printed out on the teleprinter.

The effective address search E is somewhat more useful than the W search in the PDP-11 since most addressing is relative to the PC. The nE command searches for all instructions between the two search limits that address location n. All of those found need not be actual instructions but may just have the proper values to appear to be offsets, so some caution must be used in interpreting these values. Words that contain a byte offset, a branch with a word offset, or the absolute address n are printed in the nE search.

Interrupts

ODT-11 assumes that no interrupts are desired while it is running and, after saving the current PS, sets the interrupt status to 7 so that none can occur. However, this can still lead to some problems if the teleprinter interrupt is being used in the program being debugged, since ODT uses the teleprinter as well. If the printer interrupt is enabled on entry to ODT and no interrupt is pending, exit from ODT will always generate an additional printer interrupt. But, if a printer interrupt was pending, it will occur when the user program regains control. If the teleprinter reader has just read a character upon entry to ODT, this character will be lost on exit from ODT.

Although ODT operates at priority 7, this can be changed with the $P command, which allows examination of this priority. If this location is set to 377, ODT will operate at the priority of the processor at the time of entry.

Additions to ODT

One extremely useful addition to the ODT-11 program as it stands is the nJ command, which performs an automatic JSR PC, n, and returns to ODT when done. This can be used when debugging programs that utilize FPMP11 or when calling any routine in the JSR PC mode. It can be used in conjunction with FLIP and FLOP to allow entry of floating point numbers at the teleprinter and the printout of the contents of the FLAC in decimal. For example, if the address of FLOP is 7504, the command

    *7504;J

will print out the contents of the FLAC on the teleprinter:

    34.567
    *

Similarly, if you wish to determine the value of a decimal number in FP format, calling FLIP, followed by entry of the decimal number, will convert the number and leave it in the FLAC for examination by ODT. In the following example, FLIP begins at 7712 and the number to be converted is −1.75:

*7712J −1.75 (Return)	FLIP is called and −1.75 is entered.
*7750/140340 (LF)	The FP format value of −1.75 is deter-
7752/000000 (Return)	mined by examining the contents of the FLAC at 7750.

To add the J command to the ODT source tape, a byte containing the character "J" must be added to the end of the table that begins at O.LGDR. The address of the routine JSUB must be placed at the end of the table that begins at O.LGCH, but before the table that begins at O.TL. Then the routine is simply the few lines that follow, which must be added before location O.CRET. In reassembling ODT, note that the tape is in three sections. The first contains a starting address and may be skipped if a new starting address is to be inserted, and the last contains the autostart instruction, .END O.ODT.

At the end of the O.LGDR table:

```
                          O.PROC      ;P—PROCEED
this line is added →       JSUB        ;J—PERFORM JSR PC, NNN
                          O.LGL = . − OLGDR
```

At the end of the O.LGCH table:

```
                        .BYTE 'P    ;P
this line is added →    .BYTE 'J    ;J
                        O.CLGT = . – O.LGCH
```

```
;J —COMMAND— PERFORM JSR PC, NNNN
;WHEN NNNNJ COMMAND IS GIVEN

JSUB:       TST R2          ;CHECK FOR VALID ENTRY
            BEQ O.ERR       ;ERROR IF NO NNNN ENTERED
            JSR PC, (R4)    ;PERFORM JSR
            BR O.DCD        ;GET NEW COMMAND
```


Commands In RT-11 ODT

Return	Closes open location and accepts next command.
Line Feed	Closes current location and opens next sequential location.
↑ or ∧	Opens previous location.
← or __	Indexes contents of opened location by contents of PC and opens resulting location.
>	Uses contents of open location as a relative branch and opens the referenced location.
<	Returns to sequence prior to @, > or ← and opens next location.
@	Uses contents of open location as absolute address and opens the referenced location.
r/	Opens word at location r.
/	Reopens last opened location.
r \	Opens byte at location r.
\	Reopens last opened location as byte.
!	Prints address of opened location relative to the relocation register whose contents are closest.
n!	Prints address of opened location relative to relocation register n.
$n/	Opens general register n.
$B/	Opens first word of breakpoint table.
$C/	Opens constant register.
$F/	Opens format register.
$M/	Opens first mask register.
$P/	Opens priority register.

$R/	Opens first relocation register.
$S/	Opens Status Register.
r;nA	Prints *n* bytes in ASCII format starting at location *r*, then allows input of *n* bytes from the terminal
;B	Removes all breakpoints.
r;nB	Sets breakpoint *n* at location *r*.
;nB	Removes breakpoint *n*
r;C	Stores *r* in constant register.
r;E	Searches for instructions that reference effective address *r*.
;F	Fills memory words with the contents of the constant register.
r;G	Starts execution of program at location *r*.
;I	Fills memory bytes with low-order 8 bits of the constant register.
r;O	Calculates offset from current location to *r*.
;P	Execution proceeds from breakpoint.
n;P	Execution proceeds for the next *n* instructions.
k;P	Execution proceeds from breakpoint; stops after encountering the breakpoint *k* times.
;R	Sets all relocation registers to -1.
;nR	Sets relocation register *n* to -1.
r;nR	Sets relocation register *n* to the value of *r* (default $n=0$).
R	Subtracts the relocation register whose contents are closest to but $<=$ contents of open location from contents of open location and prints the result.
nR	Subtracts contents of relocation register *n* from contents of the opened word and prints the result.
;S	Disables single-instruction mode.
;nS	Enables single-instruction mode; disables breakpoints.
r;W	Searches for words with bit patterns matching *r*.
X	Performs a Radix-50 unpack of opened location; permits storage of new Radix-50 binary number.

Single-Instruction Mode

It is occasionally useful to single-step a program one instruction at a time. Typing ";S" sets the ODT-11 single-instruction mode, which enables the T-bit in the Processor Status. It restores all breakpoints and then allows one instruction of the program to be executed each time a P command is given. Typing ";S" a second time disables this mode.

PART TWO

LABORATORY MINICOMPUTER APPLICATIONS

CHAPTER FIFTEEN

THE LPS-11 LABORATORY DATA PERIPHERAL

The LPS-11 peripheral consists of an analog-to-digital converter with eight or more channels multiplexed to it, two digital-to-analog converters to control the display, a real time clock which can tick at a number of frequencies, a digital I/O section, and a light-emitting diode (LED) numerical display. Since this material was originally written, these devices have been repackaged under the name MINC-11. The device capabilities and bit assignments are quite similar, and we will use the term LPS-11 to refer to both.

Device Addresses

Since the LPS-11 is a less common peripheral, its device addresses are variable in case other options occupy some of these addresses. The standard factory-selected device addresses are given below. They are, however, jumper selectable in increments of 40 addresses.

Device	Function	Address
ADC	Status	770400
ADC	Data/LED	770402
Clock	Status	770404
Clock	Buffer-Preset	770406
Digital I/O	Status	770410
Digital I/O	Input	770412

Device	Function	Address
Digital I/O	Output	770414
Display	Status	770416
Display	XDAC	770420
Display	YDAC	770422
External DAC	DAC	770424
Unused		770426–770434
ADC	DMA register	770436

Similarly, the interrupt vector addresses are also jumper selectable, but the standard addresses are as follows:

ADC	300	BR6
Clock	304	BR5
Digital I/O in	310	BR4
Digital I/O out	314	BR4
Display	320	BR4
Unused	324	
ADC-DMA	330	BR6

Programming the Display

The display control amounts to a Status Register and two digital-to-anaalog converters (DACs) to control the x and y deflections of the scope. The display is created by individually loading the x and y coordinates of each point and then intensifying the result by pulsing the z axis of the scope. Since neither the controller nor the scope has any memory other than the persistence of the scope phosphor, the display must be refreshed continuously. The bit assignments of the display status word are shown in Figure 15.1.

Bits	Function
15–13	Unused
12	Erase (storage scope)
11	Write through (storage scope)
10	Store (storage scope)
9	Color (of two-color scope)
8	Unused
7	Ready flag
6	Interrupt enable

Bits	Function
5	Unused
4	External delay (special scopes)
3, 2	Mode
	00 intensify with bit 0 only
	01 intensify when x is loaded
	10 intensify when y is loaded
	11 intensify when either x or y is loaded
1	Fast intensify enable
0	Intensify

Most of the preceding bits do not concern us with a typical x–y, one-color, nonstorage scope. We consider bits 15–8 as unused for our purposes. The function of bits 7, 6, and 4 is to allow the CRT (cathode ray tube) beam sufficient settling time between points being intensified. Unless some modification has been made in the LPS-11, the delay required between the loading of the point and its intensification is 21 microseconds. Once an intensify command has been given by the computer by either loading bit 0 of the status word or loading the display register specified with the mode bits, the ready flag (bit 7) drops to 0 for about 21 microseconds. It then goes back to 1 so that the next point can be loaded.

This is a tremendous waste of time with most modern high-frequency laboratory scopes, and we will not use this method in our examples. Instead, we will rely on the time it takes to complete a loop to generate the delay required to prevent streaking. Such streaking may still be observed if widely separated points (more than 10 positions) are being displayed, as in a character-oriented display of some message, since we are waiting at the wrong time. However, if the display is a spectrum of some sort, or other data acquired with the analog-to-digital converter, it is unlikely that we will need to use the delay mode at all.

The x and y display registers are 12 bits in length, occupying the 12 least significant bits of the 16-bit device word. They are both positive, unsigned numbers ranging from 0 to 7777, where the coordinates are as shown in Figure 15.2. Although the display numbers are thus unipolar, it turns out that the outputs of the corresponding analog voltages to the scope are both bipolar, so that voltage 0, 0 lies in the center of the scope,

Figure 15.1 DISTAT, display status word.

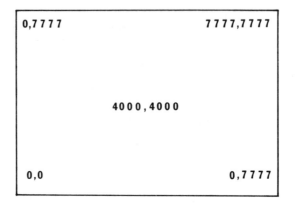

Figure 15.2 Display coordinates.

with full scale positive in the upper right corner and full scale negative in the lower left corner. It becomes apparent that this is not the best possible choice, at least digitally, and we will have to make some programming allowances to get a good bipolar display of the kind an analog scope would be expected to produce.

However, for the moment, let us assume that we wish to draw a diagonal line across the scope from lower left to upper right. To do this we start by loading both the x and the y display words with 0's, and then we increment them together. There are two other rules for writing good display programs that we will begin to observe in this first example:

1. Load the x and the y registers as close together in time as possible. Make sure that the longest possible time elapses before they are loaded again.
2. Intensify the result after the second of the two is loaded.

```
;PROGRAM TO DISPLAY A DIAGONAL LINE ON THE SCOPE
        DISTAT=170416    ;DISPLAY STATUS DEFN
        XDAC=DISTAT+2
        YDAC=XDAC+2

START:  MOV #10, DISTAT   ;FAST INTENSIFY ON Y-LOAD
        CLR XDAC                   ;X=0
        CLR YDAC                   ;Y=0
        MOV #4096., R1             ;COUNTER

LOOP:   INC XDAC                   ;ADVANCE X
        INX YDAC                   ;ADVANCE Y
        SOB R1, LOOP               ;DONE?
        BR START                   ;YES, REFRESH DISPLAY
```

This program simply sets the x and y displays to 0, increments each of them 4096 times, and then resets them to 0. It turns out that, since the output registers are only 12 bits long, incrementing a display register from 7777 will result in a 0, so that the counter is not necessary. It is good practice to leave it in, however, since there are many cases where advancing x or y circularly would lead to disastrous results.

Now let us suppose we wished to display a "spectrum" or a list of 4096 points that have been acquired or calculated and placed in successive memory locations:

```
;DISPLAY 8. POINTS ACROSS THE SCOPE
START:  MOV #8., R1              ;SET COUNTER
        MOV #DBUF, RO            ;AND POINTER
        MOV #-1, XDAC            ;INITIALIZE X

LOOP:   MOV (RO)+, YDAC          ;LOAD Y
        ADD #512., XDAC          ;ADVANCE X
L1:     TSTB DISTAT              ;DISPLAY READY?
        BPL L1
        INC DISTAT               ;NOW INTENSIFY
        SOB R1, LOOP             ;GO BACK TIL ALL 8 DONE
        BR START                 ;AND REFRESH
```

In this example, x is again initialized to -1 but is incremented in units of 1000 by adding 1000 directly to the register. After both x and y have been loaded, the program waits for the ready flag and then intensifies by incrementing DISTAT.

```
;DISPLAY 4K OF POINTS

START:  MOV #10, DISTAT          ;FAST INTENSIFY ON Y-LOAD
        MOV #-1, XDAC            ;X=-1
        MOV #DBUF, RO            ;START OF DATA
        MOV #4096., R1           ;SET COUNTER

DLOOP:  INC XDAC                 ;ADVANCE X
        MOV (RO)+, YDAC          ;LOAD Y AND INTENSIFY
        SOB R1, DLOOP            ;DONE?
        BR START                 ;REFRESH
```

This example illustrates the important point that the x and y axes should be loaded as close together in time as possible. Furthermore, the intensification should take place when the second axis is loaded. For these reasons, the x axis will always appear in line either as

```
                 ;DISTAT LOADED WITH 10
INC XDAC         ;LOAD X
MOV (R0)+, YDAC  ;LOAD Y AND INTENS
```

or

```
                 ;DISTAT LOADED WITH 4
MOV (R0)+, YDAC  ;LOAD Y
INC XDAC         ;LOAD X AND INTENS
```

In either case the x axis is loaded with the x address when it is incremented. Since we want to vary the x axis for 0 to 7777, we start by initializing it to -1. Then, when it is first incremented, the first x point becomes 0.

This method will work whenever a spectrum of adjacent points is being displayed. However, if only a few widely spread points are being displayed, streaking can be reduced by using the display ready flag.

The Analog-to-Digital Converter

The measurement of a varying input voltage and the storage of this voltage in digital form constitute one of the most important functions a laboratory computer can perform. The analog-to-digital converter (ADC) performs this function by converting a voltage in a given range to a number in a given range. The LPS-11 system ADC is a 12-bit one which can therefore represent numbers between 0 and 7777_8. The voltage range can be varied somewhat by some internal changes but is either 0 to $+5$ volts in single-input systems or -1 to $+1$ volt in multiple-input systems. The most common value is the latter, and it is used in the following discussion.

As provided, the ADC has either a single input or eight multiplexed inputs. In the full-blown LPS-11 four 10-turn potentimeters are associated with inputs 0–3. These pots control the voltage, which is read if no phone plug is in the input jack. If a plug is inserted, it cuts out the knob reading and allows the input voltage on the jack to be read. In either case a voltage of -1 volt corresponds to an ADC reading of 0, and a 1-volt input corresponds to a reading of 7777. Thus the ADC is unipolar rather than signed. This can pose a difficulty in some applications, as we will see in the next chapter.

Like other peripherals, the ADC has a Status and a Data Register. The Status Register is shown in Figure 15.3.

Bits		Function
0	ADC—start	Starts ADC under program control.
1–2	DMA	Selects DMA mode if DMA option is installed.
3	Burst mode	ADC samples continuously every 20 microseconds. Not too useful.
4	Schmitt trigger enable	Firing of ST#1 initiates one A/D conversion.
5	Clock overflow enable	ADC starts each time clock overflows.

Bits		Function
6	Interrupt enable	ADC done causes interrupt vector through address 300.
7	Done flag	Set when ADC done. Cleared when new conversion started.
8–11	Multiplexer	Loaded with numbers from 0 to 17 for the 16_{10} possible multiplexer inputs.
12–13	Unused	
14	Dual mode	Allows simultaneous acquisition of pairs of data points if this option is installed.
15	ADC—error flag	Set if an A/D conversion is started before the last one finishes or has been read, or if the multiplexer is changed during the first microsecond of the conversion time.

The conversion of an input voltage to a digital number occurs in three steps. First, the start ADC command is given. This takes a "snapshot" of the input voltage through a window a few nanoseconds wide. Then the conversion process, which takes 20 microseconds, begins. This ADC operates by a successive approximation method in which a one-half scale voltage is synthesized and compared with the input voltage, and the result used to set bit 11 of the data buffer. Then a one-fourth or three-fourths range voltage is produced and compared to generate a value for bit 10, and so forth. After 20 microseconds, the done flag of the ADC Status Register is set to 1, and the result can then be read into a register or memory. This reading should take place before a new conversion is initiated.

Let us take a very simple example of a program to read the ADC and stop with the result displayed in R0. This would proceed as follows:

```
;READ ADC INPUT 0 AND DISPLAY RESULT
        MOV #1, ADSTAT   ;START THE CONVERSION ON ADC CHANNEL 0

ADLOOP: TSTB ADSTAT      ;WAIT FOR DONE FLAG
        BPL ADLOOP       ;DONE?
        MOV ADBUF, R0    ;YES, READ ADC INTO R0
        HALT             ;AND STOP WITH VALUE IN R0
```

Generally, however, we take ADC readings more than once and apply them for such purposes as using the knobs to control the display or putting the ADC values into successive memory locations at regular intervals.

Figure 15.3 Analog-to-digital converter Status Register, ADSTAT.

Examples of this are shown in the discussion of the real time clock (p. 204), which explains the function of bit 5 of the status word.

Burst Mode

The burst mode is provided for continuous taking of data from the ADC. Data points are taken every 20 microseconds and must be read immediately from the AD buffer before a new conversion destroys them. This is equivalent to the free running feature of many oscilloscopes but has little general use.

Multiplexer

Bits 8–11 of the AD status word control which channel of the multiplexer is fed to the ADC. In the simplest system no multiplexer is present, and data must be read from channel 0. Systems with knobs have an eight-channel multiplexer; channels 0–3 are used for the knobs if no connections are made with phone plugs. An additional eight channels can also be added.

Schmitt Trigger Enable

A Schmitt trigger is a device that "fires" when a voltage crosses a preset threshold in a given direction (positive or negative slope). When the Schmitt trigger on the front panel of the LPS-11 is fired by such a voltage, one ADC conversion can be performed if bit 4 of the Status Register was previously set. In this case the ADC is started automatically, and it is not necessary to start it manually. This would simply be programmed as follows:

```
         MOV #20, ADSTAT  ;SET ST #1 ENABLE
AD1:     TSTB ADSTAT      ;WAIT FOR ADC TO FINISH WHEN ST FIRES
         BPL AD1
         MOV ADBUF, R0    ;READ RESULT
```

Dual Mode

Setting this bit causes preset pairs of channels to take data at the same instant in time through the use of two sample-and-hold circuits. They are then converted sequentially and available by two successive starts of the ADC. The pairs that can be used together require that all 16 channels be implemented in the system, as they are channels

    0 and 10
    1 and 11
    2 and 12

and so forth.

DMA Mode

The DMA mode is another option that can be installed in the LPS-11. It allows blocks of up to 4096 points to be deposited in memory without any program intervention other than setting a pointer and counter for the block. Unfortunately, *addition* of points into memory is not possible with this option, which prevents signal averaging. Furthermore, while this may save some processor overhead, the fastest data acquisition rate is little better than can be produced by methods that take points one at a time. This mode is thoroughly described in the DEC *LPS-11 Laboratory Peripheral System User's Guide*.

The LED Display

The light emitting diode (LED) display can be programmed to display any number through the use of the ADC buffer. When the ADC buffer is read, it gives the value of the current ADC conversion. However, when data are written into it, it becomes the Data Register for the LED. The bit assignments are illustrated in Figure 15.4.

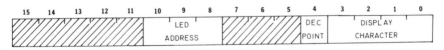

Figure 15.4   Light-emitting diode (LED) display buffer, LEDBUF.

Bits	Function
15–11	Unused.
10–8	LED addresses 0–5 define characters 1–6 from right to left.
7–5	Unused.
4	Turns on decimal point as well as character.
3–0	Character code.

Character	ASCII Code	LED Code
0	60	0
1	61	1
2	62	2
3	63	3
4	64	4
5	65	5
6	66	6
7	67	7
8	70	10
9	71	11
⊟ Test	···	12
Blank	40	13
Blank	40	14
—	55	15
Blank	40	16
Blank	40	17

The LED address is the number of the character from the right. Address 0 is the rightmost character; address 5, the leftmost character. The characters that can be loaded are the digits 0–9, the minus sign, and the blank. For all except the blank the rightmost 4 bits of the ASCII code correspond to the code loaded into the LED buffer. The decimal point is displayed, not as a separate character, but as a small dot in the lower right of the number LED. Thus setting bit 4 of the LED buffer with a given address loaded turns on that decimal point. The program given in Figure 15.5 allows the user to type in a number containing digits 0–9 and the characters minus (−) and dot (.). All other characters are treated as blanks.

Two cautionary notes are in order. First, the entire LED value must be constructed in a memory location or register and loaded all at once into the LED buffer. Second, a period in the leftmost place should automatically be preceded by a 0. In the example shown, the standard

```
;PROGRAM TO DISPLAY NUMBERS IN LED
;AS ENTERED FROM THE KEYBOARD

START:   MOV #.,  SP        ;INIT SP
         JSR PC, CRLF
RDO:     TSTB KRS           ;DON'T CHANGE DISPLAY UNTIL KB STRUCK
         BPL RDO
         CLR R1             ;LED ADDRESS
RD1:     MOVB R1, LED+1     ;LOAD LED ADDRESS
         MOVB #13, LED      ;SET EACH CHAR=BLANK
         MOV LED, LEDBUF    ;LOAD INTO DISPLAY
         INC R1             ;NEXT ADDRESS
         CMP R1, #5         ;DONE?
         BLE RD1            ;NO
         DEC R1             ;SET ADDRESS TO 5
         BR RD2B            ;DON'T READ FLAG AGAIN THIS TIME
;NOW READ CHARACTERS AND ENTER THEM IN LED
RD2:     TSTB TKS
         BPL RD2            ;IF CHAR STRUCK GO ON
RD2B:    MOV TKB, RO        ;READ IT
         BIC #177600, RO    ;CLEAR EXTRA BIT IF ON
         JSR PC, TYPE       ;PRINT IT
         CMP RO, #60        ;IS IT >=60?
         BLT DCHEK          ;NO, TRY - AND .
         CMP RO, #71        ;>71?
         BLE RD3            ;NO
         MOV #13, RO        ;YES, MAKE IT A BLANK
         BGT RD2            ;YES, IGNORE IT
RD3:     BIC #360, RO       ;CLEAR PUT ALL BUT BITS 0-3
         MOVB R1, LED+1
         MOVB RO, LED       ;LOAD ADDRESS AND CHARACTER
RD4:     MOV LED, LEDBUF    ;LOAD ACTUAL BUFFER
         DEC R1             ; ON TO NEXT CHARACTER
         BGE RD2            ;UNTIL DONE
;THEN TYPE OUT A CR-LF AND DISPLAY RESULT
         JSR PC, CRLF
         BR RDO


;CHECK FOR MINUS AND DECIMAL POINT
DCHEK:   CMP RO,#'.         ;IS IT DECIMAL POINT
         BEQ DSET           ;YES SET IT
         CMP RO, #'-        ;MINUS SIGN?
         BEQ RD3            ;YES GO LOAD IT
         BR RD2             ;OTHERWISE IGNORE CHARACTER
DSET:    CMP R1, #5         ;USE PREVIOUS LED ADDRESS UNLESS 0
         BEQ DZERO          ;DISPLAY 0. IF FIRST CHARACTER
         INC R1             ;PREVIOUS LED ADDRESS
DSET1:   MOVB R1, LED+1     ;ADDRESS
         BISB #20, LED      ;SET DECIMAL POINT
         MOV LED, LEDBUF    ;LOAD DISPLAY
         BR RD4             ;AND GO ON TO NEXT CHAR

DZERO:   CLRB LED           ;"0."
         BR DSET1           ;ADD DECIMAL POINT AND LOAD
```

Figure 15.5   LED display program.

register definitions and the routines TYPE and CRLF for printing a single character and a carriage return line feed have been omitted.

Clearly, the LED can be used to display any decimal number to six places, and one very useful application is in conjunction with the FPMP11 output routines. A calculated number can be converted to a displayed one by using the FPMP11 to convert it to an ASCII string and the LEDs to display the result.

Exercises

15.1.   Write a program to display a falling diagonal line from upper left to lower right.

15.2.   Write a program to display a cross on the scope.

15.3.   Write a program to display a horizontal line on the scope whose position can be varied with knob 1.

15.4.   Write a program to display a small cross on the scope whose vertical and horizontal positions can be varied with knobs 0 and 1.

15.5.   Write a program to display 2K of memory and halt if a "Q" is struck.

15.6.   Write a program using the FPMP11 to calculate $y = mx + b$ with m, x, and b entered at the teleprinter and the resulting y displayed in the LEDs. Display the "test pattern" if the number is too large.

15.7.   Write a program to display a region of memory defined by the Switch Register. Let the lower byte define the size of the block, and the upper byte the starting address of the block.

The Real Time Clock

The real time clock is one of the most versatile subsections of the LPS-11. It can be set to tick at virtually any rate from once every 650 seconds to once every microsecond. Although neither extreme is of much use, the wide range of values in between can make the clock useful in timing a variety of experiments. The clock status register is shown in Figure 15.6.

Bits		Function
0	Clock start	Starts clock ticking initially.
1–3	Clock range	Allows selection of eight possible clock rates.
		000: no rate
		001: 1 megahertz (note that this is faster than most instructions)
		010: 100 kilohertz

Bits		Function
		011: 10 kilohertz
		100: 1 kilohertz
		101: 100 hertz
		110: one tick for each firing of ST#1
		111: line frequency (50 or 60 hertz)
4–5	Unused	
6	Mode interrupt enable	Causes interrupt vector through location 304 whenever mode flag is set
7	Mode flag	Is set at clock overflow in modes 00 and 01, and when ST#2 fires in modes 10 and 11. Is *not cleared* by being read.
8–9	Clock mode	00: loads counter from preset and counts up to 0 *once*.
		01: repeated interval mode. Each overflow restarts count.
		10: clock is free running; firing of ST#2 loads counter into buffer and continues counting.
		11: clock runs freely, but each firing of ST#2 loads counter into buffer and *zeroes* counter.
10	MAINT ST2	Loading this bit simulates the firing of ST2. Can be used to transfer contents of counter to buffer/preset when in mode 10 or 11.
11	MAINT COUNT	When loaded with the clock counter enable turned off, this bit simulates a 1 MHz pulse to the counter. Of little general use.
12	MAINT ST1	Loading this bit simulates the firing of ST1.
13	ST#1 start	Firing of ST#1 starts clock in either mode 00 or 01.
14	ST#1 interrupt enable	Firing of ST#1 causes an interrupt.
15	ST#1 flag	Firing of ST#1 sets this flag.

In actual use the clock consists of a Status Register (CLSTAT) and a Buffer-Preset Register (CLKBUF). When the clock is turned on by setting

15	14	13	12	11	10	9	8	7	6	5	4	3	2	1	0
STI FLAG	STI INT ENABL	STI ENABL	—MAINT	BITS—		— MODE —		MODE FLAG	INT ENABL	//////	//////	——RATE	——		CLOCK ENABL

Figure 15.6   Clock Status Register, CLSTAT.

bit 0 of CLSTAT to 1, the clock in modes 00 and 01 transfers the Buffer-Preset Register to a Counter Register and starts incrementing it. When the Counter Register reaches 0, the mode flag, bit 7 of CLSTAT, is set to 1. If clock mode 00 (bits 8 and 9) was chosen, the clock then stops. If clock mode 01 was chosen, the clock overflow when the counter reaches 0 automatically restarts the counting cycle, reloading the Counter Register from the Buffer-Preset Register and counting it up to 0 again. This process continues in mode 01, the repeated interval mode, until the mode bits are changed.

When the clock counter reaches 0, it automatically sets the mode flag. It can then be tested by the usual TSTB instruction, but it should be carefully noted that reading the status word does not clear this bit, even in repeated mode. One possible programming solution is to actually clear the mode flag manually with a BIC #200, CLSTAT instruction, but this is not necessary in all cases, as we will see.

In the following example the clock is used to flash a "9" on and off in the least significant bit of the LED display once per second:

```
;PROGRAM TO FLASH A "9" ON AND OFF EVERY 1 SECOND

      START:  MOV #-1000.,  CLKBUF      ;SET CLOCK BUFFER PRESET TO -1000.
              MOV #411, CLSTAT          ;START CLOCK, 1KHZ, REPEATED INTVL
      NINE:   MOV #11, LEDBUF           ;TURN ON "9" IN RIGHTMOST DIGIT

      CLOOP:  TSTB CLSTAT               ;1 SECOND GONE BY?
              BPL CLOOP                 ;NOT YET
              BIC #200, CLSTAT          ;YES, TURN OFF CLOCK FLAG
              MOV #14, LEDBUF           ;AND TURN OFF THE "9"
      C1:     TSTB CLSTAT               ;WAIT ANOTHER SECOND
              BPL C1
              BIC #200, CLSTAT          ;TURN OFF FLAG
              BR NINE                   ;AND GO BACK AGAIN
```

Note in this example that the clock is tested twice and cleared each time. The clock should never be started until CLKBUF has been specified and should be started as close to last as possible in the initialization sequence.

One of the most important uses of mode 01 is in the acquisition of data points at regular intervals from the ADC. They are deposited in memory and displayed representing the output of some instrument or spectrometer after a given time period. In this case each clock overflow can be used to start an analog-to-digital conversion automatically without ever polling the CLSTAT mode flag. This method is infinitely preferable to one in which the clock flag is polled and the ADC started under program control because of the timing jitter introduced in a variable number of passes through the loop

```
CLOOP:    TSTB CLSTAT
          BR CLOOP
```

whenever the clock interval is not an integral multiple of the instruction time.

The clock can be used to start the ADC directly if bit 5 of the AD status word is set beforehand. In the following example, 4096 data points are acquired from the ADC at a rate of 1 each millisecond. They are stored in successive addresses. Note that all other parameters are initialized before the clock is turned on.

```
;CLOCK STARTS ADC EVERY 1 MSEC
        MOV #4096., R1   ;SET COUNTER
        MOV #DATBUF, R0  ;SET POINTER
        MOV #40, ADSTAT  ;ADC STARTS ON CLOCK OVERFLOW
        MOV #-1, CLKBUF  ;1 TICK IN COUNTER-PRESET
        MOV #411, CLSTAT         ;AND START CLOCK TICKING

;NOW TAKE 4096. POINTS
LOOP:   TSTB ADSTAT      ;ADC READY YET?
        BPL LOOP         ;NOT YET
        MOV ADBUF (R0)+  ;YES, READ ADC AND PUT IN MEMORY
        SOB R1, LOOP     ;GO BACK TIL ALL 4096 DONE
        HALT             ;THEN STOP
BATBUF: .BLKW 4096.      ;DATA GOES HERE
```

Note that in this example the clock mode flag is neither polled nor cleared. The ADSTAT done flag goes up each time the AD conversion is done and automatically is cleared each time the clock overflow starts a new conversion. The data rate limit for the process is obviously the conversion speed of the ADC, about 19 microseconds per data point at 12 bits. An internal switch allows reduction of the ADC resolution to 11, 10, 9, or 8 bits, with a decrease in conversion time to 15, 8, 6, or 4.5 microseconds.

A more sophisticated data acquisition routine, however, would display each point on the scope as it is acquired. This simply means loading XDAC with the point number (or incrementing it) and loading YDAC with the current value of the ADC. Such a program requires that XDAC be started at −1, as before, since we will be incrementing XDAC and then displaying.

```
;DISPLAY EACH POINT AS IT IS SAMPLED
     MOV  #10,  DISTAT  ;FAST INTENSIFY ON Y-LOAD
     MOV  #4096.,  R1   ;SET COUNTER
     MOV  #DATBUF,  R0   ;AND POINTER
     MOV  #40,  ADSTAT  ;ADC STARTS ON CLOCK OVERFLOW
     MOV  #-1,  CLKBUF  ;SET 1 TICK AT 1KHZ
     MOV  #-1,  XDAC    ;XDAC WILL BE 0 WHEN FIRST INCREMENTED
     MOV  #411,  CLSTAT        ;TURN ON CLOCK AND START

;DISPLAY EACH POINT AS ACQUIRED
DLOOP:  TSTB ADSTAT       ;IS ADC READY?
        BPL DLOOP         ;NOT YET
        MOV ADBUF,  (R0)  ;YES,  PUT RESULT IN MEMORY
        INC XDADC         ;ADVANCE X
        MOV (R0)+,  YDAC  ;LOAD Y,  ADVANCE MEMORY AND INTENSIFY
        SOB R1,  DLOOP    ;GO BACK TIL DONE
        HALT              ;THEN STOP
```

The preceding routine acquires 4096 points in 4.096 seconds, displaying the data on the scope as they are acquired. The display during acquisition can be helpful, especially if the acquisition rate is quite slow. A display afterward is also useful for examining the data that have been acquired.

The Schmitt Triggers

The Schmitt triggers (ST) can be used most effectively as indicators of when an analog voltage passes a certain level with a given slope. The input plugs of the LPS-11 have inputs for ST#1 and ST#2 for the two Schmitt triggers. Each also has a slope switch and a potentiometer controlling the threshold of the trigger. The threshold controls allow continuous adjustment of the threshold to any point on a signal between -5 and $+5$ volts. When a voltage passes the preset threshold having the slope specified with the slope switches, the Schmitt trigger will fire.

If the triggering signal is connected to ST#1, it can be used for the following purposes:

1. Cause one increment of the clock Counter Register if the rate bits are set to 111.
2. Start the clock ticking in either mode 0 or 1 if bit 13 of CLSTAT is set.
3. Cause an interrupt through location 304 if bit 14 of CLSTAT is set.
4. Set bit 15 of CLSTAT for program testing.
5. Initiate one ADC sample if bit 5 of ADSTAT is set.

Schmitt trigger number 2 is associated with a free running clock mode in which its firing will cause the current value of the clock counter to be loaded back into the buffer. This can be accomplished in two ways, depending on the setting of mode bits 8 and 9.

Mode 10   The clock counter is free running. When ST#2 is fired, the contents of the counter are transferred to the CLKBUF, and the counter continues counting. The mode flag (bit 7) is set each time ST#2 fires, but is not cleared by a TSTB instruction.

Mode 11   The clock counter is free running as in mode 10. When ST#2 is fired, the contents of the counter are transferred to the buffer CLKBUF and the counter is zeroed. Counting continues. The mode flag is set by the firing of ST#2. This is an excellent method of measuring the time between firings of the Schmitt trigger for the production of a time interval histogram.

Let us take two examples of the use of the Schmitt triggers. First, we will use ST#1 to initiate a scan of 4.096-second data acquisition, as in our preceding data acquisition example. To do this, we simply allow ST#1 to start the clock, based on some external signal that determines when the data are ready. Then acquisition proceeds as before. In this case we will not start the clock by setting bit 0; rather, we will allow ST#1 to start it. The clock overflow will then start the ADC and begin the acquisition process.

```
;ACQUIRE DATA POINTS AFTER FIRING OF SCHMITT TRIGGER
      MOV  #4, DISTAT    ;INTENSIFY ON X-LOAD
      MOV  #4096., R1    ;SET COUNTER
      MOV  #DATBUF, R0   ;SET POINTER
      MOV  #-1, XDAC     ;X=-1
      MOV  #-1, CLKBUF   ;ONE COUNT AT 1 KHZ
      MOV  #40, ADSTAT   ;CLOCK OVERFLOW STARTS ADC
      MOV  #20410, CLSTAT        ;ENABLE ST#1 TO START CLOCK

;NOW WAIT FOR ST#1 TO GO HIGH AND FOR AD TO START
;ADC FLAG WON'D GO HIGH UNTIL ST HAS STARTED
;AND CLOCK HAS STARTED ADC

ALOOP:  TSTB ADSTAT      ;ADC READY?
        BPL ALOOP        ;NOT YET
        etc.
```

In our second example we will assume that we start the clock ticking at 100 hertz and then wish to measure the time intervals between which 40 successive firings of ST#2 occur. This can be accomplished in mode 11, which transfers the contents of the counter back to the CLKBUF each time ST#2 fires. The 40 numbers are stored in 40 successive memory locations and can be displayed in some form or printed out after the run is over.

```
;MEASURE TIME INTERVALS BETWEEN WHICH 40 SUCCESSIVE
;FIRINGS OF ST#2 OCCUR.

        MOV  #40.,  R1      ;INTERVAL COUNTER
        MOV  #DBUF,  R0     ;SET POINTER
        MOV  #1413,  CLSTAT          ;START CLOCK RUNNING, MODE 11, 100HZ

C1:     TSTB CLSTAT         ;MODE FLAG IS ET WHEN ST2 FIRES
        BPL  C1             ;NOT YET
        BIC  #200,  CLSTAT          ;CLEAR CLOCK FLAG
        MOV  CLKBUF,  (R0)+         ;STORE INTERVAL IN MEMORY
        SOB  R1,  C1        ;GO BACK TILL 40 DONE
        HALT
```

It should be noted that the repeated interval mode is one of the few where running the clock at 1 megahertz makes sense, since it can then measure small intervals very accurately. However, the range of the numbers must be carefully controlled so that all intervals lie within $2^{16} - 1$ or 65,535 times the clock period.

Use of Interrupts

By far the most efficient way of acquiring data and at the same time maintaining a continuous display is to allow the ADC to interrupt the display whenever a new point is ready and to let the main program be the display. When all points are done, the ADC can be disabled from further interrupts by clearing the AD status word. In the program given in Figure 15.7 the display is maintained continuously as a background program, while in the foreground the ADC takes 2K of points every 7 milliseconds and adds them into memory.

Digital I/O Section

The digital I/O section of the LPS-11 consists of a 16-bit Output Register, a 16-bit Input Register, and two relays. The digital I/O has been manufactured in two rather different forms, called the LPSDR and the LPSDR-A. The LPSDR, the older version, allows the input or output of 16 signals, but interrupts can be caused only upon the receipt of a "ready" signal, indicating new data on the input lines. In the LPSDR-A, however, interrupts can be caused when any selected input lines are set. Interrupts can be caused during output only by the receipt of a "data accept" signal from whatever external device is receiving the signal. This last feature is of little general utility but may be valuable in some special situations.

```
;DISPLAY 2K OF MEMORY AND ACQUIRE 2K OF POINTS
ADVECT=300          ;ADC INTERRUPT VECTOR ADDDRESS
BR6=300             ;BUS REQUEST LEVEL 6
ADSTAT=170400       ;ADC DEVICE ADDRESS
ADBUF=ADSTAT+2      ;ADC BUFFER ADDRESS
CLSTAT=ADBUF+2      ;CLOCK STATUS REGISTER
CLKBUF=CLSTAT+2     ;CLOCK BUFFER-PRESET
DISTAT=170420       ;DISPLAY STATUS REGISTER
XDAC=DISTAT+2       ;X-AXIS DAC
YDAC=XDAC+2         ;Y-AXIS DAC

.ASECT
.=1000
START:  MOV #., SP          ;INITIALIZE THE SP
        MOV # ADSERV, @#ADVECT  ;SET ADC INTERRUPT VECTOR
        MOV #BR6, @#ADVECT+2    ;BUS REQ 6
        MOV #140, ADSTAT       ;START ADC ON CLOCK-INTERRUPT WHEN DONE
        MOV #4, DISTAT         ;FAST INTENS ON X-LOAD
        MOV #-7, CLKBUF        ;7x 1 KHZ
        MOV #2048., R3         ;COUNTER FOR ADC
        MOV # DBUF, R2         ;POINTER FOR ADC
        MOV # 411, CLSTAT      ;START CLOCK AND SAMPLING

;CLOCK IS NOW RUNNING. POINTS WILL BE TAKEN
;BY INTERRUPT ROUTINE ADSERV EVERY 7 MSEC.

;MEANWHILE DISPLAY THE DATA AREA
DISPLA: MOV #2048., R1      ;SEPARATE COUNTER FOR DISPLAY
        MOV # DBUF, R0      ;DISPLAY POINTER
        MOV # -2, XDAC      ;INIT X AT -2
;INNER DISPLAY LOOP
D1:     MOV (R0)+, YDAC     ;LOAD Y WITH EACH POINT
        ADD #2, XDAC        ;ADVANCE X
        SOB R1, D1          ;LOOP BACK TILL DONE
        BR DISPLA           ;AND REFRESH DISPLAY

;ADC INTERUPT SERVICE ROUTINE
ADSERV: MOV ADBUF, (R2)+    ;PUT EACH POINT INTO MEMORY
        DEC R3              ;DONE WITH SCAN?
        BNE AEXIT           ;NO
        CLR ADSTAT          ;YES, DISABLE ADC INTERRUPTS
        CLR CLSTAT          ;AND TURN OFF THE CLOCK
AEXIT:  RTI                 ;EXIT FROM THE INTERRUPT

DBUF:   .BLKW 2048.         ;DATA STARTS HERE

.END START
```

Figure 15.7   Program to display 2K of memory and acquire 2K of points.

Relays

The most readily understood and easily used of the LPSDR functions are the relays. These are connected to two banana plugs on the front panel of the LPS-11; when they are closed, a switch closure is made between the two posts of that relay. This switch closure can be used to raise and lower a plotter pen, or turn a plotter or other device on or off. The relays are controlled by bits 0 and 8 of the Status Registers shown in Figure 15.8. If the bits are set, the relay switch is closed; if they are zero, the

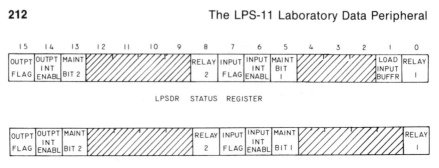

Figure 15.8   LPSDR and LPSDR-A, digital I/O Status Registers, IOSTAT.

relays are opened. The relays are capable of switching a 5-ampere resistive load at 115 volts or 2.5 amperes at 230 volts. The relays should not be used for other than resistive loads, as the manufacturer warns that voltage spikes resulting from inductive loading can cause deterioration of the contacts. Furthermore, the relays shoudl not be expected to perform at exceedingly high speeds, as their contact bounce may last as long as 20 milliseconds.

Status Register

The Status Registers of the LPSDR and LPSDR-A differ slightly in capabilities, as shown in Figure 15.8. The principal difference is that the Input Register can be read at will by setting bit 1 of the LPSDR Status Register, whereas the LPSDR-A can be read only when a DATA READY signal reaches the input connector.

Bits	Function
0	Set to turn on relay 1; clear to turn off.
1	LPSDR *only*. Set to read Input Register.
2–4	Unused.
5	Maintenance bit—causes input interrupt.
6	Input interrupt enable. Causes interrupt when input flag sets.
7	Input flag. Set when NEW DATA READY signal is received at pin 2 of input connector.
8	Set to turn on relay 2; clear to turn off relay 2.
9–12	Unused.
13	Causes output interrupt for maintenance purposes.
14	Setting of output flag causes an interrupt.
15	Sets when "external data accept" is received at pin 3 of output.

General Description of Levels of Input and Output Registers

The Input and Output Registers utilize standard TTL logic conventions, with the exception that a logical 1 is a ground (low) and a logical 0 is +3 volts (high). These registers are protected at an overvoltage of ±20 volts dc, after which a fusible resistor opens to protect the rest of the circuit.

Pin Assignments of the Input and Output Connectors

The Input and Output Registers are accessed at the rear of the LPS-11 through two 25-pin connector plugs marked "Input Register" and "Output Register." The pin assignments of these two registers are similar and are shown schematically in Figure 15.9. The pin assignments are given for the male plug as viewed from the rear or the female connector on the LPS-11 as viewed from the front.

The pin assignments for the 16 input and output lines are entirely self-explanatory. However, two additional important signals in both the input and the output connector warrant some explanation.

When data is sent to the output register with a MOV or BIS instruction, a signal called INT NEW DATA READY is sent as an indication that the data on the output lines have just changed. This signal, which is on pin 2 of the *input* connector, is a 1-microsecond pulse to ground, which can be used to clock some external flip-flop or other external device. If the program must know that the external device has accepted the data and that the Output Register can now be changed, the external device can send a signal back to the computer on pin 3 of the Output Register. This signal must be a pulse of 1-microsecond or greater duration that, when sent, will set the output flag. If the interrupt is enabled by setting bit 14 of the IOSTAT Status Register, an interrupt can also occur.

Input data can be sent to the Input Register by setting to ground the lines that are to represent *ones* (negative logic). These are not read into the Input Register, however, until either (*a*) a "NEW DATA READY" pulse is received at pin 2, or (*b*) the programmer sets bit 1 of the LPSDR Status Register IOSTAT (this will not work with the LPSDR-A). When one of these events has occurred, the data at the input connector are read into the Input Register and the input ready flag is set. If the input interrupt is enabled by setting bit 6 of IOSTAT, an interrupt is also generated. The reading of the data by either method also causes the generation of an INPUT DATA ACCEPT pulse at pin 3 of the input connector, which can be sent to the external device.

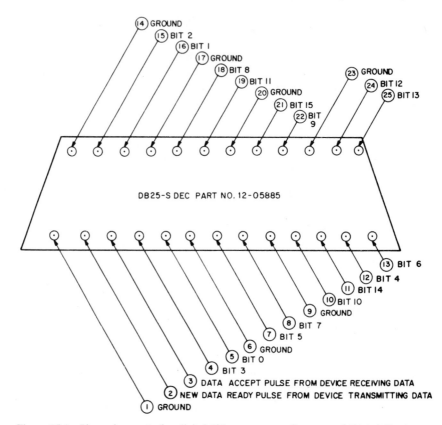

Figure 15.9 Pin assignments for digital I/O connectors. Courtesy of Digital Equipment Corporation.

It should be carefully noted that, at least in the LPSDR, the NEW DATA READY signal at the output connector is sent by pin 2 of the *input* connector, and, conversely, the signal indicating that new data are ready at the input connector is received by pin 2 of the *output* connector.

Programming Examples

Let us consider some examples of how the LPSDR can be used.

The following program opens and closes relay 1 every few hundred milliseconds, depending on the processor speed. It is useful for raising and lowering a plotter pen a few times to shake the pen and get the ink started.

```
;SHAKE THE PEN 5 TIMES
PENSHAK: MOV #5, R1              ;COUNT PEN SHAKES
SHK10:  CLR R0          ;SET R0 TO BE DECREMENTED 65,536 TIMES
        BIS #1, IOSTAT          ;RELAY1 LOWERS PEN
DOWN:   DEC R0                  ;WASTE 65536 CYCLES
        BNE DOWN
        BIC #1, IOSTAT          ;RAISE PEN BY TURING OFF RELAY
UP:     DEC R0                  ;COUNT DOWN 65536 CYCLES AGAIN
        BNE UP
        SOB R1, SHK10           ;GO BACK 5 TIMES
        RTS PC                  ;AND EXIT
```

The following program reads the Input Register of the LPSDR (only) by setting bit 1 of the I/O Status Register.

```
BIS #2, IOSTAT      ;INITIATE READ OF INPUT REGISTER
MOVE INPUT, R0      ;READ INPUT REGISTER INTO R0
```

The following program fragment reads the Input Register only when the input flag has been set.

```
DI: TSTB IOSTAT     ;NEW DATA AT INPUT REGISTER?
    BPL DI          ;NO, KEEP LOOKING
    MOV INPUT, R0   ;YES, READ INTO R0
```

The following program fragment sends data to the Output Register without checking on whether previous data have been accepted:

```
MOV #25146, OUTPUT
```

CHAPTER SIXTEEN

SIGNAL AVERAGING OF LABORATORY DATA

Very often laboratory experiments produce signals so small that they are overwhelmed by the noise of the system. This usually means that the sensitivity of the technique is insufficient to examine the sample in question. However, if the signal is there, no matter how small, and if the noise is random, there is a good chance that the weak signal can be improved by the technique of signal *averaging*.

There are several requirements for accurate signal averaging:

1. It must be possible to take repetitive scans of the spectrum or signal in such a way as to ensure that exactly the same region is being scanned each time.
2. The computer must receive this signal at a constant rate and with some sort of trigger to ensure that the same information goes into the same memory locations each time.
3. The signal must have only random noise distributed around 0 volts; or, if there is any dc offset, it must be balanced out in the amplifiers or in the computer.
4. The analog-to-digital converter must represent 0 volts as zero counts and be bipolar in its conversion about 0 volts. In other words it must represent positive voltages as positive numbers and negative voltages as negative numbers. This last condition is not met by the ADC of the LPS-11, but the lack can be compensated for by software, as we will see.

216

If these four conditions are satisfied, the signals from successive scans may be coadded into memory. Under these conditions the signal will grow linearly with the number of scans, N:

$$\text{signal} = k_1 N$$

Meanwhile, if the noise is indeed random, containing no coherent contributions, the noise will coadd less efficiently, growing at a rate proportional to the square root of the number of scans:

$$\text{noise} = k_2 N^{1/2}$$

Finally, the total signal-to-noise ratio after N scans can be represented by

$$\frac{\text{signal}}{\text{noise}} = \frac{k_1 N}{k_2 N^{1/2}} = K N^{1/2} \tag{16.1}$$

Thus we can say that the signal-to-noise ratio (S/N) grows as the square root of the number of scans.

Note that so far we have said nothing about averaging. In fact, the *average* of a number of scans is seldom taken, since it would require dividing the sum through by the number of scans. This average would contain no information not present in the sum; the only difference would be that the magnitude of the resulting numbers would be smaller. This offers no particular advantage, since we can easily program the computer to scale the data down to any given range before displaying them. However, the division of the sum by N has the significant disadvantage that it prevents the experimenter from further data acquisition and summation, since the data have been divided down to a small fraction of their final range. If averaging were allowed to commence at this point, the scans that were added into the scaled sum would be weighted statistically by a factor much greater than they deserved. Thus, no matter how often the term signal *averaging* is used, what is meant is signal *summation*.

To allow proper signal averaging in the PDP-11 we must convert the result produced by the LPS-11 ADC into the bipolar form, as required by condition 4. Recall that the LPS-11 ADC returns a value of 0000 for -1 volt and a value of 7777 for $+1$ volt. Therefore, 0 volts returns a value of 4000. If we did nothing to this value, we would fill up memory to overflowing in 32 scans, even with a 0-volt input. Clearly, a 0-volt input should have no effect on the contents of memory, and we thus subtract 4000_8 from each ADC reading before continuing with averaging.

A typical ADC interrupt routine for performing this task is shown below:

```
;SET UP ADC AS BEFORE:
        MOV  $ADINT,  @$300     ;INITIALIZE INTERRUPT VECTOR
        MOV  #340,  @#302       ;NEW PS FROM HERE
        MOV  # 140,  ADSTAT     ;AD STARTS FROM CLOCK- INTERRUPTS WHEN DONE
        MOV  #DISBUF, R4        ;SET AD POINTER
        MOV  #4096, R5          ;POINT COUNTER
        MOV  # -370, CLKBUF     ;CLOCK BUFFER-PRESET
;TURN ON THE CLCK AND START ACQUISITION
        MOV  #405, CLSTAT       ;CLOCK ON, 100KHZ, REPTD INTERVAL
```

(ADC interrupts the background routine that resides here)

```
        ;ADC INTERRUPT ROUTINE
ADINT:  MOV  ADBUF,  R3         ;PUT ADC VALUE INTO R3
        SUB  # 4000, R3         ;CONVERT TO BIPOLAR VALUE
        ADD  R3,  (R4)+         ;ADD INTO MEMORY
        DEC  R5                 ;CHECK COUNTER- ARE WE DONE?
        BEQ  AD10               ;STOP AVERAGING AT END OF SCAN
        RTI                     ;RETURN TO BACKGROUND ROUTINE

AD10:   CLR  CLSTAT             ;STOP AVERAGING AND CONTINUE DISPLAY
        RTI
```

Similarly, it is necessary to bias the display before loading the YDAC, since the data in memory are properly signed, but the y-display axis puts out -1 volt when loaded with a 0 and $+1$ volt when loaded with 7777. As a result it would be necessary to add 4000 to a single scan stored in memory in bipolar fashion. To increase versatility, however, it is desirable to view more than one scan when summed into memory. In this case the data in memory must be shifted by some amount before being loaded into the YDAC. The shifting of the data before the load into the YDAC is done in a single instruction, using the ASH instruction multiple shifts. The amount that the ASH instruction shifts is contained in another word that can be set by some external command.

```
        ;VARIABLE EXPANSION BIPOLAR DISPLAY ROUTINE
        ;USING THE LOCATION 'VSHIFT' SET EXTERNALLY
        ;CAN BE USED DURING ADC INTERRUPT CONTROLLED ACQUISITION

DISPLA: MOV  #DISBUF, R0        ;SET DISPLAY POINTER
        MOV  #4096., R1         ;AND COUNTER
        MOV  #6, DISTAT         ;FAST INTENSIFY ON X-LOAD
        MOV  #-1, XDAC          ;SET X TO -1

        ;INNER DISPLAY LOOP
DLOOP:  MOV  (R0)+, R2          ;GET EACH SIGNED DATA POINT
        ASH  VSHIFT, R2         ;SHIFT BY AMT IN LOCN VSHIFT
        ADD  #4000, R2          ;ADD 4000 TO BIAS DISPLAY
        MOV  R2, YDAC           ;LOAD Y
        INC  XDAC               ;ADVANCE X
        SOB  R1, DLOOP          ;GO BACK TILL SCAN COMPLETE
        BR   DISPLA             ;AND REFRESH THE DISPLAY
```

Memory Overflow in Signal Averaging

Signal averaging by summation has the disadvantage, however, that memory overflow will eventually occur. Both the noise and the signal are growing at a finite rate, and eventually one of them will cause a computer memory word to overflow. It is commonly assumed that this memory overflow occurs quite quickly, causing the subsequent loss of information. In fact, many people believe that, if we represent the word length of w and the ADC length by d, then memory overflow will occur after 2^{w-d} scans:

$$TS = 2^{w-d} \qquad (16.2)$$

where TS = total number of scans. In fact, if the signal being averaged has a low S/N, the number of scans which can be obtained is significantly higher than equation 16.2 indicates. For example, let us suppose that we are averaging a signal having an initial S/N of $10:1$. Given a 12-bit ADC and a 16-bit word, we will assume that the memory channels containing the most intense information contain 10/11 signal and 1/11 noise. A 12-bit 4096-count ADC will then contain 3724 counts of signal and 372 counts of noise. After four scans there will be

$$4(3724) = 14{,}896 \quad \text{counts of signal and}$$
$$\sqrt{4(372)} = \underline{744} \quad \text{counts of noise, or}$$
$$15{,}640 \quad \text{total counts in memory}$$

The total memory capacity used is 15,640 counts, whereas equation 16.2 would have predicted that 4(4096) or 16,384 counts would have been used. Clearly the assumptions embodied in equation 16.2 are false.

Let us now look at a general equation for the number of counts in memory after the total number of scans. The ADC contains d bits or 2^d counts, and these counts contain either signal S or noise N. The total number of counts attributable to signal is given by

$$\text{counts of signal} = 2^d \cdot \frac{S}{S + N}$$

Similarly, the noise in the ADC is given by

$$\text{counts of noise} = 2^d \cdot \frac{N}{S + N}$$

The total number of scans we can take is determined by how fast the computer word fills up. The signal will grow at a rate proportional to TS, and the noise at a rate proportional to $\sqrt{TS}$. When their sum is equal to the total number of counts in the word, 2^w, the computer word is full and no more scans can be taken; thus, setting 2^w equal to the sum of the signal and the noise, we have

$$2^w = \frac{[TS(S) + \sqrt{TS}(N)] \cdot 2^d}{S + N}$$

Now, since we generally refer to signal to noise as S to 1, we set $N = 1$ and then have

$$2^w = \frac{[TS(S) + \sqrt{TS}] \cdot 2^d}{S + 1} \qquad (16.3)$$

We can solve equation 16.3 for TS by a simple substitution. First we write the equation in quadratic form:

$$TS(S) + \sqrt{TS} - 2^{w-d}(S + 1) = 0$$

Then, letting $L = \sqrt{TS}$, we can solve for L or for $\sqrt{TS}$, giving

$$\sqrt{TS} = \frac{-1 + [1 + 4S(S + 1)2^{w-d}]^{1/2}}{2S} \qquad (16.4)$$

Note that equation 16.4 does indeed reduce to 16.2 for $S \gg 1$.

It is apparent from equation 16.4 that the total number of scans is dependent not only on w and d but also on the signal-to-noise ratio of the input signal. When S is very large, we can take 2^{w-d} scans without overflowing memory, but TS can be very much larger for very small S. Furthermore, since TS is dependent on 2^{w-d}, the larger $w - d$, the more scans that can be taken. This means not only if w were larger but also if d were smaller! In fact, when the only signal to be averaged is a very small one, a 1-bit ADC is sufficient. This, however, does not hold true for large-dynamic-range signals, as we shall see later. A complete discussion of these points is given in reference 1.

We now must deal with the appropriate action to be taken when we detect imminent memory overflow. The possibilities are as follows:

1. Go on averaging anyway.
2. Average in multiple precision.
3. Divide down memory and the ADC and go on.
4. Use a normalized averaging algorithm to prevent overflow.

These possibilities are dealt with in the following sections.

Continuation of Averaging After Memory Overflow

In some cases continuing to average is a perfectly acceptable alternative. The decision depends entirely on the nature of the signal that is being averaged and the treatment that the data will receive afterward. If the data consist of some large peaks containing irrelevant information in one portion of the spectrum, and of other, smaller peaks requiring further averaging in another portion of the spectrum, it does not matter if the large peak overflows memory several times while the small peak is gaining in strength through further averaging.

However, if the data of interest lie on top of or on a shoulder of the large peak, then memory overflow clearly cannot be permitted. If it occurred, the data of interest would overflow with it. If, on the other hand, the data being averaged are to be manipulated afterwards by a Fourier transform to find either the power spectral density (PSD), the frequency domain spectrum, or some correlation function, memory overflow cannot be permitted as it will render the results of the transform totally meaningless. Taking a Fourier transform of data containing peaks that have overflowed memory is tantamount to taking a Fourier transform of a step function or a square wave. This introduces additional noise and harmonics throughout the spectrum, causing it to lose large amounts of information.

Averaging in Multiple Precision

Another possibility is the collection of data in more than one word per data point. With the byte addressing capabilities of the PDP-11 it is also possible to collect data in half-words, thus allowing one-and-one-half precision if desired. This method ensures that the parameter w in equation 16.4 will not be satisfied as soon as if only single-precision averaging took place, but eventually any size memory word will indeed be filled up.

Double precision has the further disadvantage of reducing the total number of data points available for data storage. Thus, while an 8K PDP-11 could easily accommodate a 4K program and a 4K single-precision data array, only a 2K double-precision array could be accommodated. In addition, most minicomputers, the PDP-11 included, have few methods of successfully handling multiple-precision data reduction. Multiple-precision programming becomes much harder to manage, and multiple-precision multiplication and division generally cannot be done in fixed point or, to any significance, in floating point either. Thus one of the first steps after accumulating a multiple-precision average is scaling the data to single precision before further data reduction. This markedly reduces the number of significant figures, of course, but makes data reduction much easier. Since this scaling step is usually introduced before further processing, it is reasonable to consider methods of maintaining data in single precision and still preventing overflow.

Double-Precision Instructions.   The PDP-11 has the ability to carry out multiple-precision addition with great simplicity using the ADC and SBC instructions. These add or subtract the C-bit from the register specified.

Add carry	b055ar	ADC (B)	Add C-bit into destination.
Subtract carry	b056ar	SBC (B)	Subtract C-bit from destination.
Sign extend	0067ar	SXT	Fill destination with 0's if N-bit is 0 and with 1's if N-bit is 1.

Simple unsigned double-precision summation could be carried out as follows:

```
          CLR R0          ;HIGH ORDER WORD
          CLR R1          ;LOW ORDER WORD
LOOP:     ADD (R2) +,R1   ;ADD EACH INTO R1
          ADC R0          ;ADD EACH CARRY INTO R0
          SOB R3,LOOP
```

Signed multiple-precision addition is somewhat more complex, however. Consider the addition of a single-precision value, such as one from an analog input, to a double-precision average. There are four cases to consider:

1. Input positive, no carry.
2. Input positive, carry occurs.
3. Input negative, no carry.
4. Input negative, carry occurs.

Examining some examples, we see the following:

042316	012612	042316	177302
	+ 24		1623
042316	012636	042317	001125
(1) Input +, no carry		(2) Input +, carry occurs	
high word is unaffected		high word is incremented	

042316	012612		042316	000003
	177704			177704
042316	012516		042315	177707

(3) Input −, carry occurs   (4) Input −, no carry
    high word is unchanged   high word is decremented

We see from the examples that the carry bit is added into the high word if the single-precision word is positive, and ignored if the single-precision word is negative. Furthermore, the high word is decremented if the single-precision word is negative and no carry occurs, but left unchanged if a negative word is added and there is a carry. We can design an algorithm to carry out this signed addition if we use the ADC and SXT (sign extend) instructions:

1. Get each single-precision word.
2. Sign extend it into a temporary location.
3. Add temporary location to high word.
4. Add single-precision word into low word of sum.
5. Add carry into high word.

This is illustrated as an ADC interrupt service routine:

```
;ADC SERVICE ROUTINE FOR DOUBLE PRECISION ACQUISITION
;R3 AND R4 ARE POINTERS
ADINT:  MOV RO, -(SP)     ;SAVE RO
        MOV R1, -(SP)     ;AND R1
        MOV ADBUF, R1     ;READ ADC
        SUB #4000, R1     ;MAKE IT BIPOLAR
        SXT RO            ;AND MAKE RO -1 OR 0
        ADD RO, (R3)      ;ADD SIGN INTO HIGH WORD
        ADD R1, (R4)+     ;ADD READING INTO LOW WORD
        ADC (R3)+         ;AND CARRY INTO HIGH WORD
;RETURN FROM INTERRUPT
        DEC R5            ;CHECK COUNTER
        BEQ AQDONE        ;DONE?
        MOV (SP)+, R1     ;RESTORE REGISTERS
        MOV (SP)+, RO
        RTI               ;AND RETURN FOR MORE POINTS

ADONE:  ;end of scan-take appropriate action...
```

Memory Scaling on the Detection of Imminent Overflow

If we are averaging signals and suddenly discover that memory overflow is imminent, we can simply divide the signal in memory by 2 or 4 and continue averaging. This method is statistically defensible, however, only if the input is divided down by the same amount. Then, when averaging continues, the input reading is scaled to the right before addition into memory by the same amount as was the entire average.

This scaling is best performed only between scans, as the amount of time required might otherwise cause the loss of a data point. The following simple program tests for memory being within 7777 counts of being full:

```
;TEST FOR IMMINENT MEMORY OVERFLOW
OVTEST:  MOV  #DISBUF, R4   ;SET POINTER TO ARRAY
         MOV  SIZE, R5      ;SET COUNTER TO ARRAY SIZE
OV10:    MOV  (R4)+, R3     ;GET EACH DATA POINT
         BGE  OV20          ;IS IT POSITIVE?
         NEG  R3            ;NO, CALC ABS VALUE
OV20:    CMP  R3,#70000     ;IS OVERFLOW IMMINENT?
         BGE  OV30          ;YES, DIVIDE DOWN BY 2
         SOB  R5, OV10      ;NO, KEEP LOOKING
         RTS  PC

;OVERFLOW DETECTED- DIVIDE MEMORY DOWN BY 2
OV30:    MOV  #DISBUF, R4   ;SET POINTER
         MOV  SIZE, R5      ;AND COUNTER
OV40:    ASR  (R4)+         ;DIVIDE EACH POINT BY 2
         SOB  R5, OV40      ;GO BACK UNTIL DONE
         DEC  ASHIFT        ;ADD 1 TO SHIFTS DURING ACQUISITION
         CMP  ASHIFT, #-12. ;QUIT IF WE HAE REACHED 12 SHIFTS
         BLE  OV50          ;NO MORE ACQUISIION POSSIBLE
         RTS  PC            ;ELSE EXIT

;ABORT ACQUISITION IF NO MORE SHIFTS POSSIBLE IN 12-BIT ADC
OV50:    HALT

ASHIFT:  0        ;NUMBER OF SHIFTS DURING ACQUISITION
```

This program divides memory through by 2 and then subtracts 1 from location ASHIFT. If ASHIFT becomes greater than 12 right shifts, the program halts, since no further scaling is possible. The ASH instruction in the signal averaging routine then operates in conjunction with location ASHIFT to shift each data point down before its addition into memory. Recall that a right shift requires a negative number in ASHIFT. The routine for averaging with shifting is

```
;ADC INTERRUPT SERVICE ROUTINE WITH SHIFTING OF INPUT DATA
ADINT:   MOV  ADBUF, R3     ;GET EACH DATA POINT
         SUB  #4000, R3     ;MAKE NUMBER BIPOLAR
         ASH  ASHIFT, R3    ;SHIFT R3 ASHIFT PACES RIGHT
         ADD  R3, (R4)+     ;ADD INTO MEMORY
         DEC  R5            ;DONE?
         BEQ  AD10          ;YES, SHUT OFF ADC
         RTI                ;NO, CONTINUE

AD10:    ....     ;STOP ACQUISITION AT END OF SCAN
```

Normalized Averaging

It is also possible to prevent memory overflow by the use of the normalized averaging algorithm. Let us consider the equation for obtaining a true average of signals in memory. If we represent the average after n scans by A_n and each scan by S_i, then

$$A_n = \frac{S_1 + S_2 + \cdots + S_{n-1} + S_n}{n} \qquad (16.5)$$

Similarly, the average after $n - 1$ scans is given by

$$A_{n-1} = \frac{S_1 + S_2 + \cdots + S_{n-1}}{n - 1} \qquad (16.6)$$

Multiplying equation 16.5 by n and 16.6 by $n - 1$, and subtracting equation 16.6 from 16.5, we have

$$nA_n - (n - 1)A_{n-1} = S_n$$

Subtracting A_{n-1} from both sides, dividing by n, and rearranging gives

$$A_n = \frac{S_n - A_{n-1}}{n} + A_{n-1} \qquad (16.7)$$

Equation 16.7 is the expression for the so-called normalized averaging procedure. It amounts to the subtraction of the current average from each scan and the addition of this difference divided by the current number of scans with the previous average. In actual fact, the division is usually accomplished in signal averagers and computers by right shifting the quantity $S_n - A_{n-1}$ by the number of bits corresponding to the first power of 2 greater than or equal to n. Thus the equation becomes

$$A_n \leq \frac{S_n - A_{n-1}}{K(n)} + A_{n-1} \qquad (16.8)$$

where $K(n)$ is the first power of 2 greater than or equal to n.

Although this technique will indeed prevent memory overflow, it does not prevent signal underflow, and the number of scans that can be taken is limited to 2^d, where d is the word length of the ADC as before. As soon as n, the number of scans, becomes greater than 2^{d-1}, the value of K is raised to 2^d, effectively reducing all differences to 0. Furthermore, the dynamic range of the signal is decreased with each increase of K, so that the technique is ill-suited to measurements involving a wide dynamic range.

To write a program to perform normalized averaging, we simply set the scan counter COUNT to 1 and let another location NSHIFT represent the next higher power of 2 than the number of scans completed. NSHIFT is incremented to one greater shift value whenever the left shifting of a 1 NSHIFT times causes a number less than the current sweep count. For example, after one scan, COUNT will be 1 and NSHIFT will be 1. If we set a register equal to 1 and then shift it left NSHIFT ($=$ 1) times, we have 2. The number of scans (one) is still less than 2, so no change is made in NSHIFT. However, after three scans, COUNT will be 3 and NSHIFT will still be equal to 1. Therefore shifting 1 once to the left will produce 2, which is less than three scans. At this time, we increase the value of NSHIFT to 2.

In actuality, the value of NSHIFT must normally be negative to cause a right shift, and it must be negated to use as a left shift during the test of NSHIFT versus COUNT. The values of NSHIFT and COUNT are as follows for the first 17 scans:

COUNT	1	2	3	4	5	6	7	8	9	10	11	12	13	14	15	16	17
NSHIFT	-1	-1	-2	-2	-3	-3	-3	-3	-4	-4	-4	-4	-4	-4	-4	-4	-5

The program to perform normalized averaging is outlined below.

```
;NORMALIZED AVERAGING
;START BY SETTING UP COUNTERS AND SHIFTS
        MOV #1, COUNT    ;SET SCAN COUNTER TO 1
        MOV #1, NSHIFT   ;SET SHIFT COUNTER FOR NORMALIZATION
        MOV #DISBUF, R4  ;SET ARRAY POINTER
        MOV SIZE, R5     ;AND COUNTER

;         :
;         :

;ADC INTERRUPT SERVICE ROUTINE
ADINT:  MOV ADBUF, R3    ;GET LATEST DATA POINT
        SUB (R4), R3     ;S(N)-A(N-1)
        ASH NSHIFT, R3   ;(S(N)-A(N-1)/K(N)
        ADD R4, (R4)+    ;ADD INTO MEMORY
        DEC R5           ;DONE WITH SCAN?
        BEQ ADONE        ;YES
        RTI              ;ELSE EXIT SERVICE ROUTINE
;NORMALIZE AT END OF SCAN
ADONE:  INC COUNT        ;COUNT SCANS
        CLR ADSTAT       ;TURN OF ADC
        CLR CLSTAT       ;AND CLOCK
        MOV NSHIFT, R4   ;GET CURRENT SHIFT VALUE
        NEG R4           ;MAKE POSITIVE FOR LEFT SHIFT
        MOV #1, R3       ;PUT 1 IN R3 FOR SHIFTING LEFT
        ASH R4, R3       ;SHIFT R3 NSHIFT PLACES LEFT
        CMP R3, COUNT    ;IS R3>COUNT?
        BLT AOUT         ;NO, JUST EXIT
        DEC NSHIFT       ;YES, INCREASE # OF SHIFTS
AOUT:   MOV # DISBUF, R4        ;RESET POINTER
        MOV SIZE, R5     ;AND COUNTER
        MOV #xxxx, ADSTAT     ;RESET ADC
        MOV #xxx,CLSTAT  ;AND CLOCK
        RTI              ;AND BEGIN NEW SCAN
```

Schmitt Trigger Start of Each Scan

Regardless of the method of averaging, the use of the Schmitt trigger is the most common way to signal the computer that a new scan is to be initiated. In this case Schmitt trigger 1 can be used to turn on the clock, and the end of each clock period can then start the ADC. Once the clock is running, further triggers will be ignored until the clock is turned off again. This is accomplished as follows:

```
MOV #140, ADSTAT     ;AD DONE CAUSES INTERRUPT
MOV #20404, CLSTAT   ;CLOCK STARTS ON NEXT ST #1,
                     ;100KHZ, REPT
```

in the initialization routine and by a

```
CLR ADSTAT    ;DISABLE FURTHER CONVERSIONS
CLR CLSTAT    ;AND TURN OFF CLOCK
```

at the end of each scan.

Signal Averaging with High-Dynamic-Range Signals

Sometimes signal averaging is not as simple as in the cases described earlier. When a very small signal is the only signal, a 1-bit digitizer is sufficient to detect and digitize it. The data will eventually average to fill memory in the usual way. However, if the spectrum contains a strong signal of lesser interest, such as a solvent line in spectra obtained from solutions, the small signal of interest may be much more difficult to extract.

The exact degree of difficulty depends on the necessity of avoiding memory overflow. Sometimes it is perfectly permissible to allow the large peak or signal to overflow memory, as long as the small signal does not also overflow memory. In other cases, as when the small signal is riding on top of the large signal or when the resulting average is to be correlated or Fourier transformed, memory overflow cannot be permitted.

The ability to detect a large-dynamic-range signal is related to the resolution of the ADC. A successive approximation ADC will set the least significant bit for a signal equal to one-half of the range of the least significant bit, or $0.5(1/2^d)$. The ADC will not detect a signal any smaller than this. Thus the 12-bit PDP-11 digitizer will detect a signal equal to 1 part in 8192 or 1 in 2^{d+1}. Thus, if we assume that we have a large signal which fills the ADC on each scan, the small signal we wish to average must be 1/8192 of that or greater, or it cannot be detected at all. Furthermore, the total number of scans that can be taken without overflowing memory is now just 2^{w-d}, since the large signal is so large that no averaging is necessary. The fact that the small signal of interest requires averaging is unimportant. Therefore, using single precision and a 12-bit ADC, the PDP-11 or any other 16-bit computer can obtain only 2^{16-12} or 2^4 or 16 scans. Furthermore, the normalized averaging algorithm described in the preceding section cannot be used, as the dynamic range of the signal is attenuated by the division process after the second scan. The stepping down of the ADC described earlier will not work either, because the smaller number of bits in the ADC will no longer detect sufficient dynamic range.

The only method that will allow sufficient dynamic range for more averaging without memory overflow is the double-precision method. Unfortunately, this means that all further operations on the data must also be performed in double precision, or the advantage will be lost.

The problems and utilities of signal averaging will be appreciated only after the reader has written and tested some programs that allow the display and memory to be manipulated.

Exercises

16.1   Write a program to acquire 4096 data points over a 2-minute interval, displaying the entire spectrum whenever a data point is not being acquired.

16.2   Modify Exercise 16.1 so that the data rate or sweep time can be specified by a software command. Be sure to check for the zero case, zero loaded into the clock buffer and data out of range.

16.3   Modify your acquisition program to scale down the ADC each time overflow is detected, and print this scaling factor on the teleprinter. Each scan should start only when ST#1 fires or when a G command is typed.

16.4   Modify your signal averaging program to contract and expand the display by C and X commands from the teleprinter while the averaging is going on.

16.5   Modify your signal averaging program to take and display an arbitrary number of data points, at any data rate, where both the scan time and the number of points are obtained by teleprinter command. (*Hint:* What value would you add to the XDAC each time? How could you obtain it?)

Timing in Signal Averaging

We generally consider the problem of timing in signal averaging as simply a question of how long our data persist and the adjustment of the scan time to encompass all of them. This is a useful simple assumption, but in some cases it cannot be considered to constitute a complete statement of the problem. For example, suppose that we are observing some evoked response signal that contains both high- and low-frequency components. If this response contains a 1–2 hertz sine wave and another signal of, say, 50 hertz, we must be sure that we are sampling the data often enough to pick up details from both signals. According to the sampling theorem, we must sample a sine wave twice per cycle to define it unambiguously. Thus we must sample the 50-hertz sine wave, which has a period of 0.02 second, at least once every 0.01 second. This requires a sampling rate of 100 hertz.

Conversely, the highest frequency that can be accurately represented is that which is one-half the sampling frequency. This is just another way of saying that, if we do not sample the data at least twice per cycle of the highest frequency, these data cannot be correctly interpreted later. This highest frequency that we can accurately represent is called the *Nyquist*

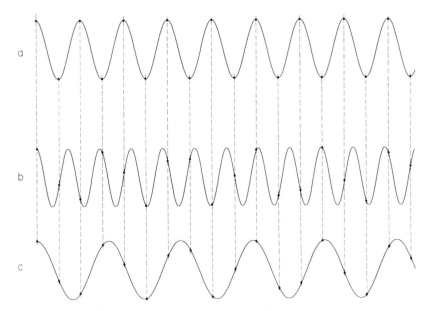

Figure 16.1 Sampling near the Nyquist frequency. (*a*) Sampling at the Nyquist frequency. (*b*) Sampling at frequency $N + \Delta f$, (*c*) Sampling at frequency $N - \Delta f$.

frequency. The consequences of not sampling at a high enough rate are shown in the figures. Figure 16.1a shows a sine wave sampled at just the Nyquist frequency, N. Figure 16.1b shows a frequency somewhat higher than the Nyquist frequency, which we will represent as $N + \Delta f$. Figure 16.1c shows a third frequency $(N - \Delta f)$ which is just the same amount below N as the frequency in Figure 16.1b was above N. Each of these three frequencies is sampled at just the Nyquist frequency, N.

It is clear from either tracing or inspection that the data which the computer "sees" at the frequencies $N + \Delta f$ and $N - \Delta f$ are identical. In fact, the points have exactly the same amplitude in the two cases, leading to the misidentification of a frequency $N + \Delta f$ as being $N - \Delta f$. This phenomenon is known as "fold-back," or "aliasing," and can be avoided only by correct selection of the sampling frequency.

Reference

1.   J. W. Cooper, *Computers and Chemistry* **1**, 51 (1976).

CHAPTER SEVENTEEN

METHODS OF DISPLAYING AND PLOTTING DATA

Variable Width Display

As soon as data have been acquired or averaged into the memory of a computer, the question always arises of how to display them properly. Clearly, the entire spectrum can be displayed point by point, as we have demonstrated in the preceding chapters. However, it is often desirable to be able to expand the data in both the horizontal and vertical directions as efficiently as possible. As we shall see, the algorithms for vertical expansion are trivial, and those for horizontal expansion rather complex.

One of the most convenient ways of expanding a set of data points is through the use of parameters knobs such as those available on the LPS-11. These knobs allow control of multiplexer channels 0–3 if no phone plug input is inserted in the corresponding input jack. When a phone plug is inserted, it simply switches off the input from the knobs. The knobs are merely 10-turn potentiometers (pots) which place a resistance across a voltage so that full scale left corresponds to −1 volt or a digital reading of 0 and full scale right corresponds to +1 volt or a reading of 7777. We will use two of these knobs to control the position and the expansion of the displayed data.

Let us suppose that we want to use knob 1 to control the size of the display or the number of points to be displayed. If we symbolize the reading of knob 1 by k_1 and the maximum unipolar ADC reading by 2^d, where d is the ADC resolution, the fraction of points that we are to display is given by $k_1/2^d$. The number of points is then calculated by multiplying the total number of points in the array, SIZE, by this fraction, giving the variable size, VSIZE:

$$\text{VSIZE} = \text{SIZE} \left(\frac{k_1}{2^d} \right)$$

The variable starting address, VSA, is to be read from knob 0, whose reading we will symbolize by k_0. Now we will want to be able to shift the starting point of the display so that the first point through the last point of the array can be centered on the screen for expansion. Thus the starting address will vary from $-\text{SIZE}/2$ to $+\text{SIZE}/2$, having a total range of SIZE. Now we must make the reading of k_0 bipolar to calculate numbers in this range, so that the range of starting addresses will be

$$\text{VSA range} = \frac{(k_0 - 2^{d-1})\text{SIZE}}{2^d}$$

The actual starting address will then be obtained by adding on the address of the beginning of the array to be expanded, SA, giving:

$$\text{VSA} = \text{SA} + \frac{(k_0 - 2^{d-1})\text{SIZE}}{2^d}$$

The preceding equation will expand the display and vary its size as desired. However, the expansion will be about the first point displayed, so that those to the left of the first point are continually lost. It is therefore desirable to expand the data about the center of the display. At any given time the display will be showing VSIZE points, and those not shown will be the quantity (SIZE − VSIZE). Since we want to put the expansion point in the center, we simply put half of these points on either side of the displayed area by adding half of this difference to the VSA, giving

$$\text{VSA} = \text{SA} + \frac{(k_0 - 2^{d-1})\text{SIZE}}{2^d} + \frac{(\text{SIZE} - \text{VSIZE})}{2}$$

Although this equation as given is correct in the abstract, modifications are necessary for the PDP-11. First, the PDP-11 uses *byte* addressing, so that adjacent words are two addresses apart. Thus the range and size terms must be doubled. Second, the 10-turn pots that comprise the parameter input knobs are designed so that full scale counterclockwise produces a 0 and full scale clockwise a 7777. Therefore, the display will move to the right (decreasing VSA) when the knob is turned to the left, and vice versa. This is easily corrected by negating the range term. Finally, the value of d for the PDP-11 is 12. These three changes are incorporated in the final equations:

$$VSIZE = SIZE \left(\frac{k_1}{4096}\right)$$

$$VSA = SA + \frac{(k_0 - 2048)SIZE}{2048} + (SIZE - VSIZE)$$

Now, to display a variable number of points evenly across the scope, we will need to advance the x axis of the display by some variable amount per data point. Since we must refresh the display as rapidly as possible, we cannot calculate the number of x steps per point in floating point but rather must rely on integer division. This presents some special problems, as we see in the following example.

Suppose that we have read the knobs before this scope scan and find that VSIZE = 2051 points. Since the XDAC has 4096 steps in it, we then calculate the number of steps per point by dividing

$$\frac{4096}{2051} = 1 \text{ in integer division}$$

Thus we would advance x by one step for each new data point and end up with a display 2051 points wide, or slightly more than halfway across the scope. Now 4096/2051 really equals 1.997, and it is this lost fraction that prevents the x axis from being a full 4096 steps wide. We can recover some of this range by taking advantage of the fact that the computer word is usually longer than the XDAC and using these remaining bits to contain some fractional data. The 16-bit word of the PDP-11 can contain unsigned numbers as high as 65,535. This number can be converted to the value loading into the x axis by shifting it four places to the right, or dividing by 16. Now let us divide

$$65,536/2051 = 31$$

We will call this value, 31, the *shifted x increment*, SXINC. In advancing our x axis now, we will keep a running sum, adding SXINC each time, and then shift it four places to the right before loading into the XDAC. For example, the first x point will appear at 0 and the second one at 31/16 or 1 as before. But the third point will appear at 62/16 = 3 rather than the 2 that ordinary integer division would have produced. The accompanying table shows some of the x values for the 2051 points:

Point No.	x by Integer Division	x by SXINC Method
0	0	0
1	1	1
2	2	3
3	3	5
4	4	7
10	10	19
1024	1024	1984
2051	2051	3973

In this example the display thus fills 3973/4096 (97%) instead of 2051/4096 (50%) of the scope, a marked improvement. Now 65,536 was chosen as the large number by which the shifted x increment was calculated because the entire running sum of the x-axis values could then be contained in a single 16-bit word. If double precision were used instead, the value of x at any point could be calculated to much greater accuracy and rounded to single precision by a double-precision shift instruction. However, the necessary double-precision arithmetic adds more time to the calculation and thus more flicker to the display for the small improvement in linearity. Therefore, we define the shifted x increment as

$$\text{SXINC} = \frac{2^w}{\text{VSIZE}}$$

in general, where w is the computer word length and

$$\text{SXINC} = \frac{1\,000000}{\text{VSIZE}}$$

for the 16-bit computer having the facility for double-precision division.

The necessity for multiple shifts and fast multiplication and division again forces us to the conclusion that the capable laboratory computer must contain multiply and divide hardware. A program utilizing these features and the equations given above is shown in Figure 17.2. The flowchart (Figure 17.1) is of general utility for any computer language. The bracketed numbers in Figure 17.2 refer to the box numbers in the flowchart in Figure 17.1.

Discussion of Flowchart for Variable Display

Boxes 1 and 2. Knob 1 is read, and the variable size parameter is calculated by multiplying k_1 by the total size of the array to be expanded,

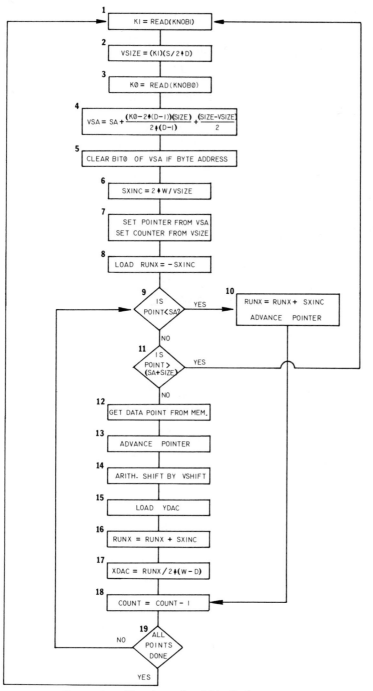

1 KI = READ(KNOBI)

2 VSIZE = (KI)(S/2↑D)

3 K0 = READ(KNOB0)

4 $VSA = SA + \dfrac{(K0 - 2 \uparrow (D-1))(SIZE)}{2 \uparrow (D-1)} + \dfrac{(SIZE - VSIZE)}{2}$

5 CLEAR BIT0 OF VSA IF BYTE ADDRESS

6 SXINC = 2↑W/VSIZE

7 SET POINTER FROM VSA
SET COUNTER FROM VSIZE

8 LOAD RUNX = −SXINC

9 IS POINT < SA? —YES→ **10** RUNX = RUNX + SXINC
ADVANCE POINTER

NO

11 IS POINT > (SA+SIZE) —YES→

NO

12 GET DATA POINT FROM MEM.

13 ADVANCE POINTER

14 ARITH. SHIFT BY VSHIFT

15 LOAD YDAC

16 RUNX = RUNX + SXINC

17 XDAC = RUNX/2↑(W−D)

18 COUNT = COUNT − 1

19 ALL POINTS DONE —NO→

YES

Figure 17.1 Flowchart of variable display program.

divided by the largest value which can be read in the ADC. Note that k_1 is positive and unipolar.

Boxes 3 and 4. Knob 0 is read and converted to a variable starting address by the equation shown. Note that the second and third terms must be doubled for the PDP-11 or other machines utilizing byte addressing. Knob 0 is assumed to be unipolar positive and is converted to bipolar by subtracting 2^{d-1}.

```
            ;VARIABLE DISPLAY OF DATA
      ;VSIZE AND VSA ARE CALCULATED FROM THE READINGS OF KNOBS 0 AND 1


DISO:    MOV #., SP
         MOV #401, ADSTAT           ;READ MUX 1, KNOB 1        [1]
      ;VSIZE=KSIZE*(ADC)/4096.                                 [2]
            JSR PC, ADREAD   ;READ IT
            MUL KSIZE, RO    ;KSIZE*ADC
            ASHC #-12., RO   ;DIVIDE BY ADC RESOLUTION, 12-BITS
            MOV R1, VSIZE    ;THIS IS THE VARIABLE SIZE
            BNE D05          ;PREVENT ZERO POINTS
            INC VSIZE        ;MAKE IT 1


      ;NOW GET THE VARIABLE STARTING ADDRESS
      ;VSA= SA+(ADC-2048.)*SIZE/2048.   + (SIZE-VSIZE)
D05:        MOV #1, ADSTAT   ;SET MUX TO KNOB 0                 [3]
            JSR PC, ADREAD   ;AND READ VALUE INTO RO
            SUB #4000, RO    ;MAKE BIPOLAR
            NEG RO           ;REVERSE SENSE OF KNOB 0
            MUL KSIZE, RO    ;(ADC-4000)*SIZE                  [4]
            ASHC #-11., RO   ;DIVIDE RO+R1 BY 2048. TO ALLOW FOR BYTE ADDRESS
ING
            MOV KSIZE, R2    ;CALC SIZE-VSIZE
            SUB VSIZE, R2
            ADD R2, R1
            ADD SA, R1       ;ADD ON STARTING ADDRESS OF BLOCK
            BIC #1, R1       ;PREVENT ODD ADDRESS              [5]
            MOV R1, VSA      ;AND SAVE

      ;SHIFTED X-INCREMENT SXINC=1;000000/VSIZE                [6]
            MOV #1, RO       ;1
            CLR R1           ;000000
            DIV VSIZE, RO    ;17 BITS/VSIZE=XADV PER POINT *4
            MOV RO, SXINC    ;SHIFTED X-INCREMENT

      ;NOW DISPLAY RESULT
            MOV VSA, RO      ;GET VSA                          [7]
            MOV VSIZE, R1
            MOV #6, DISTAT
            MOV SXINC, R3    ;R3 KEEPS RUNNING XDAC SUM SHIFTED LEFT 4
            NEG R3                                             [8]
D10:        CMP RO, SA       ;IS VSA<SA?                       [9]
            BLT XADV         ;YES, ADVANCE X WITHOUT DISPLAYING
            CMP RO, PMAX     ;IS RO> LAST POINT?               [11]
            BGE D20          ;YES, STOP ANC CHECK TTY
            MOV (RO)+, R4    ;PUT ADDRESS IN R4                [12,13]
            ASH VSHIFT, R4   ;AND SHIFT                        [14]
            ADD #4000, R4    ;BIAS BEFORE DISPLAY
```

Figure 17.2   Listing of variable display program.

```
;NOW ADVANCE X
          ADD SXINC, R3    ;ADD SHIFTED X-INCREMENT TO R3   [16]
          MOV R3, R2       ;COPY FOR SHIFTING
          ASH #-4, R2      ;AND SHIFT 4 PLACES RIGHT
          MOV R4, YDAC     ;LOAD Y FIRST                    [15]
          MOV R2, XDAC     ;LOAD 0 AND INTENSIFY            [17]
          SOB R1, D10      ;DONE?                           [18,19]
D20:      TSTB TKS         ;TTY STRUCK
          BPL DISO         ;NO
          MOV TKB, R0
          BIC #-200, R0    ;CLEAR OUT PARITY BITS
          CMP R0, #'C      ;CONTRACT?
          BEQ CONTR        ;YES
          CMP R0, #'X      ;X?
          BNE DISO         ;NO
          INC VSHIFT       ;YES, EXPAND
          BR DISO
CONTR:    DEC VSHIFT       ;CONTRACT-ONE MORE RIGHT SHIFT
          BR DISO          ;AND CONTINUE


;ADVANCE SHIFTED X SUM WITHOUT LOADING INTO DISPLAY    [10]
XADV:     ADD SXINC, R3    ;ADVANCE SHIFTED X-SUM--NO DISPLAY
          ADD #2, R0       ;ADVANCE R0
          SOB R1, D10      ;GO ON WITH DISPLAY
          BR DISO          ;OR RESTART IF NO LUCK

;READ ADC- MULTIPLEXER AND START LOADED EXTERNAL TO ROUTINE
ADREAD:   TSTB ADSTAT      ;ADC READ
          BPL ADREAD       ;NO
          MOV ADBUF, R0    ;YES, READ IT
          BIC #1, R0       ;CLEAR LSB OF ADC TO PREVENT NOISE
          RTS PC

PMAX:     SAPNT+8192.      ;ADDRESS OF LAST POINT
VSHIFT:   0                ;VERTICAL DISPLAY SHIFT
KSIZE:    4096.
SXINC:    0                ;SHIFTED X-INCREMENT
VSIZE:    0
VSA:      0                ;VARIABLE STARTING ADDRESS
SA:       SAPNT            ;DISPLAY BUFFER STARTS HERE
```

Figure 17.2 (*Continued*)

Box 5. Since the PDP-11 utilizes byte addressing at odd addresses, we must clear bit 0 of the VSA to ensure that only words will be addressed.

Box 6. The shifted x increment, SXINC, is calculated here by dividing the range of the computer word by the variable size.

Boxes 7 and 8. The pointer and counter are initialized. The running sum of the x-axis points is set to $-$ SXINC, so that the first point will be at 0.

Boxes 9 and 10. A test is made to see whether the calculated address now contained in POINT is greater than the starting address of the array. If it is not, the running sum of x increments is advanced but this sum is *not* loaded into the XDAC, thus allowing a dark screen at the beginning of the display for points less than SA. Control is then transferred to the bottom of the loop, where the counter is decremented.

Box 11. The address POINT is tested to see whether it is above the last point of the array, given by SA + SIZE. If it is, the display loop is terminated and control returns to box 1.

Boxes 12–15. A legal data point is obtained from memory, shifted, and loaded into the YDAC.

Boxes 16 and 17. The running x sum, RUNX, is advanced and then shifted into position by four right arithmetic shifts.

Boxes 18 and 19. The counter is decremented, and if non-zero, the loop continues from box 9. Otherwise the knobs are again read for a new display at box 1.

Variable Vertical Display

Up to this point we have expanded the display vertically by the use of the arithmetic shift instruction (ASH) in conjunction with the Switch Register or some memory location containing the number of shifts to be performed. This method has the advantage that it can be used even during data acquisition, where it would be impossible to read a knob without missing several data points. During acquisition a teleprinter interrupt can be enabled, allowing service whenever convenient to accept a C or X command.

However, during data reduction, it is desirable to be able to expand or contract the data and even invert them without being limited to simple powers of 2. This can, of course, be effected by using the parameter knobs to generate a multiplier for the vertical direction as well as for VSA and VSIZE. This method has the disadvantage of being slower, however. It can cut down the scope refresh rate by a factor of 2 over the ASH method. The simple program given in Figure 17.3 illustrates the facility with which the display can be multiplied by a positive or negative number taken from a bipolar knob.

Cursor Display

It is often desirable to display on the screen an intensified point that can be moved around the data and used to interrogate them as to the value of the point or its time or frequency. This can easily be done using the knobs of the LPS-11 to control the position of the cursor point.

If, however, there are about 4096 points, the noise of the knobs or of the ADC itself may cause a value to be returned that is not time invariant, and the cursor thus will appear to jump around on the display. Since this

```
;VARIABLE MAGNITUDE DISPLAY
START:  MOV #.. SP
        MOV #1001, ADSTAT       ;READ KNOB 2
        JSR PC, ADREAD          ;WAIT AND READ INTO R4
        SUB #4000, R4           ;MAKE BIPOLAR
        MOV #DSTART, R0         ;INITIALIZE POINTER AND COUNTER
        MOV #4096., R1
        MOV #4, DISTAT          ;INTENS ON X-LOAD
        MOV #-1, XDAC           ;LOAD X

LOOP:   MOV (R0)+, R2
        MUL R4, R2              ;MULT EACH POINT BY KNOB READING
        ASHC #-11, R2           ;SHIFT TO SINGLE PRECISION
        ADD #4000, R3           ;BIAS FOR DAC
        MOV R3, YDAC            ;LOAD Y
        INC XDAC                ;ADVANCE X
        SOB R1, LOOP            ;GO BACK TIL DONE
        BR START               ;REFRESH

;SUBROUTINE TO READ ADC IN R4
ADREAD: TSTB ADSTAT            ;ADC DONE?
        BPL ADREAD             ;NO
        MOV ADBUF, R4          ;YES
        RTS PC                 ;EXIT
DSTART: 0
.END START
```

Figure 17.3   Listing of variable magnitude display program.

is obviously an undesirable situation, a cursor display that utilizes a coarse and a fine control from two different knobs, neither of which uses the least significant bits of the knob reading, is described here.

Such a cursor routine simply reads the knobs, converts them into an *x*-axis value, and then obtains the *y* value from the corresponding *y* address. In the case of PDP-11, of course, the *y* address must allow for the fact that words are *two* addresses apart because of the byte addressing scheme. The cursor point is then intensified 64 times so that it appears markedly brighter than the other points. This is an arbitrary value chosen empirically and may be changed, if desired, for other numbers of points.

The cursor 12-bit value is positive and unipolar and is obtained by combining bits 4–11 of the coarse knob with the right shifted bits 4–7 of the fine knob. All other bits are masked out by shifting and the BIC instruction. A routine for the display of 4096 points with a cursor that moves from point to point is shown in Figure 17.4.

Plotting of Data on an *x–y* Plotter

One task that seems extremely simple but turns out to be quite involved is the plotting of data on an *x–y* or *y–d* plotter. The simplest method is one in which the scope leads are run in parallel to the plotter and the display slowed down by an appropriate delay, so that the data are presented

```
;CURSOR DISPLAY USING COARSE AND FINE CONTROLS
;KNOB 2 IS COARSE - BITS 2-11
;KNOB 3 IS FINE - BITS 0-3 TAKEN FROM BITS 6-10 OF READING

START:    MOV #4096., R1        ;DISPLAY COUNTER
          MOV #DSTART, R0       ;POINTER
          MOV #-1, XDAC         ;INIT X DAC
          MOV #10, DISTAT       ;INTENS ON Y-LOAD

LOOP:     INC XDAC              ;ADVANCE X
          MOV (R0)+, YDAC       ;LOAD Y
          SOB R1, LOOP          ;GO BACK TILL 1 REFRESH DONE
          MOV #1001, ADSTAT     ;READY TO READ KNOB 2
          JSR PC, ADREAD        ;READ ADC
          BIC #3, R0            ;SAVE LESS NOISY BITS, 2-11
          MOV R0, R1            ;SAVE IN R1
          MOV #1401, ADSTAT     ;READY TO READ KNOB 3
          JSR PC, ADREAD        ;READ ADC
          ASH #-6, R0           ;SHIFT RIGHT 4 PLACES
          BIC #177740, R0       ;USE ONLY BITS 0-3
          ADD R1, R0            ;AND COMBINE COARSE AND FINE
          MOV R0, XDAC          ;THIS IS X POSITION
          ASL R0                ;X 2 FOR BYTE ADDRESSING
          ADD #DSTART, R0       ;ADD ON ADDRESS OF FIRST DATA POINT
          MOV #100, R1          ;INTENSIFY 64 TIMES
CURSE:    MOV (R0), YDAC        ;LOAD Y SPOT TO INTENSIFY
          DEC R1
          BNE CURSE             ;64 INTENSIFICATIONS
          BR START              ;AND REFRESH

;POSITIBE UNIPOLAR ADC ROUTINE
ADREAD:   TSTB ADSTAT           ;WAIT TILL DONE
          BPL ADREAD
          MOV ADBUF, R0         ;READ INTO R0
          RTS PC                ;AND RETURN
DSTART:   0                     ;DATA TO DISPLAY STARTS HERE
.END
```

Figure 17.4   Listing of cursor display program.

quite slowly, a point at a time, to the plotter. Unfortunately, this method leads to great problems in pen skipping when two adjacent points are rather far apart. Furthermore, the line between the points is left to the slewing rate of the plotter itself. It may be straight, curved, or nonexistent, depending on the response of the plotter to a rapid change in input voltage.

The next most obvious way to plot out data is to draw a straight line between adjacent points, first in the x direction and then in the y direction. This can lead to a perfectly adequate plot when a large number of points are involved, but a "New York skyline" type plot results for fewer numbers of points. Once this phenomenon has been observed, it becomes apparent that an interpolating plot routine must be written. Furthermore, although we have made some rough approximations to full scale while displaying a variable number of points using the variable display routine, the plot must fill *exactly* full scale regardless of how many points are plotted. In other words, the full range of the XDAC must be used whether we are plotting 113 or 4096 points, and adequate interpolation is required in both the x and y directions to prevent unaesthetic-looking spectra.

Let us suppose again that we have to plot 2051 data points full scale on an x–y plotter, using the full range of the XDAC. If we perform an integer calculation regarding the number of points to plot, we will find that we should advance x by one step for each point plotted, leading to a plot only 2051 x steps wide, as before. However, if we calculate the quantity 4096/2051 in floating point, we find that it equals 1.997. For the relatively slow plot routine, it is perfectly acceptable to actually calculate the number of steps to the next point in floating point and then convert it to fixed point. This method is slow, but plotting is generally a slow process anyway because of pen responses and recorder slewing rates.

The floating point method is unduly complicated, however, and requires an enormous amount of memory space just to perform the interpolation. The fixed point method we will describe here (Figures 17.5 and 17.6) takes advantage of the fact that, instead of calculating the number of steps per point, it is much more accurate to calculate the total number of steps to get to the *next* point. This method involves no cumulative errors and can easily be implemented in fixed point.

The number of steps to the next point is represented by the variable STTODO, and it is calculated before plotting each point by the equation

$$\text{STTODO} = \frac{(\text{PTDONE})(\text{TOTSTP})}{\text{TOTPNT}} - \text{STDONE}$$

where   PTDONE is the number of points already plotted in the spectrum,
TOTSTP is the total number of steps across the plot,
TOTPNT is the total number of points,
STDONE is the number of steps already done in the plot.

Note that there is no requirement that either TOTSTP or TOTPNT be a power of 2, and hence a number of points less than full scale may be used. This is useful in plotting the data on a small section of the page. Furthermore, it is quite possible to plot more points than steps using this algorithm.

Detailed Description of Plot Routine Flowchart

Boxes 1 and 2. The pointer, counter, and total number of steps are initialized from the calling routine. The total number of steps done is set to 1, the number of points done to 1, and the value of the XDAC to 0.

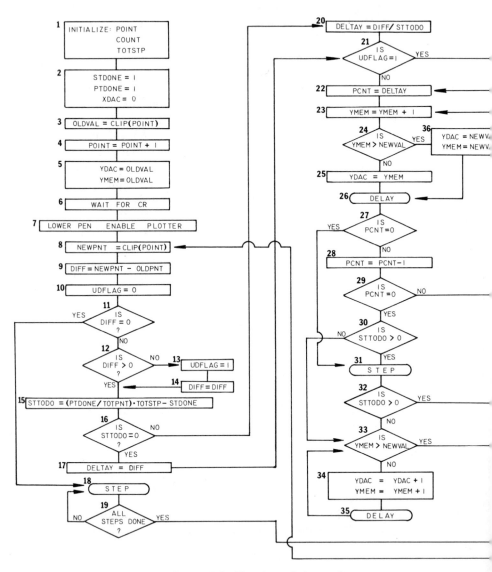

Figure 17.5 Flowchart of plot routine.

Boxes 3–7. The location OLDVAL is set to the clipped value of the number pointed to by POINT. Clipping amounts to performing any right or left shifting as specified by VSHIFT and then adding 4000. If the result is greater than 7777 or less than 0, the value is replaced by 7777 or 0. The pointer is then advanced, and the first value of the YDAC is loaded from

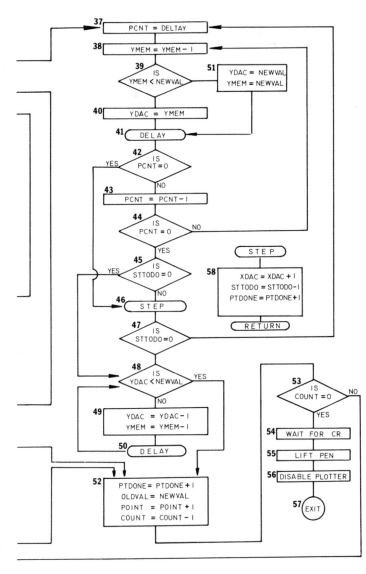

Figure 17.5 (*Continued*)

OLDVAL. A copy of the YDAC is also kept in a memory location called YMEM so that the results can be checked without changing the YDAC. The program then waits for a Return to be typed before proceeding, so that the user can turn on the plotter. Once the plotter is on, the program closes one relay to enable the plotter and another to lower the pen.

```
.TITLE PLOT
;PLOT SUBROUTINE
;CALLED BY
;       JSR R3, PLOT
;       ADDR                    ;ADDRESS OF ADDRESS OF ARRAY TO BE PLOTTED
;       SIZE                    ;ADDRESS OF NUMBER OF POINTS IN ARRAY
;       STEPS                   ;NUMBER OF PLOTTER STEPS
;LOCATION VSHIFT IS ASSUMED TO CONTAIN THE NUMBER OF VERTICAL SHIFTS
;TO BE APPLIED TO EACH POINT BEFORE PLOTTI
NG
;IN THE USUAL ASH VSHIFT, RN FORM

PLOT:   MOV @(R3)+, POINT       ;FIRST ADDRESS                       [1]
        MOV @(R3), TOTPNT       ;TOTAL POINTS
        MOV @(R3)+, COUNT       ;ALSO SET COUNTER
        DEC COUNT               ;ONE LESS POINT
        MOV (R3)+, TOTSTP       ;TOTAL STEPS
        MOV R3, -(SP)           ;SAVE R3 ON STACK
;INIT 1ST POINT AND WAIT FOR CR
        MOV #12, DISTAT         ;INIT DISPLAY
        CLR STDONE              ;STEPS DONE                          [2]
        MOV #1, PTDONE          ;SET POINTS DONE TO 1
        CLR XDAC                ;X=0
        MOV @POINT, R2
        JSR PC, CLIP            ;TEST FOR CLIPPING IN R2
        MOV R2, OLDVAL          ;SET UP FIRST POINT AS OLDVAL        [3]
        ADD #2, POINT           ;ADVANCE POINTER TO NEXT POINT       [4]
        MOV R2, YDAC            ;LOAD Y                              [5]
        MOV R2, YMEM            ;AND MEMORY COPY
        CLR STDONE              ;STEPS DONE
        MOV #1, IOSTAT          ; (ENABLE OUTPUT) WITH 1 RELAY       [6]
        JSR PC, CRWAIT          ;WAIT FOR CR
        BIS #400, IOSTAT        ;LOWER PEN WITH 2ND RELAY            [7]

PLT15:  MOV @ POINT, R2         ;GET EACH POINT
        CLR UDFLAG              ;UP-DOWN FLAG                        [10]
        JSR PC, CLIP            ;RETURN FULL SCALE IF TOO BIG
        MOV R2, NEWVAL          ;LIMITING VALUE OF Y                 [8]
        SUB OLDVAL, R2          ;CALCULATE DIFFERENCE
        BPL PL20                ;UP?                                 [12]
        INC UDFLAG              ;1=DOWN                              [13]
        NEG R2                  ;ABSOLUTE VALUE                      [14]
PL20:   MOV R2, DIFF           ;SAVE DIFFERENCE                     [9]
;CALCULATE STEPS TO DO TO GET TO THIS POINT
;STTODO= (PTDONE/TOTPNT)*TOTSTP-STDONE                               [15]
        MOV PTDONE, R4          ;POINTS DONE TO DATE
        MUL TOTSTP, R4          ;PTDONE*TOTSTP
        DIV TOTPNT, R4          ;(PTDONE*TOTSTP)/TOTPNT
        SUB STDONE, R4          ;(PTDONE*TOTSTP)/TOTPNT -STDONE
        MOV R4, STTODO          ;SAVE AS STEPS TO DO
PL21:   TST STTODO
        BNE PLT25               ;STEPS NON-ZERO                      [16]
        MOV DIFF, DELTAY        ;NO STEPS HERE                       [17]
        BR PLT27                ;SO PLOT WITHOUT STEPPING
```

Figure 17.6   Listing of plot routine.

Boxes 8–14. The value of the next point to be plotted (NEWPNT) is obtained from memory and clipped as described earlier. The value DIFF is calculated by subtracting the clipped value from the OLDVAL, and UDFLAG is set to 1, meaning plotting is to be down if DIFF is negative. The value of DIFF is negated if negative. If DIFF is 0, control goes to box 18.

```
;DELTAY = DIFF/STTODO
PLT25:  CLR R4
        MOV DIFF, R5              ;LOW WORD OF DIVIDEND
        BEQ PLT70                ;NO VERTICAL CHANGE, JUST PLOT X[11]
        DIV STTODO, R4           ;DIFF/STTODO
        MOV R4, DELTAY           ;QUOTIENT INTO DELTAY
;PLOT UP OR DOWN STTODO TIMES BY DELTA Y
;MAKING LAST STEP= NEWVAL
PLT27:  TST UDFLAG               ;UP?                              [21]
        BNE DOWN                 ;DOWN

;GO UP
UP:     MOV DELTAY, R1           ;COUNT INCREMENTS                 [22]
PLT30:  INC YMEM                 ;MEMORY COPY OF YDAC              [23]
        CMP YMEM, NEWVAL         ;IS MEM>NEWVAL?                   [24]
        BGT PLT35                ;YES SUBSTITUTE NEWVAL
        MOV YMEM, YDAC           ;NO, LOAD YDAC                    [25]
PLT37:  JSR PC, DELAY            ;WAIT                             [26]
        TST R1                   ;0 YET?                           [27]
        BEQ PLT38                ;YES DON'T DRAW MORE
        DEC R1                   ;SUBTRACT                         [28]
        BNE PLT30                ;DO MORE                          [29]
        TST STTODO               ;ANY STEPS?                       [30]
        BEQ PLT40                ;NO THIS TIME
PLT38:  JSR PC, STEPS            ;TAKE A STEP                      [31]
        DEC STTODO               ;COUNT THEM
        BNE UP                   ;CONTINUE UNTIL ALL DONE          [32]
PLT40:  CMP YDAC, NEWVAL         ;                                 [33]
        BGE NXTPNT               ;GO ON IF REACHED
        INC YDAC                 ;IF LESS MAKE EXACTLY EQUAL
        INC YMEM                 ;                                 [34]
        JSR PC, DELAY            ;                                 [35]
        BR PLT40

;GO DOWN
DOWN:   MOV DELTAY, R1           ;                                 [37]
PLT50:  DEC YMEM                 ;                                 [38]
        CMP YMEM, NEWVAL         ;                                 [39]
        BLT PLT55                ;IF ALREADY LESS SUBST NEWVAL
        MOV YMEM, YDAC           ;OTHERWISE LOAD YDAC              [40]
PLT57:  JSR PC, DELAY            ;DIDDLE                           [41]
        TST R1                   ;0 ALREADY?                       [42]
        BEQ PLT58                ;YES, TAKE STEPS
        DEC R1                   ;SUBRACT 1                        [43]
        BNE PLT50                ;GO BACK TILL ALL DONE            [44]
        TST STTODO               ;ANY STEPS?                       [45]
        BEQ PLT60                ;NOT HERE
PLT58:  JSR PC, STEPS            ;                                 [46]
        DEC STTODO               ;COUNT THEM
        BNE DOWN                 ;UNTIL ALL DONE                   [47]
PLT60:  CMP YDAC, NEWVAL         ;                                 [48]
        BLE NXTPNT               ;ALL DONE?
        DEC YDAC                 ;NO, MAKE EXACTLY EQUAL
        DEC YMEM                 ;                                 [49]
        JSR PC, DELAY            ;                                 [50]
        BR PLT60
PLT55:  MOV NEWVAL, YDAC         ;LOAD NEWVAL                      [51]
        MOV NEWVAL, YMEM         ;MAINTAIN CORRESPONDENCE
        BR PLT57

PLT35:  MOV NEWVAL, YDAC         ;                                 [36]
        MOV NEWVAL, YMEM         ;COPY
        BR PLT37
```

Figure 17.6 *(Continued)*

```
;NO CHANGE-PLOT STRAIGGHT LINE
PLT70:  JSR PC, STEPS           ;                       [18]
        DEC STTODO              ;                       [19]
        BNE PLT70

;NOW GET NEXT POINT
NXTPNT: INC PTDONE              ;COUNT POINTS DONE       [52]
        MOV NEWVAL, OLDVAL      ;UPDATE OLDVAL
        ADD #2, POINT          ;AND ADVANCE POINTER
        DEC COUNT
        BEQ PLT80               ;                       [53]
        JMP PLT15

;CLIPPING ROUTINE SETS OUTPUT TO FULL SCALE IF # IS >12 BITS
CLIP:   CLR SIGN               ;ZERO SIGN FLAG
        TST R2                 ;TAKE ABSOLUTE VALUE OF POINT
        BGE CL10
        NEG R2
        INC SIGN               ;SET SIGN FLAG
CL10:   MOV R2, R3            ;COPY INTO R3
        CLR R2                 ;ZERO R2 FOR DP SHIFT
        ASHC VSHIFT, R2        ;NOW SHIFT
        TST R2                 ;R2 SHOULD BE ZERO IF NO OVERFLOW
        BNE OVRFLW             ;THERE IS OVERFLOW
        BIT #170000, R3       ;ARE UPPER 4 BITS SET?
        BNE OVRFLW             ;YES, OVERFLOW HAS OCCURRED
        MOV R3, R2            ;COPY BACK INTO R2
CL20:   TST SIGN               ;RESTORE SIGN
        BNE CL30
        NEG R2
CL30:   ADD #4000, R2         ;AND ADD BIAS
        RTS PC                 ;EXIT
OVRFLW: MOV #3777, R2         ;FULL SCALE POSITIVE
        BR CL20                ;RESTORE SIGN AND EXIT
PLT80:  JSR PC, CRLF          ;                       [54]
        CLR IOSTAT            ;TURN OFF BOTH RELAYS    [55,56]
        JSR PC, CRWAIT        ;WAIT FOR CR
        MOV (SP)+, R3         ;RESTORE R3
        RTS R3                 ;RETURN                  [57]

;WAIT FOR CR
CRWAIT: .TTYIN R1             ;GET CHARACTER
        BIC #-200, R1        ;CLEAR OUT PARITY BIT
        CMP R1, #CR
        BNE CRWAIT           ;WAIT FOR CR
        JSR PC, CRLF         ;ELSE PRINT CRLF
        RTS PC               ;AND EXIT

;DELAYS
DELAY:  MOV DLSET, I         ;GET NUMBER TO COUNT DOWN
DLA10:  DEC I                 ;DECREMENT IT
        .TTINR R3            ;ANY KEY STRUCK?
        BCC DLA20            ;NOPE, KEEP COUNTING DOWN
        BIC #-200, R3       ;CLEAR PARITY
        CMP R3, #'F         ;F-FAST
        BEQ FAST
        CMP R3, #'S         ;S-SLOWER
        BEQ SLOW
        CMP R3, #'Q         ;Q-QUIT
        BEQ DLA30
DLA20:  TST I                 ;SEE IF WE'VE COUNTED TO 0 YET
        BNE DLA10            ;NO, KEEP LOOPING
        RTS PC               ;ELSE EXIT
```

Figure 17.6 (*Continued*)

```
DLA30:   CMP (SP), #STP5          ;POP EXTRA ITEM
         BNE DLA35                ;IF INTERRUPT WAS FROM "STEPS"
         TST (SP)+                ;POP
DLA35:   TST (SP)+                ;POP OFF SUBROUTINE RETURN
         BR PLT80                 ;AND GO TO EXIT ROUTINE

FAST:    ASR DLSET                ;DIVIDE DOWN DLSET
         BNE DLA20
         INC DLSET                ;PREVENT 0
         BR DLA20

SLOW:    ASL DLSET                ;MULTIPLY UP DLSET
         BNE DLA20
         MOV #100000, DLSET       ;PREVENT 0 AT HIGH END
         BR DLA20

;STEP TAKER
STEPS:   INC XDAC                 ;                              [58]
         JSR PC, DELAY
STP5:    INC STDONE
         CMP XDAC, #7777          ;BE SURE X DOESN'T OVEFLOW
         BEQ STP10                ;IT WILL
         RTS PC

STP10:   TST (SP)+                ;POP OFF STEPS RETURN ADDRESS
         BR PLT80                 ;AND EXIT

;CONSTANTS
DLSET:   400
STDONE:  0       ;STEPS DONE
STTODO:  0       ;STEPS TO DO
OLDVAL:  0       ;LAST POINT
NEWVAL:  0       ;LATEST POINT
YMEM:    0       ;MEMORY COPY OF Y
DELTAY:  0       ;Y-STEPS PER X-STEP
DIFF:    0       ;NEWVAL-OLDVAL
TOTSTP:  0       ;TOTAL STEPS
TOTPNT:  0       ;TOTAL POINTS
PTDONE:  0       ;POINTS DONE TO DATE
VSHIFT:  0       ;VERTICAL DISPLAY SHIFT
SIGN:    0       ;1 IF POINT IS NEGATIVE

POINT:   0
COUNT:   0
UDFLAG:  0       ;UP-DOWN FLAG
```

Figure 17.6 (*Continued*)

Boxes 15–17. The number of steps to plot to the next point is calculated using the equation shown. If this value, STTODO, is 0, the number of y increments per step, DELTAY, is set equal to DIFF.

Boxes 18 and 19. If DIFF was 0, there is to be no change in the y-axis plotting, and steps are taken one at a time until they are all done for this point. Control then transfers to the end of the loop, where the pointers and counters are modified at box 52.

Boxes 20 and 21. If DIFF was nonzero, the quantity DELTAY is calculated, where DELTAY is the amount of y change per x step. If UDFLAG is 1, the plot will be down and control is transferred to box 37. Otherwise the plot is up.

Boxes 22 and 37. A counter, PCNT, is set equal to DELTAY each time through the loop.

Boxes 23–26 and 38–41. The value of the memory copy of the YDAC is incremented (decremented) by 1 and then compared with the ultimate goal, NEWVAL. As long as YMEM $\neq$ NEWVAL, the value of YMEM is loaded into the YDAC. The program then delays for a variable time by counting memory cycles.

Boxes 36 and 51. If YMEM = NEWVAL, no more *y* steps are to be taken, and YDAC and YMEM are set equal to NEWVAL each time through the loop so that no further vertical excursions occur.

Boxes 27 and 42. The value of PCNT is tested. This is the counter which determines whether there are more vertical steps to be taken. If it is 0, there are no more steps and control is transferred to box 31 or 46, where the *x* axis is advanced. If it is now 0, it is by division truncation at box 20.

Boxes 28–29 and 43–44. The number of vertical steps taken is counted by decrementing PCNT. The value is decremented and again tested for 0. If it was not 0 above but is now 0, then it had a nonzero value and the number of *y* steps, DELTAY, is now completed. Control is transferred back to box 23 or 38 to take more *y* steps if PCNT is still not 0.

Boxes 30–32 and 45–47. If PCNT has just become 0, we now need to test to see whether there are more steps to do. If there are not, control is transferred to box 33 or 48. Otherwise one *x* step is taken which also

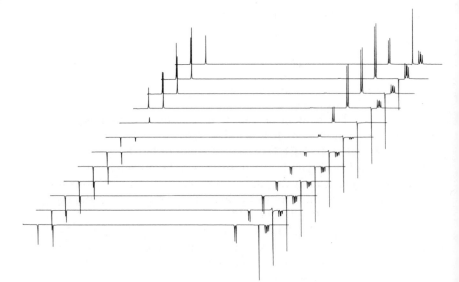

Figure 17.7   Stacked plots of an inversion recovery T_1 experiment on adenosine.

decrements STTODO. The value of STTODO is then tested; and if it is not yet 0, another DELTAY vertical steps are taken by going back to box 22 or 37.

Boxes 33–35 and 48–50. When all the calculated steps have been taken in both directions, it is still necessary to test to see whether YMEM has reached NEWVAL, since division truncation may have left a few more vertical steps to do. These are taken in boxes 34 and 39. Control from both columns then moves to box 52.

Boxes 52 and 53. Here the number of points done is incremented, the value of OLDVAL copied from NEWVAL, the pointer advanced, and the counter decremented. Then the counter is tested for 0. If it is 0, the plot is done. If it is not 0, control returns to box 8.

Boxes 54–57. Before exiting, the routine waits for a Return in case the plotter enable and pen lift are not controlled by relays, then issues the commands to disable the plotter, so that the returning display will not cause the pen to scribble all over the plot, and lifts the pen. The subroutine then exits.

Subroutine STEP

Box 58. The routine advances the XDAC by one, decrements STTODO, and increments PTDONE and STDONE. It then exits.

Subroutine CLIP. This subroutine, not shown, shifts the data point to the right or left according to the value of VSHIFT and then adds 4000 to display the result. It then tests to see whether the shifted point is off scale by being greater than 7777 or less than 0. If the point is less than 0, the value is replaced by 0. If greater than 7777, the value is replaced by 7777.

Continuous Line Display

By assigning 2K or 4K of memory as a display *buffer*, it is possible to display a spectrum or a portion of one with the "dots" connected. This can be accomplished by modifying the plot routine in Figure 17.6 to calculate new data points and then store them in the display buffer region of memory. The display routine then draws vertical lines between one point and the next, skipping some *y* points to keep the score refresh rate high.

Stacked Plotting

Plots of several spectra can be stacked and offset as in Figure 17.7 by making the number of plotter steps less than full scale and starting at a slightly offset x point for each new spectrum. The y axis is offset by adding a constant to each point before plotting it.

Plotting Examples

Figure 17.8 shows a spectrum containing 4K of data points, plotted full scale. Figure 17.9 shows the expansion of the left portion of the spectrum plotted using the interpolating technique detailed above. Figure 17.10 shows, for comparison, the "New York skyline" plot obtained when a small portion of a spectrum is expanded and plotted without interpolation.

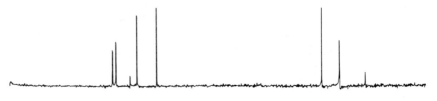

(b)

Figure 17.8   Full plot of carbon-13 spectrum of 3-ethylpyridine.

Figure 17.9   Expanded plot of two lines of spectrum in Figure 17.8.

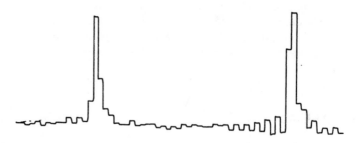

Figure 17.10   Expanded plot of same region as in Figure 17.9 without interpolation illustrating the "New York skyline" effect.

CHAPTER EIGHTEEN

DETECTION AND PRINTING OF PEAKS

It is often desirable, after acquiring data, to print out some information regarding the spectra, usually the height and time (or frequency) of the peaks. Whereas it is very easy to detect the peaks in a spectrum visually, detecting peaks by computer program is a nontrivial task.

Detection of peaks requires, first, the establishment of a mean noise value or threshold below which all data are assumed to be noise. This can be done by displaying a horizontal line on the spectrum and allowing the operator to adjust it, using a knob to define the cutoff point. This method is not very accurate, however, since it requires quite a high level of display resolution to select the threshold. In another commonly used method, the user enters at the teleprinter a value that defines the threshold numerically. This is prone to even more error if the data magnitudes vary widely. Furthermore, data having a rolling baseline are quite difficult to analyze by either of these methods.

It is also possible for the computer to detect the mean noise by some sort of calculation procedure and arrive at the threshold without any operator intervention. Unfortunately this can lead to the program making some totally unwarranted assumptions for spectra having a wide standard deviation but a number of important small peaks.

The method that has been found to be the best compromise between these approaches is essentially a combination method, in which the mean noise is calculated and peaks are then picked that are larger than the mean noise. The intensity of a peak is then compared with a user-adjustable threshold or minimum intensity, and only peaks larger than this threshold are actually printed out.

The mean noise method that is most easily implemented in the small minicomputer utilizes the sum of the absolute values of the differences of adjacent points divided by the number of points. In actual practice peaks are selected that are two to four times greater than the mean noise.

Once the mean noise has been calculated, a program for picking peaks can be written. However, there are several cases for which the program must allow.

1. Peak rising from the baseline and descending back to it.
2. Peak rising from the baseline and descending only partway back to it before another overlapping peak begins.
3. Peak starting above the baseline, overlapping with another, and either descending or not descending to the baseline.
4. Negative peaks, which descend below the baseline and return to it.

In essence, then, the peak picking routine must detect both positive and negative peaks and overlapping ones with minimum error. The first point, last point, and maximum must be returned to the calling program for interpretation and printout.

Flowchart Discussion

A flowchart of a peak printout program is shown in Figure 18.1. In essence the program (Figure 18.2) detects a peak if its absolute value: (1) rises above the baseline by an amount greater than the mean noise, (2) falls below the mean noise, or (3) falls by an amount greater than the mean noise and then rises again by an amount greater than the mean noise.

As written, the program does not correctly detect a peak whose maximum suddenly plunges from a large positive value to a large negative value, although this capability could easily be added with a sign flag.

The flowchart is interpreted as follows.

Box 1. The mean noise is calculated as the sum of the absolute values of the differences between adjacent points divided by the number of points.

Box 2. The pointer is initialized to the start of the array.

Boxes 3–5. Each point is tested for having an absolute value greater than the mean noise. Points are rejected until one rises above the mean noise criterion.

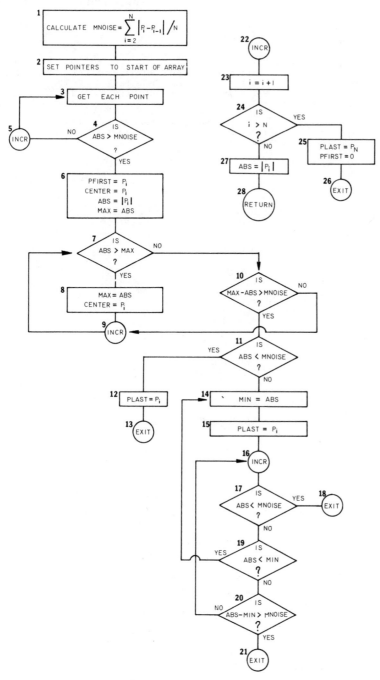

Figure 18.1 Flowchart of peak picking routine.

253

```
.GLOBL PFIRST, PLAST, CENTER,PEASET,PEAPCK
;PEASET SUBROUTINE
;CALLED BY
;       JSR R3, PEASET
;       SA              ;ADDRESS OF WORD CONTAINING BLOCK START ADDRESS
;       SIZE            ;ADDRESS OF WORD CONT. SIZE OF BLOCK
;THIS ROUTINE SETS PFIRST, PLAST, LSTPNT AND CALLS MNCALC
;TO CALCULATE THE MEAN NOISE

PEASET: CLR PFIRST              ;NO PEAK FOUND IF PFIRST=0
        MOV @(R3), PLAST        ;SET PLAST TO START OF BLOCK
        MOV @(R3)+, LSTPNT      ;SET LAST POINT
        MOV @(R3)+, SIZE        ;SET SIZE
        ADD SIZE, LSTPNT        ;LAST POINT
        ADD SIZE, LSTPNT        ;TWICE FOR BYTE ADDRESSING
        JSR PC, MNCALC          ;CALCULATE MEAN NOISE
        RTS R3                  ;AND EXIT

;;MEAN NOISE IS CALCULATED ON SIZE POINTS STARTING AT PLAST
;THE MEAN NOISE IS THE SUM OF THE ABSOLUTE VALUES OF THE DIFFERENCES
;OF ADJACENT POINTS
;CALCULATE MEAN NOISE
MNCALC: MOV SIZE, R1            ;COUNTER                 [1]
        MOV PLAST, R0           ;POINTER
        TST (R0)+               ;2ND POINT
        CLR R4                  ;HIGH SUM
        CLR R5                  ;LOW SUM
        DEC R1                  ;COUNTER-1
MN10:   MOV (R0)+, R2           ;GET EACH POINT
        SUB -4(R0),R2           ;GET DIFFERENCE FROM LAST ONE
        BPL MN20                ;TAKE ABSOLUTE VALUE
        NEG R2                  ;NEG IF -
MN20:   ADD R2, R5              ;ADD INTO RUNNING SUM
        ADC R4                  ;ADD ON IF CARRY
        SOB R1, MN10            ;GO BACK TILL DONE

;NOW DIVIDE DOWN BY NUMBER OF POINTS
        ASHC #5, R4             ;ADJUSTABLE MULTIPLIER ON NOISE
        DIV SIZE, R4
        MOV R4, MNOISE          ;SAVE MEAN NOISE
        RTS PC                  ;AND EXIT
```

Figure 18.2   Listing of peak picking routine.

Box 6. The constants PFIRST, CENTER, ABS, and MAX are initialized. ABS is the absolute value of the current point, MAX is the largest ABS found so far, CENTER is the address of MAX, and PFIRST is the address of the first point of the peak.

Boxes 7–9. Each time a new point is examined, its absolute value, ABS, is compared with MAX. If ABS is greater than MAX, MAX is changed to ABS and CENTER is changed to the address of ABS. Box 9 calls for the subroutine INCR, which increments the pointer, sets a new ABS, and tests for the end of the spectrum.

Box 10. As soon as a data point is found whose value is less than MAX, control is transferred to box 10. Here a test is made to see whether the difference between MAX, the peak maximum, and the current point is greater than the mean noise, MNOISE. If it is, the peak has fallen in

```
;PEAK PICKING ROUTINE
;CALLED AFTER PEASET BY
;         JSR PC, PEAPCK
;RETURNS FIRST AND LAST PEAK ADDRESSES IN PFIRST AND PLAST.
;MAXIMUM ADDRESS IN PEAK IS FOUND IN CENTER
;THE Z-BIT IS 0 IF A PEAK WAS FOUND
;AND A 1 IF NO MORE PEAKS WERE FOUND.
;THUS AN APPROPRIATE CALLING PROCEDURE WOULD BE
;         JSR R3, PEASET         ;SET POINTERS AND CALC MEAN NOISE
;         DATA                   ;ADDRESS OF START OF DATA
;         SIZE                   ;SIZE OF DATA BLOCK
;
;         JSR PC, PEAPCK         ;CALL PEAK PICKER
;         BEQ STOP               ;QUIT IF NO MORE PEAKS FOUND
;         :
;         :
;         :
;SIZE:    4096.
;DATA:    20000


;PEAK PICKER- ASSUMES THAT PFIRST, LSTPNT AND MNOISE HAVE BEEN PRIMED
;RETURNS WITH NEXT PEAK > MEAN NOISE

PEAPCK: MOV PLAST, R0   ;START PICKING HERE        [2]

;DIDDLE UNTIL A POINT > MNOISE IS FOUND
PICK00: MOV (R0), R2                               [3]
        BPL PICK10      ;ABS VALUE
        NEG R2
PICK10: SUB MNOISE, R2  ;POINT- MEAN NOISE      [4]
        BGT PICK20      ;GO ON IF > MEAN NOISE
        JSR PC, INCR    ;MOV ON TO NEXT POINT   [5]
        BR PICK00       ;GO BACK UNTIL ONE > MNOISE IS FOUND
PICK20: MOV R0, PFIRST  ;THIS IS FIRST POINT      [6]
        MOV R0, CENTER  ;CENTER POINT OF PEAK
        MOV (R0), R2    ;ABS
        BPL PICK30
        NEG R2
PICK30: MOV R2, ABS     ;ABS VALUE AND MAX SET
        MOV R2, MAX

;IS ABS> MAX?                                      [7]
PICK40: CMP ABS, MAX    ;CHECK FOR CHANGE IN LINE
        BLT PICK50      ;NOT GREATER
        MOV ABS, MAX    ;SAVE NEW MAX            [8]
        MOV R0, CENTER  ;CENTER ADDRESS
PICK45: JSR PC, INCR    ;GO ON TO NEXT POINT     [9]
        BR PICK40       ;KEEP LOOKING FOR RISE IN LINE

;SEE IF MAX-P(I)>MNOISE                              [10]
PICK50: MOV MAX, R4
        SUB ABS, R4     ;HAS LINE FALLEN BELOW MEAN NOISE?
        CMP R4, MNOISE
        BGT PICK60      ;YES, LINE IS NOW FALLING
        CMP ABS, MNOISE ;IS ABS< MNOISE?         [11]
        BGE PICK45      ;NO
```

Figure 18.2 (*Continued*)

```
;LINE HAS FALLEN BELOW MEAN NOISE                    [12]
;THIS IS THEREFORE THE END OF THE PEAK
EXIT:    CMP RO, PFIRST  ;IS LAST AND FIRST POINT THE SAME?
         BNE EXIT1       ;NO
         TST (RO)+       ;YES, ADVANCE BY 1
EXIT1:   MOV RO, PLAST   ;SET PLAST AND EXIT
         TST PFIRST
         RTS PC          ;                    [13]

;LINE IS NOW DECREASING-- SAVE PLAST EACH TIME
PICK60:  MOV (RO), PLAST                      [15]
         MOV ABS, R3     ; STORAGE OF MIN POINT [14]
PICK70:  JSR PC, INCR    ;NEXT POINT          [16]
         CMP ABS, MNOISE ;IS ABS > MNOISE?    [17]
         BLT EXIT        ;NO, LINE IS OVER    [18]
         CMP ABS, R3     ;NO, IS ABS < MIN POINT [19]
         BLT PICK60      ;YES, LINE IS STILL FALLING

;HAS LINE RISEN BY AMOUNT > MEAN NOISE  AFTER FALLING?
         SUB ABS, R3     ;LAST ABS- THIS ABS    [20]
         NEG R3          ;REVERSE SIGN
         CMP R3, MNOISE  ;IS DIFF > MNOISE?
         BLT PICK70      ;NO
         BR EXIT         ;YES END OF LINE       [21]

;SUBROUTINE TO INCREMENT POINTER AND SET NEW ABS
INCR:    TST (RO)+       ;                     [22,23]
         CMP RO, LSTPNT  ;> LAST POINT?        [24]
         BEQ IEXIT       ;YES
         MOV (RO), ABS;NO, SET NEW ABS         [27]
         BPL INCR10
         NEG ABS         ;ABS VALUE
INCR10:  RTS PC          ;                     [28]
IEXIT:   TST (SP)+       ;POP SUBROUTINE RETURN
         CLR PFIRST      ;NO PEAK FOUND IF 0   [25]
         BR EXIT         ;                     [26]


PFIRST:  0
CENTER:  0
PLAST:   0
MNOISE:  0
LSTPNT:  0
MAX:     0
ABS:     0
SIZE:    0
COUNT:   0
.END
HZPPT:   .FLT2 2.418
DSTART=20000
```

Figure 18.2 (*Continued*)

intensity by an amount greater than the mean noise, and this defines the start of the downward slope of the peak. If the point has not fallen from MAX by an amount greater than the mean noise, the fluctuation is assumed to be just noise and new data points are sought by returning to box 9 and thus to box 7.

Box 11. If the peak is now falling in intensity, control passes to box 11, where a test is made to see whether the current data point, stored in ABS, is less than the mean noise. If it is, the peak has dropped below the baseline noise and this defines the end of the peak. If so, PLAST, the address of the last point in the peak, is set to the current value of the pointer, P_i, and the program exits from the peak picking subroutine. If the point is not in the baseline region, control passes to box 14.

Boxes 14–18. When the program reaches this point, the peak is declining in intensity but has not fallen back to the baseline yet. We must now continually remember the minimum value that the peak reaches, so that we can detect the case of a new peak forming without the valley between the two reaching the baseline noise. Therefore, in boxes 14–16, we set MIN and PLAST and go on to the next point by calling INCR.

In box 17 we test again to see whether the absolute value of the current point is less than the mean noise. If it is, we exit at box 18. If it is not, we go on to box 19.

Box 19. Here we test to see whether the current data point is less than the minimum, remembered as MIN. If it is, we reset the value of MIN by going back to box 14. If the current point is greater than MIN, the peak may be rising to form a new peak. This case is tested for in box 20.

Box 20. Here is the crucial test of a multiplet whose splitting does not return to the baseline. If the value of the current point is greater than MIN by an amount greater than the mean noise, we decide that this means a new peak is starting and exit. If the current data point has not risen above MIN by an amount greater than the mean noise, we ascribe this rise to random noise fluctuations and go back to box 16 to test the next point.

Subroutine INCR

Boxes 22–28. This subroutine advances the value of P_i, the address of the point being tested, and then sets the absolute value of the current data point into ABS. It also is the place where a test is made to make sure that the peak being tested does not lie outside a preselected range. If the point P_i is incremented beyond the last point, exit from the peak picking routine occurs with PFIRST equal to 0. When PFIRST is 0, the driver routine accepts this as an indication that no more peaks are to be found.

Subroutines PEAPCK and MNCALC. In the program shown in Figure 18.2, the subroutine PEAPCK assumes that the starting address and the last point of the spectrum have been set into PEAPCK before it is called for the first time. Each time it exits, it returns a PFIRST, PLAST, and CENTER address of the next peak after the current setting of PLAST. It also sets PFIRST to 0 and exits when all points in the spectrum have been examined without finding further peaks. The subroutine MNCALC calculates the mean noise as the sum of the absolute values of the differences of adjacent points.

Interpolation Between Points

After a routine such as PEAPCK has been used to pick out peaks from the data, the question arises as to how to report the data to the user. If the peaks are relatively broad, this will not be a problem, but there is some difficulty in calculating and reporting the position (in units such as frequency) and the intensity of peaks which are defined by only a few points, such as are present in carbon-13 nmr data. These peaks may have only 3–10 points per peak, and clearly in such cases, the actual peak maximum and center position are not represented by any of the points on the peak.

While it is somewhat controversial to perform any "interpolation" because the data has not been measured that accurately, we report here for your consideration an algorithm for the parabolic interpolation between the top three points of a peak. The derivation of this simple algorithm is due to Dr. Bruno Guigas and Dr. Elaine Braun-Keller of Bruker, and acknowledgement is given here for helpful discussions with them.

Suppose we have three points near the top of a peak, including its apparent maximum, having intensities $I1$, $I2$, and $I3$, having x-positions $x1$, $x2$, and $x3$.

To fit a parabola to three points, we plug the points into the equations

$$Y = A x^2 + Bx + C$$

and obtain three simultaneous equations in A, B and C. We can simplify this process by transforming the coordinates so that $x2$ is the origin, giving points $(1, I_3)$, $(0, I_2)$, and $(-1, I_1)$. Then our three equations are

$$I_3 = A (+1)^2 + B (+1) + C$$
$$I_2 = A (0)^2 + B (0) + C$$
$$I_1 = A(-1)^2 + B(-1) + C$$

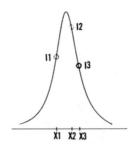

Figure 18.3 Coordinates of points near the top of a peak.

so we have

$$I_3 = A + B + C$$
$$I_2 = C$$
$$I_1 = A - B + C$$

or

$$I_3 = A + B + I_2$$
$$I_1 = A - B + I_2$$

Adding and subtracting, we have

$$A = (I_1 + I_3 - 2I_2)/2$$
$$B = (I_3 - I_1)/2$$
$$C = I_2$$

To find the maximum of the function

$$Ax^2 + 2Bx + C = 0$$

we take the first derivative and solve for x, giving

$$x = -B/2A$$

The maximum peak position thus occurs at $-B/2A$, a fractional distance from the highest point. This fraction can be calculated directly from the values of the intensities I_1 through I_3 given earlier.

Integration of Peaks in a Spectrum

The simplest way of calculating and reporting the integral under a peak, once it has been detected, is to sum the contents of all memory points contained within the peak. Given that the width of each channel is the same (one sampling interval), we can say that this sum represents a good approximation of the integral. By this summation we are really calculating the sum of the areas of the rectangles, which are a first approximation of the curve described by the data points shown. Thus, as shown in Figure 18.4, this integral is the one given by the *rectangular rule*. This is obviously a fairly crude approximation of the integral, especially in cases where the number of data points on the peak is relatively small.

The next best approach to the integration of the peak in memory is the trapezoidal integral, which is sketched in Figure 18.5. Here the areas of the small triangles shown on top of the rectangles are also included in the integral.

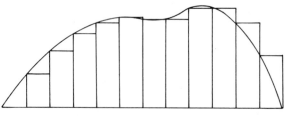

Figure 18.4   Rectangular integral shown as sums of rectangles drawn under the curve.

Let us consider (Figure 18.6) the area of one rectangle, given by f_1x where f_1 is the height of the left-hand side and x the width of one channel:

$$I_r = f_1 x$$

Then we can form the trapezoidal approximation to the integral by adding on the area of the triangle on top of the rectangle:

$$I_t = f_1 x + \tfrac{1}{2}(f_2 - f_1)x$$

or, rearranging,

$$I_t = \frac{2f_1 + f_2 - f_1}{2} = \frac{f_1 + f_2}{2}$$

Now the trapezoidal integration of an entire peak in this way leads to the formula

$$I_t = \frac{f_1}{2} + f_2 + \cdots + f_{n-1} + \frac{f_n}{2}$$

since all the $f/2$ terms add up except the first and the last. Trapezoidal integration is generally a much more accurate method of performing integration than the rectangular method and requires little additional programming.

One of the most accurate methods of integration, however, is Romberg integration, discussed in several numerical analysis textbooks. In essence,

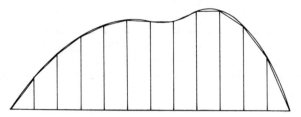

Figure 18.5   Trapezoidal integration, where the areas of the small triangles are added to the area of the rectangular integral.

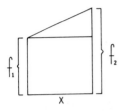

Figure 18.6 Coordinates of single rectangle and triangle used in rectangular and trapezoidal integration.

this method relies on the fact that the amount of error inherent in two trapezoidal integrations using different step sizes can be used to calculate a better integral than is available from either one of the step sizes alone. Unlike the classical case of series approximation, however, we cannot reduce the step size arbitrarily to as small an x-axis increment as is needed for a good integral because we are dealing with signal-averaged data having a finite number of points in the spectrum. However, we can increase the step size by a factor of 2 or 4, as needed, for Romberg integration.

In the simplest case, Romberg integration amounts to taking the trapezoidal integral of the peak using every point, then taking it again using every other point, and combining them as follows:

$$I_n = \text{integral using every point}$$
$$I_{2n} = \text{integral using every other point}$$
$$I_{\text{Romberg}} = I_n + \frac{I_n - I_{2n}}{3}$$

This simple extrapolation is sufficient for many cases involving small numbers of points. However, it can be expanded further by also calculating the trapezoidal integral for every fourth point and every eighth point and adding in further correction terms. The general formula is

improved value =

$$\text{more accurate} + \left(\frac{1}{4^k} - 1\right)(\text{more accurate} - \text{less accurate})$$

where k represents the step size as follows:

k	Step
0	Every point
1	Every second point
2	Every fourth point
3	Every eighth point

Exercises

18.1   Write a program to print out the peaks in a real or synthetic spectrum, acquired using your signal averaging program (and the knobs if no data are available), that prints out the center of each peak in time units and the contents of the memory channel, and then stops and intensifies the entire peak from PFIRST to PLAST while displaying the entire spectrum.

18.2   Write a program that picks peaks as in Exercise 18.1 and interpolates to find the actual center of each peak.

18.3   Write a program to perform rectangular, trapezoidal, and one order of Romberg integration, and print out the results for comparison on a series of picked peaks.

18.4   Write a program to plot out any selected knob-expanded region of the spectrum or to display it with the dots connected in a 4K display buffer.

CHAPTER NINETEEN

SIMPLE DIGITAL CIRCUITS

In Chapter 2, we showed the logic symbols for the AND, inclusive OR and exclusive OR. In this chapter, we look at some of the simple integrated circuit packages that are available to perform these functions. The discussion in this chapter follows the general outline of that given by Cazes.[1] Readers are also referred to more extensive texts on scientific uses of digital circuits by Perone and Jones,[2] Wilkins et al.,[3] and Arnold.[4]

We will start by considering one more type of logic element: the NAND. The NAND gate performs a logical AND and complements (or *inverts*) the result. The truth table for the NAND is

	0	1
0	1	1
1	1	0

We introduce the NAND gate here because it is the fundamental building block that can be used for nearly all common computer circuits. Its logic symbol is given in Figure 19.1. Note that this symbol is like the ordinary AND except for the small circle at the output side. This circle represents an *inversion* of the signal, so that whatever was true, or high, is now low or false, and vice versa.

NAND gates are available in a number of configurations: The simplest is the SN7400 series integrated circuit chip, which contains four NAND gates in a single 14-pin chip. Since most of the integrated circuits (or IC's) we will be dealing with are in similar packages, it is important to under-

263

A —[D)o— A·B
B —   **Figure 19.1**   The NAND gate.

stand the conventions for connecting up these chips. The SN7400 chip
and all other 14-pin chips operate only when connected to power, in this
case +5 volts, and to ground. Since by convention, all unconnected pins
of a chip have a +5-volt potential once power is applied, connection to
ground is essential for the chip to operate properly.

If you look at a 14-pin IC, you will see that it has a small indentation
on one side, and some numbers printed on it. The numbers on a 7400-
series chip should contain the chip number, usually among some other
letters and numbers that tell which manufacturer made it and what year
and week it was made in.

Using the indentation, we can always recognize pin 1. If you hold the
chip so that the indentation is at the left, pin 1 is directly below the
indentation and the pins are numbered from 1 to 14, going around the
chip in a counterclockwise direction (see Figure 19.2).

Looking at one of these NAND gates more closely, we see that we
could connect signals (either +5 for logical 1 or ground for logical 0) to
pins 1 and 2 and observe their logical NAND at pin 3, assuming that the
usual power connections (pins 7 and 14) had been made.

A number of *breadboarding* devices exist for plugging in a few IC's
and wiring up a few pins with some solid hookup wire. We will assume
that you can obtain one of these, such as the E&L Instruments Digi-
designer, or equivalent products made by Heathkit or Radio Shack, to
hook up some of the experiments in this chapter.

Here, we suggest that you plug an SN7400 IC into your breadboarding
device and

1. Connect +5 volts to pin 14.
2. Connect ground to pin 7.
3. Connect a lamp to pin 3.
4. Connect a switch to pin 1.
5. Connect another switch to pin 2.
6. Observe the state of the lamp for each of the four possible switch
   positions.

The Inverter

Another important building block for digital circuits is the *inverter*. This
device simply inverts 1 (+5 volts) to 0 (ground) and 0 to 1. We can make

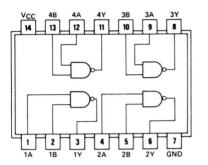

Figure 19.2 Pin assignments of a SN7400 integrated circuit.

a simple inverter, by tying together the two inputs of a NAND gate (see Figure 19.3) or we can actually use the inverter chip, such as is available in the SN7404, which contains six inverters from pin 1 to 2, 3 to 4, 5 to 6, 9 to 8, 11 to 10 and 13 to 12 (see Figure 19.4).

Exercise

19.1 Construct an AND gate using only an SN7400 chip.

Making a Simple Adder

Now let us consider what happens when we want to perform binary addition in a computer. The task becomes one of examining the results of binary addition:

$$
\begin{array}{cccc}
0 & 0 & 1 & 1 \\
\underline{0} & \underline{1} & \underline{0} & \underline{1} \\
0\ 0 & 0\ 1 & 0\ 1 & 1\ 0
\end{array}
$$

As we see, the sum of two binary digits produces two outputs: the *sum* and the *carry*. Now let's rewrite these results in truth table form for the sum and carry:

Sum:

	0	1
0	0	1
0	1	0

Carry:

	0	1
0	0	0
0	0	1

 **A** ———▷○— Ā **Figure 19.3**   Using a NAND gate for an inverter.

We see immediately that the sum of 2 bits produces a result that is identical to the exclusive OR, and the carry a result identical with the logical AND. We can easily construct a simple adder where the sum is A XOR B and the carry A AND B, which are written in Boolean logic notation as

$$\text{sum} = A \oplus B$$
$$\text{carry} = A \cdot B$$

This circuit is shown in Figure 19.5.

We call this a half-adder rather than a full adder because it produces a carry output but has no carry input coming in, so it cannot be cascaded to form a complete adder for many bits.

Exercise

19.2   Connect up the binary half-adder shown in Figure 19.6 and verify that it works. Be especially careful that each chip you use has both +5 and ground connected. Using the SN7486 chip, design a simpler circuit that contains XOR gates.

The Simple RS Flip-Flop

You can make a simple memory system that can be set and reset (cleared) using two NAND gates connected so that each output is an input to the other. The two inputs are called Set and Reset and the two outputs Q and Q-not, symbolized by $\overline{Q}$.

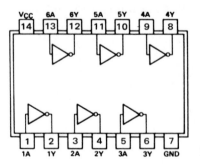

Figure 19.4   The SN7404 inverter.

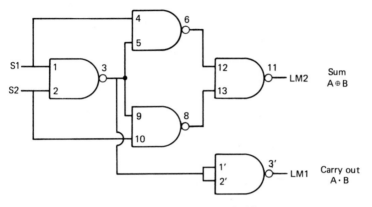

Figure 19.5 The binary half-adder.

One input sets Q to 1 and $\overline{Q}$ to 0, the other Resets Q to 0 and $\overline{Q}$ to 1. Thus, at any time after the event represented by pressing switch 1, we can observe the remembered value by examining the output Q, or its complement $\overline{Q}$.

Exercise

19.3 Connect up the simple RS flip-flop and see what happens for switch 1 = 1 or switch 2 = 1. What happens when they are both 1? What happens when they are both 0?

The Clocked RST Flip-Flop

It often happens that we want to remember what happened at a particular instant using a flip-flop. If we add NAND gates to the simple RS flip-flop,

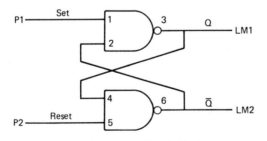

Figure 19.6 A simple RS flip-flop.

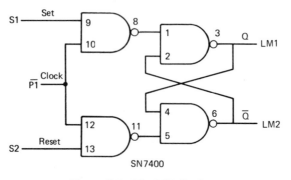

Figure 19.7   The RST flip-flop.

we have made what is called an RST flip-flop, which only changes state when a clock pulse changes from 0 to 1. This is shown in Figure 19.7.

Construct the circuit shown in Figure 19.7 and vary switches controlling the inputs S1 and S2 and observe the result when P1 momentarily becomes 1. What happens when S1 and S2 are both 1? Both 0? What happens if the pulse change for the clock is from 1 to 0 instead of 0 to 1?

The Data Latch

As with the simple flip-flop, the RST flip-flop can have an indeterminate state if the set and reset inputs are not complementary. This is remedied by allowing only one input to the flip-flop and connecting it through an inverter to the complementary input. Such a circuit is called a *data latch*.

Exercise

19.4   Construct a data latch from an inverter and an RST flip-flop and try it out.

The JK Master-Slave Flip-Flop

The data latch just described could still lead to an undetermined state if the states of R and S inputs change *while* the clock pulse is high. The JK master-slave flip-flop is actually two flip-flops tied together with the first one (called the master) connected to the inputs and the second (called the slave), connected to the master.

The data at the inputs is transferred to the master when the clock pulse changes from 0 to 1 and the data in the master is copied into the slave when the clock pulse falls from 1 to 0. The most important advantage of the JK master-slave flip-flop is that setting the two inputs always has a definite effect. The inputs are now renamed J and K. The action of the JK flip-flop is defined for all four possible combinations of inputs.

1. If both J and K are 1 at the clock pulse, the flip-flop will change to its other state.
2. If both J and K are 0 at the clock pulse, the flip-flop will not change from its current state.
3. If J and K are complements, they will be copied into the outputs Q and $\overline{Q}$ at the time of the clock pulse.

While it is possible to build up a JK-type flip-flop from two SN7400's and one SN7410 four-input NAND, the circuit is fortunately available in more convenient single packages such as the SN7476, which contains two such flip-flops. This IC is a 16-pin package with pins 1–8 on the bottom row and pins 16–9 along the top. Power connections are at pins 5 and 13 as shown in Figure 19.8.

In addition to the J and K inputs, there are also Preset and Clear inputs, which unconditionally set the state of Q to 1 or 0, respectively, when they are *grounded*. No clock pulse is needed in conjunction with Preset or Clear.

Exercise

19.5   Try plugging in the SN7476 chip and testing out these properties.

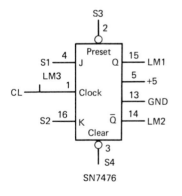

Figure 19.8 The JK master-slave flip-flop connections for one of two in an SN7476 package.

The Binary Counter

The most useful characteristic of the JK flip-flop is the precise definition of the action of the flip-flop when the previously indeterminate states of 00 and 11 are encountered. This can be used to construct a binary counter. In this counter, both inputs of the flip-flop are tied to 1 so that every clock pulse causes a change in the state of the outputs. The Q output of each flip-flop is used to clock the next flip-flop causing it to change each time the lower flip-flop changes from 1 back to 0, or "overflows." The circuit shown in Figure 19.9 will allow a four-stage binary counter which will display its current value in the four lamps. To watch it work, it is useful to have a low-frequency square wave generator to generate the pulses to cause the counting. Alternatively, you can just use a pushbutton that goes high when depressed.

Exercise

19.6 Construct the circuit in Figure 19.9 SN7476 chips. Does it count up or down? What simple change will cause it to count in the other direction?

The Shift Register

A shift register can be constructed from four JK flip-flops, using the usual property of copying the data from inputs to outputs when the clock pulse

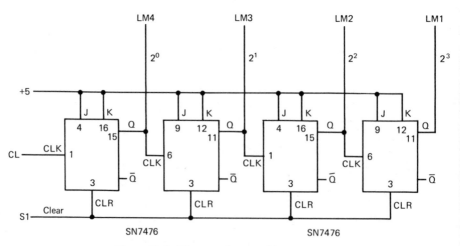

Figure 19.9 The asynchronous binary counter.

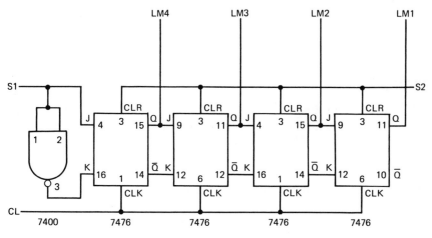

Figure 19.10   The Shift Register.

occurs. The initial data must be entered to the shift register with an inverter in the first stage to keep the inputs complementary. Then, each time a clock pulse occurs, the data are transferred to the next stage of the shift register. As shown in Figure 19.10 the circuit shifts data out the end of the register to be lost. Alternatively, this circuit could be modified to allow for rotation of the bits.

Interfacing to the PDP-11

The PDP-11 LPS-11 lab peripheral contains a digital I/O section which can be used to communicate 16 bits of parallel data from the computer to the "outside world." However, both the digital input and output connectors utilize *negative logic*, so that a binary 1 is equivalent to a ground state and a logical 0 to +5 volts. Thus you may need to utilize inverters or complement statements in your programs to correlate these data with your positive logic circuits.

The 16 bits of output data are sent to the Output Register by MOVing a 16-bit word into the Output Register. This is how the MOV occurs: a 1-microsecond pulse is sent on the OUTPUT READY line, which on the LPS-11 is actually pin 2 of the *input* connector. This line is normally high, and the pulse is a 1-microsecond GND pulse. The bits that are 1's are set to GND on the output lines.

Data are received at the input register upon receipt of the DATA READY pulse (GND) of 1 microsecond or greater length at the EXT

DATA READY line, which is (on the LPS-11) pin 2 of the *output* connector.

Exercise

19.7   Write a program to read the Input Register each time a pulser button is depressed on your breadboard, and print out the result as 1's and 0's on the terminal.

Detecting Imminent Memory Overflow

The problem of incipient memory overflow during signal averaging is a great one, since in some cases the data may not overflow without destroying valuable information, such as the shape of sine waves. This is discussed in Chapter 16. Memory can be said to be *almost* full when all of the upper bits are 1's. Memory is almost full of *signed numbers* when the sign bits are 0 and several lower bits are on OR when the sign bits are 1 and several lower bits are off.

Exercises

19.8   Design a circuit to detect whether imminent memory overflow is present in a 16-bit word sent to the output register of the PDP-11. If the overflow is going to occur shortly, then the circuit should send a 1 to the Input Register. Otherwise, the Input Register input should be 0. Write a program to test this circuit.

19.9   Design a circuit that will allow you to detect imminent overflow if any one of a series of words sent to the output register is almost full. Send back a 1 after all of them have been sent. Write a program to send such an array of data words containing just one overflow case. You will need to use a flip-flop to "remember" the last case each time. Submit the circuit, the program and a demonstration of the result for credit.

References

1.   J. Cazes, *Experimenting with Digital Circuits*, E&L Instruments, Derby, Connecticut, 1977.

2. S. P. Perone and D. O. Jones, *Digital Computers in Scientific Instrumentation*, McGraw-Hill, New York, 1973.
3. C. L. Wilkins, S. P. Perone, C. E. Klopfenstein, R. C. Williams, and D. E. Jones, *Digital Electronics and Laboratory Computer Experiments*, Plenum, New York, 1975.
4. James T. Arnold, *Simplified Digital Automation with Microprocessors*, Academic Press, New York, 1979.

CHAPTER TWENTY

THE USE OF THE FOURIER TRANSFORM IN THE LABORATORY

In this chapter we will discuss the Fourier transform as it applies to nuclear magnetic resonance (nmr) and to infrared (ir) spectroscopy. We will draw many of our examples from these fields to try to make the Fourier transform a believable and useful process. Then we will discuss the mathematics of the transform and close with general hints on the writing of such transforms, including an example of a real and of an inverse real transform.

Fourier Transform Nmr

In the conventional swept nmr experiment, referred to as continuous wave or cw nmr, a spectrum is obtained by slowly sweeping the radio frequency while holding a magnetic field constant. For proton (^{1}H) nmr, nmr signals can be observed at 60 megahertz and 14 kilogauss and at proportionally higher frequencies and fields. In such an experiment we sweep slowly through the spectrum at a rate determined by *slow-passage conditions*,[1] a maximum rate that is dictated by the nuclear resonance phenomenon. This is typically about 1 hertz per second for high-resolution spectra. Thus proton nmr (pmr) spectra can be obtained in about 1000 seconds over the full 1000-hertz frequency range. Naturally spectra occurring in a narrower frequency range will require less time to acquire.

However, should the signal be so weak that signal averaging is required, the amount of time per scan remains 1000 seconds. Even if we sacrifice some resolution and sweep at 100 seconds per scan, we find that signal averaging remains a very slow process. Suppose, however, that we sweep one half of a spectrum with one oscillator and at the same time sweep the other half with another oscillator. Then we take two detectors related to these two oscillators and run their outputs into two inputs of a computer or signal averager, assigning half the computer's memory to each input. We can then scan the entire spectrum twice as fast without losing any resolution, since each oscillator scans only half the spectrum.

The logical extension of this experiment is to have one oscillator for each frequency we wish to observe, each oscillator tied to a detector and eventually to a particular channel of an ADC, each channel of which, in turn, is connected to a single point in our computer or signal averager. Then we could signal average data extremely rapidly, even after allowing for the fact that continuous irradiation will lead to saturation. This thought experiment is known as the *multichannel nmr spectrometer*.

Clearly, such an experiment would be prohibitively expensive as well as physically impossible. However, it turns out that an analog to this experiment can be found in the pulsed-Fourier nmr experiment.[2]

To understand how pulsed-Fourier nmr can resemble the multichannel nmr spectrometer, we need to discuss a classical treatment of the magnetic resonance phenomenon. If we place in a magnetic field a sample containing nuclei that resonate, the nuclear magnetic moments, which can be regarded as stemming from a spinning charged particle, tend to align themselves with the applied magnetic field H_0. Since they are not quite perfectly aligned individually, they tend to *precess* around the applied field at a frequency in the radiofrequency range, called the *Larmor frequency*[1] (Figure 20.1). The resultant of the entire ensemble of nuclei,

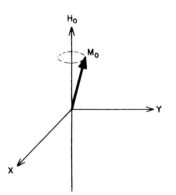

Figure 20.1 Magnetization of nuclei in magnetic field, precessing at Larmor frequency.

however, is perfectly aligned with H_0, since the various vectors are not in phase with each other.

During the swept nmr experiment the magnetic component of the applied field can be treated as a rotating magnetic field in the x–y plane, which, when it corresponds in frequency with the Larmor frequency of a particular type of nuclei, causes magnetic resonance to occur. The resulting energy absorption can be measured and plotted out as a function of frequency.

In the pulsed-Fourier experiment, a short duration single-frequency pulse, lasting perhaps a few microseconds, is applied and causes the tipping of all nuclei whose chemical shift ($\omega - \omega_0$) is less than $\gamma H_1/2\pi$, where H_1 is the power of the rf pulse. This rf power is generally chosen to excite the entire chemical shift range uniformly for a particular nucleus. If the rf pulse width is properly chosen, all of the nuclei in this range are tipped into the x–y plane, where they begin to precess at their respective Larmor frequencies (Figure 20.2).

By placing a detector coil in the x–y plane, we can observe these excited nuclei as they whiz by at their respective Larmor frequencies, and thus we are observing the entire spectrum at once. This is analogous to the multichannel nmr experiment and presents an advantage over swept nmr, known as *Fellget's advantage*.[3]

Thus we can observe all frequencies at once. If we consider what we observe, however, by looking at the y axis conceptually as the vectors whiz by, we realize that we will see a signal that grows in intensity as the vector reaches the $+y$ axis and then decreases in intensity as the vector moves on. The signal will have a negative amplitude as the vector reaches

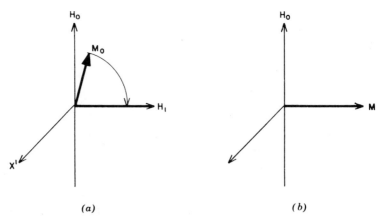

(a)   (b)

Figure 20.2 Tipping of magnetization M_0 by 90-degree pulse.

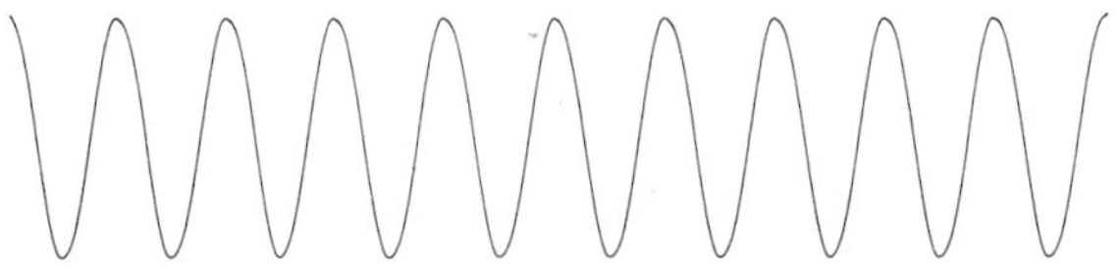

Figure 20.3   Sine wave induced by magnetization M rotating in x–y plane.

the $-y$ axis and then will again start to increase as the vector begins to become positive. What we have described, as can be shown by elementary trigonometry, is a sine wave whose frequency is the precession frequency. This might look as shown in Figure 20.3. Now it turns out that there is no real reason for the nuclei to stay in the x–y plane after their excitation, and they slowly decay back to their equilibrium positions along the z axis according to two relaxation processes called *spin-lattice relaxation* (T_1) and *spin-spin relaxation* (T_2). These are illustrated as a slow return to the z axis and a fanning out of the individual vectors in Figure 20.4. They result in an exponentially decaying sine wave, as shown in Figure 20.5, which is a plot of absorption versus time and can be converted to a plot of absorption versus frequency by inspection, as in Figure 20.6.

Now let us consider a more complex system consisting of two absorption frequencies. In the time domain it would appear as a set of coadded sine waves, but in the frequency domain as a pair of lines, as shown in Figure 20.7.

In these cases conversion from the time to the frequency domain can be accomplished by inspection. In the case of a real spectrum such as the carbon-13 nmr spectrum of 3-ethylpyridine shown in Figure 20.8,

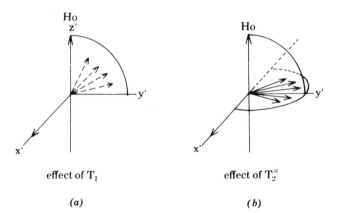

effect of T_1

effect of T_2^*

(a)

(b)

Figure 20.4   (*a*) Relaxation by spin-lattice relaxation mechanism, T_1. (*b*) Relaxation by spin-spin relaxation mechanism, T_2.

Figure 20.5 Decaying sine wave produced by observing excited magnetization M decay according to spin-spin relaxation and inhomogeneity effects.

however, the time domain spectrum is so much more complex than the frequency domain one that conversion is a nontrivial task.

Fourier Transform Infrared Spectroscopy

The usual chemists' method of measuring infrared spectra consists of a rather slowly swept frequency scan through the range of ir frequencies, producing a plot of absorption versus frequency much as in nmr. In general, scanning the data much more rapidly produces a great reduction in resolution.

However, the Michelson interferometer[4] provides a very efficient method of obtaining spectra more rapidly. In this experiment a beam of light containing a broad band of infrared frequencies is passed through a beam splitter. One half of the split beam goes to a fixed mirror and then out to the sample. The other half of the split beam goes to a mirror which can be moved fairly rapidly over a path several centimeters long. The beam is reflected from this mirror and out to the sample as well. If we were to observe the light just as it impinged on the sample, we would find that various light frequencies underwent constructive and destructive interference as a function of the moving mirror's position. The sample then absorbs some of these frequencies, and the interferogram of the light that is not absorbed is observed by a detector. These constructive and destructive interference patterns can be Fourier transformed with appropriate phase correction to produce a conventional ir spectrum. The details

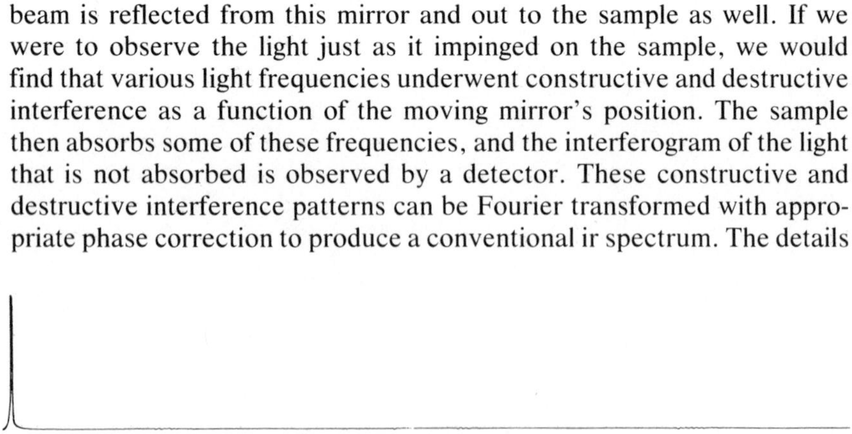

Figure 20.6 Single peak produced by frequency domain plot of the data in Figure 20.5.

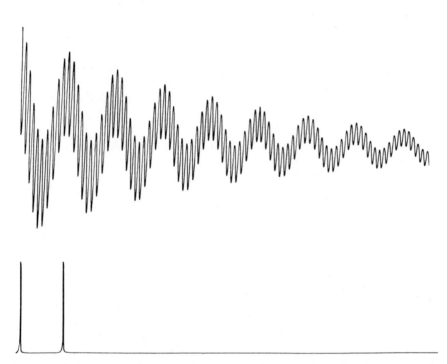

Figure 20.7 Two coadded decaying sine waves, such as might be produced by a two-peak nmr spectrum in the time domain. Same data plotted in the frequency domain.

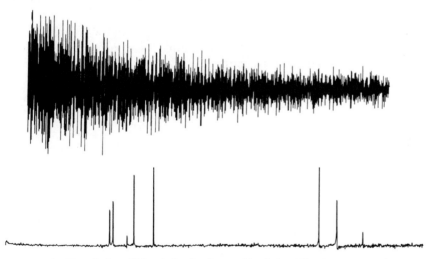

Figure 20.8 Top: Carbon-13 free-induction decay of 3-ethylpyridine. Bottom: Fourier transform of above fid.

of instrument design are discussed in an article by Griffiths, Foskett, and Curbelo.[4]

The principal reasons for the use of ir interferometery are somewhat different from those for Fourier transform nmr since weak sample absorptions are less of a problem in infrared. The main applications include

ir emission spectroscopy
Emission of remote sources, including astronomical bodies
Reflection spectroscopy
Rapidly reacting species
Gas chromatograph-infrared

and others where rapid measurement may be essential. Many of these are dealt with in a book by Griffiths.[5]

Mathematics of the Fourier Transform

The method for converting between the time and the frequency domain is known as the *Fourier transform*. If we have an array of data points acquired in a time domain spectrum, we can convert them to the frequency domain by means of the equation

$$A_r = \sum_{k=0}^{N-1} X_k \exp\left(\frac{-2\pi irk}{N}\right), r = 0, \ldots, N - 1 \qquad (20.1)$$

where A_r is the rth point in the frequency domain, and X_k the kth point in the time domain. The X_k's may be complex and the A_r's are almost always complex. We can rewrite this as

$$A_r = \sum_{k=0}^{N-1} (X_k) W^{rk}, r = 0, \ldots, N - 1 \qquad (20.2)$$

where

$$W = \exp\left(\frac{-2\pi i}{N}\right) \qquad (20.3)$$

This equation defines a closed continuous function, and Fourier analysis is usually, in fact, performed as integrals of continuous functions rather than summations. However, for the Fourier transform of a discrete number of data points, the small digital computer can perform the preceding summations to obtain the requisite transform. It turns out that this can be a very slow and time-consuming process, if the equation is implemented as shown. It amounts to N^2 multiplications and several times that many additions, where N is the number of data points in the spectrum.

Since this process can be so slow, regardless of the language it is coded in, the Fourier transform was for years the bottleneck in data reduction, with transforms taking 5–30 minutes and large chunks of the computer budget. In 1965, however, Cooley and Tukey[6] reported an algorithm for the calculation of the transform that reduced the number of multiplications to $N \log_2 (N)$, thus allowing a considerable saving in computer time. For example, whereas a 4096-point conventional Fourier transform required $4096^2 = 16.7$ million multiplications, the Cooley–Tukey method required only $4096 \log_2 (4096) = 4096(12) = 49{,}152$ multiplications.

This method, which seems a bit arcane at first, can be best understood by considering the following derivation as given by Cochran et al.[7] Let us consider a time series, X_k, consisting of N samples. We first divide it into two functions, Y_k and Z_k, each having $N/2$ points. We will let Y_k consist of the even-numbered points and Z_k of the odd-numbered points:

$$Y_k = X_0, X_2, X_4, \ldots$$

$$Z_k = X_1, X_3, X_5, \ldots$$

We can write these more compactly as

$$\left.\begin{array}{l} Y_k = X_{2k} \\ Z_k = X_{2k+1} \end{array}\right\}, k = 0, 1, 2, \ldots, \frac{N}{2} - 1$$

Each of the new functions, Y_k and Z_k, has discrete Fourier transforms, given by

$$B_r = \sum_{k=0}^{(N/2)-1} Y_k \exp\left(\frac{-4\pi irk}{N}\right) \tag{20.4}$$

$$\left.\vphantom{\sum}\right\}, r = 0, 1, 2, \ldots, \frac{N-1}{2}$$

$$C_r = \sum_{k=0}^{(N/2)-1} Z_k \exp\left[\frac{-2\pi ir(2k+1)}{N/2}\right] \tag{20.5}$$

The transform that we really are after, however, is A_r, which we can write in terms of the odd- and even-numbered points as

$$A_r = \sum_{k=0}^{(N/2)-1} \left\{ Y_k \exp\left(\frac{-4\pi irk}{N}\right) + Z_k \exp\left(\frac{-2\pi irk(2k+1)}{N/2}\right) \right\} \tag{20.6}$$

or

$$A_r = \sum_{k=0}^{(N/2)-1} Y_k \exp\left(\frac{-4\pi irk}{N}\right)$$

$$+ \exp\left(\frac{-2\pi ir}{N}\right) \left[\sum_{k=0}^{(N/2)-1} Z_k \exp\left(\frac{-4\pi irk}{N}\right)\right]$$

(20.7)

where $r = 0, 1, 2, \ldots, N - 1$ in equations 20.6 and 20.7.

Equation 20.7 can be rewritten, using the definitions of B_r and C_r, as

$$A_r = B_r + \exp\left(\frac{-2\pi ir}{N}\right) C_r, \qquad r = 0, 1, 2, \ldots, \frac{N}{2} - 1 \quad (20.8)$$

For values of r greater than $N/2$, the discrete transforms B_r and C_r repeat periodically. Thus, if we substitute $r + N/2$ for r above, we get

$$A_{r+N/2} = B_r + \exp\left(\frac{-2\pi i(r+N/2)}{N}\right) C_r, \qquad 0 \le r < \frac{N}{2}$$

$$= B_r + \exp\left(\frac{-2\pi ir - 2\pi iN/2}{N}\right) C_r$$

$$= B_r + \exp\left(\frac{-2\pi ir}{N} - \pi i\right) C_r \qquad (20.9)$$

$$= B_r + \exp\left(\frac{-2\pi ir}{N}\right) (-1)C_r$$

$$= B_r - \exp\left(\frac{-2\pi ir}{N}\right) C_r$$

Now, using equation 20.3, we can rewrite equations 20.8 and 20.9 as

$$A_r = B_r + W^r C_r \qquad (20.10)$$

$$0 \le r < \frac{N}{2}$$

$$A_{r+N/2} = B_r - W^r C_r \qquad (20.11)$$

Equations 20.10 and 20.11 show that the first and last $N/2$ points of a transform, A_r, can be obtained from the transforms of two $N/2$-point arrays, Y_k and Z_k.

If we have a method for obtaining the N-point transform, A_r, in the classical sense as given in equation 20.1, we can obtain it instead by the two $N/2$-point transforms, B_r and C_r. This is important because, whereas obtaining A_r directly requires N^2 multiplications, the method utilizing equations 20.10 and 20.11 requires $2(N/2)^2$ or $N^2/2$ multiplications.

Now we can extend this argument to as many subdivisions as we like. We could obtain D_r and E_r from the four smaller arrays, S_k, T_k, U_k, and V_k, each having $N/4$ points, and so forth. These reductions can be carried out as long as the remaining number of points is divisible by 2. In the most common limiting case, $N = 2^n$, and we can make n such reductions, arriving eventually at a one-point function. The discrete transform of a one-point function is, naturally, the sample itself. Thus the Cooley–Tukey Fourier transform algorithm becomes merely a point shuffling process, consisting entirely of complex multiplications and additions. Furthermore, the number of multiplications is reduced from N^2 to $N \log_2 (N)$.

Such a transform is shown in the signal flow graph of Figure 20.9 for an eight-point transform. Lines start at a point in one column and go to a point in the next column. Where lines intersect, we add the data together. Numbers written along side lines denote multipliers. Clearly half the multiplications expected are eliminated. Moreover, half the remaining multiplications are eliminated through Euler's formula

$$e^{-i\pi} = -1 \tag{20.12}$$

Now let us look more closely at this graph. The frequency domain points, A_i, end up in order, but the time domain points, X_i, start out of

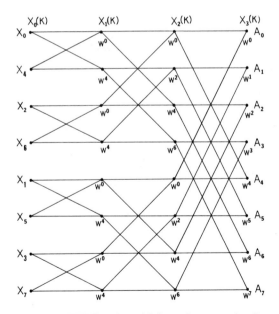

Figure 20.9   FFT flowchart, bit inversion occurring first.

order. The order $X_0, X_4, X_2, X_6, \ldots$ seems a peculiar one and impossible to predict until we examine the indices as binary numbers.

Position	Index	Binary Index
0	0	000
1	4	100
2	2	010
3	6	110
4	1	001
5	5	101
6	3	011
7	7	111

The order of the indices can be deduced from examination of the last column. The indices are the bit-reversed or *bit-inverted* representations of their positions in the list. Thus position 1 (001) is bit inverted to find the index 4 (100), and X_4 is placed in this position.

It is also possible to write the Cooley–Tukey Fourier transform, or fast Fourier transform (FFT), so that the X_i's start out in order. In this case the frequency domain points, A_i, end up in the bit-inverted order, as shown in Figure 20.10.

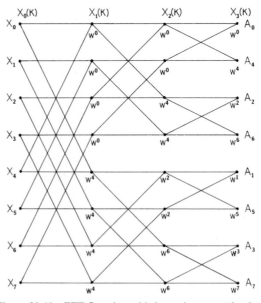

Figure 20.10   FFT flowchart, bit inversion occurring last.

There are other ways of looking at the Cooley–Tukey method. The classical Fourier transform can be considered the multiplication of the data by all possible sine waves from 0 to the Nyquist frequency. The Cooley–Tukey method can then be regarded as a method of taking advantage of the harmonics of sine waves and their integral multiples. The Cooley–Tukey method can also be considered as recognition of the fact that

$$W^k = W^{k \bmod N}$$

Finally, the Cooley–Tukey FFT algorithm can be elegantly derived by use of matrix algebra.[8]

Practical Details of the Transform Process

We now know in theory how a Fourier transform is to be written. However, although this is usually where articles on the subject leave the reader, it may be well to point out some simple mathematical manipulations that make the whole process much more simple than it first appears.

We have only briefly considered the form of the W terms. We know that we must multiply each point in each column by some power of W and that, by equation 20.3,

$$W = \exp\left(\frac{-2\pi i}{N}\right)^4 = \exp\left(\frac{-2\pi i}{N}\right) \tag{20.3}$$

First let us consider W^0. This simple case is clearly just 1. Powers of W that are factors of 4 are also simplified when $N = 8$ since

$$W^4 = \exp\left[\left(\frac{-2\pi i}{8}\right)^4\right] = e^{\pi i} = -1$$

by Euler's formula (20.12).

However, all the remaining powers of W can be simplified to a more tractable form by recalling the more general form of Euler's formula:

$$e^{iy} = \cos(y) + i\sin(y) \tag{20.13}$$

Thus some simple odd power of W such as W^3 becomes

$$W^3 = \exp\left(\frac{6\pi i}{8}\right) = \cos(0.75\pi) + i\sin(0.75\pi) = -0.707 + 0.707i$$

We have reduced the calculation of some exponential to a simpler calculation involving a sine and cosine. We can simplify this calculation even further, however, by including in our program a sine *look-up table*. Then we need not even evaluate π, but instead use the coefficient of π as an index to the look-up table.

The next trick that makes the *in place* transform possible is the fact that points can be operated on in pairs, so that the new value always is placed where the old one was at such a time that the old value is no longer needed. Examining the flowchart for the eight-point transform again, we see that every point in every column after the first is calculated from the values of two points in the preceding column, and that these two points are used in the calculation of only one other point. For example, the point $X_1(1)$ is calculated from $X_0(1)$ and $X_0(5)$; furthermore, the only other point that utilizes the values of $X_0(1)$ and $X_0(5)$ is $X_1(5)$. Thus $X_1(1)$ and $X_1(5)$ constitute a pair whose values are calculated at the same time. It turns out that these point pairs always occur, regardless of the size of the transform, and that the distance between them is halved on each successive pass through the transform.

Let us first derive the form in which the equations are most simply calculated during the transform. Each point, $X_i(k)$, in any column is the result of two previous points, $X_{i-1}(k)$, and each of the X_i's is in fact a *complex* number consisting of two data points. We will see later how to circumvent this restriction when only real data are involved.

Now let us consider the two points $X_i(m)$ and $X_i(n)$. We will simplify the notation by letting

$$X1 = X_{i-1}(m), \qquad X2 = X_{i-1}(n)$$

and

$$X1' = X_i(m), \qquad X2' = X_i(n)$$

Then, looking at our flow diagram, we find that in general

$$X1' = X1 + X2W^y, \qquad X2' = X1 + X2W^z$$

where y and z are two different exponents of W. However, this can be further simplified by recognizing that the exponents y and z are always $N/2$ apart and that

$$W^y = -W^{y+N/2} \qquad (20.14)$$

Then we can write

$$X1' = X1 + X2W^y \qquad (20.15)$$

$$X2' = X1 - X2W^y \qquad (20.16)$$

This not only simplifies the number of W^y terms to be calculated, but also reduces the number of multiplications by one half.

We will now expand equations 20.15 and 20.16, using 20.13 and letting

$$X1 = R1 + iI1, \qquad X2 = R2 + iI2$$

where $R1$ and $I1$ are the real and imaginary coefficients of the complex number $X1$, and i is the square root of -1.

Expanding, we have

$$R1' + iI1' = (R1 + iI1)$$
$$+ (R2 + I2)(\cos y + i \sin y) \qquad (20.17)$$
$$R2' + iI2' = (R1 + iI1)$$
$$- (R2 + iI2)(\cos y + i \sin y) \qquad (20.18)$$

Collecting terms and rearranging, we have

$$R1 = R2 \cos y + I2 \sin y + R1 \qquad (20.19)$$
$$R2' = -R2 \cos y - I2 \sin y + R1 \qquad (20.20)$$
$$I1' = -R2 \sin y + I2 \cos y + I1 \qquad (20.21)$$
$$I2' = R2 \sin y - I2 \cos y + I1 \qquad (20.22)$$

Note that only $R2$ or $I2$ is multiplied by $\sin y$ and $\cos y$ and that the $R1$ and the $I1$ are always added. This simplifies the programming process considerably, since we need only throw four values up in the air, perform four multiplications, and catch the results on the way down, requiring only a small amount of temporary storage. In the actual transform process we simply start with the widest possible distance between $X1$ and $X2$ and progressively narrow it on each pass. Then for the next pass we divide the data into two groups or "cells" and move through them until we have exhausted each cell. In each successive pass there are more smaller cells, until each cell consists of only two points. When we have completed this cell, we are done except for the bit-invert shuffling of the data points. Note that if we shuffle the data into bit-inverted order first, we start with the smallest possible cell size and continue until we have the largest possible cell size, as shown in Figure 20.9, Furthermore, the exponents of W are not bit-inverted and in fact can be incremented directly within each cell.

One last programming trick concerns the first pass through the data, which will always consist of only additions, since all the W's are raised to powers that make the result either 1 or -1. For the first pass, we can write a separate program segment if we wish to make the transform as fast as possible. The relevant equations will be

$$R'(m) = R(m) + R(n) \tag{20.23}$$

$$I'(m) = I(m) + I(n) \tag{20.24}$$

$$R'(n) = R(m) - R(n) \tag{20.25}$$

$$I'(n) = I(m) - I(n) \tag{20.26}$$

The Sine Look-Up Table

It is necessary to calculate a sine look-up table at some point before beginning the transform. It can be part of the program file, or it can be calculated upon start-up. The table need not be anywhere near as long as the spectrum to be transformed; usually 512 or 1024 points with interpolation is sufficient for data lengths as large as a minicomputer can handle—up to, say, 64K. The sine function varies periodically from 0 to 1, to 0, to −1, and back to 0. Each of these four quadrants is symmetric with the first quadrant along one plane or another, and thus only the first quadrant need be stored.

Since the sine function is always less than or equal to 1, the data in memory must use a fractional rather than an integer representation. In the most common one, bit 15 remains the sine bit, and bits 14–0 all represent fractional bits, with the binary point thus lying between bit 15 and bit 14. Although the integer 1 cannot be represented exactly in this way, the number 77777 is quite close to 1, which would really be represented as (1, 000000) in two words, since the sine bit is in bit 15. This format for the integer 1 is quite adequate, considering the larger errors that will be generated by successive round-off as the transform progresses.

To generate such a look-up table, it is merely necessary to divide the maximum input value by the number of points in the table and then multiply by the index of the data point before calling the sine function in the floating point package. We know that the sine of 90° = 1.000. sine functions often accept data in radians as well, so that the sine of ($\pi/2$) radians is 1.000; also, the sine function in FPMP11, as well as in the floating point packages of a number of other computers, returns a value of 1.00 for an input of $\pi/2$, so that input to FPMP11 is in units of radians. Therefore, we make the first point of the sine table 0 and then calculate the sine of $\pi/2$ (1/512, 2/512, 3/512, . . . , 511/512) for a 513-point look-up table. We set the last point of the table to fixed point 1.0000, or 77777.

Binary Fractions

Now that we have the sine look-up table generated in fractional form, we need to consider how we are going to handle the time domain data themselves. The simplest way to handle these data so that they all will be in the same format is to consider all of them as fractional as well, again with the binary point between bits 15 and 14. This data representation has some implications that we need to think out carefully before proceeding.

Suppose that we consider a simple 4-bit computer with multiplication hardware or software that simulates multiplication. If we multiply together any two 4-bit numbers, the hardware will, by definition, produce an 8-bit result. For example, the multiplication of 2×2 will occur as $0010 \times 0010 = 0000\ 0100$, producing 4 as expected. Similarly, if we multiply together two numbers that yield a double-precision result, such as 4×4, we will get $0100 \times 0100 = 0001\ 0000$. Now, in fractional multiplication, we always expect the binary point to reside at the left side of the number if the two numbers multiplied together are less than 1, as they always will be when the binary point precedes the most significant data bit. Thus, if we multiply 0.5×0.5, we expect to get 0.25 for an answer.

With this same multiplication hardware, 0.5 will be represented as 0100, and we have just shown that $0100 \times 0100 = 0001\ 0000$, while the fractional answer that we want is $0010\ 0000 = 0.010\ 0000 = 0.25_{10}$. Therefore, we can develop the following rule for using integer multiplication hardware for fractional multiplication:

> When performing fractional multiplication, the product must be multiplied by 2 or shifted one place left to produce a properly scaled binary fraction.

Note also that the single-precision product of a binary multiplication comes from the high-order word of the double-word product.

Although fractional division is not used very often because of the difficulty in controlling the magnitude of the result, we include it here for completeness. In our 4-bit computer, we assume that we can load a double word with the dividend and a single word with the divisor. Suppose that we were to divide 32 by 4. We would expect an answer of 8, and the computer would indeed provide it as follows:

$$\frac{32}{4} = \frac{0010\ 0000}{0100} = 1000$$

assuming that the number is unsigned.

Now, if we wish to divide 0.25 by 0.5, using the same divide hardware, we discover that we have set up the same problem:

$$\frac{0.25}{0.5} = \frac{0.010\ 0000}{0.100} = 1000$$

The answer we expect, 0100 or 0.5, can be generated only by shifting the product to the right. Our division rule then is:

Before performing fractional division, the entire dividend must be shifted once to the right.

Shifting the quotient after, instead of before, division can often lead to difficulties with the sign bit and is not recommended.

The Sine Look-Up Routine

We know from examining the sine function that all four quadrants are in some way symmetrical with the first one. Thus we have stored only the first quadrant of the sine table. The remaining quadrants are calculated using the following rules:

Quadrant	Input Value	Sine
1	$0 - \pi/2$	$\sin y$
2	$\pi/2 - \pi$	$\sin(-y)$
3	$\pi - 3\pi/2$	$-\sin y$
4	$3\pi/2 - 2\pi$	$-\sin(-y)$

The numbers we want to take the sine of are the exponents of W, since from equations 20.3 and 20.13 we know that

$$W^y = \cos\left[2\pi\left(\frac{y}{N}\right)\right] + i\sin\left[2\pi\left(\frac{y}{N}\right)\right] \qquad (20.27)$$

Thus y/N becomes the position of the sine look-up table and is related with appropriate scaling to the actual address in the table where the sine is found. The sine table consists of, say, 513 sine points ranging from 0 to 1. From the preceding table, we know how to obtain the sines in other quadrants, and we need only arrive at the most efficient format for the exponents y.

We might simply use values of y varying from 0 to 512 to represent values from 0 to $\pi/2$. Then we would have values from 0 to 2048 representing the numbers in all four quadrants. This method will return all the desired sines but leaves no room for interpolation. We can easily interpolate in our look-up table, however, if we let the input values range from 0 to 16383 in the first quadrant, thus allowing the entire 16-bit word as a value of y. Then the lower 4 bits will be interpolated bits, and they will indicate what fraction of the difference between two values is to be added on. This is more easily seen in the accompanying table.

Input Value (Octal)	Bit 15	Bit 14	Return Sine
0–37777	0	0	$\sin y$
40000–77777	0	1	$\sin(-y)$
100000–137777	1	0	$-\sin y$
140000–177777	1	1	$-\sin(-y)$

Thus we simply write a program that negates y if bit 14 was set and negates the sine if bit 15 was set. After noting bits 14 and 15, we note the contents of bits 0–3 for interpolation purposes, shift y four places to the right, and look up the sine in that address.

Interpolation amounts to multiplying the difference between this sine and the next one by the value of bits 0–3 divided by 2^4. This is most easily accomplished by multiplying this difference by the number held in bits 0–3 and shifting four places to the right. This shifted sum is then added to the lower sine in the look-up table. A program for accomplishing these things is shown in Figure 20.11.

Scaling During the Transform

One important factor that must be considered in these integer operations in the transform is the problem of numerical scaling. We are always multiplying together fractions, so that their product will always be less than 1 and thus will fit in 16 bits, but the additions that are performed during each pass through the transform will, in general, cause the size of the data to increase continually. Therefore, we must continually examine the data and assure ourselves that no memory overflow will take place soon.

It is not sufficient to allow the data to be held between half scale and full scale, as the operations shown by equations 20.19–20.22 can cause the data to more than double. Therefore, we continually test the data to see whether bits 14 and 13 are set, and shift the entire data array one place to the right if bit 13 is on, and two places to the right if bit 14 is on. This prevents one pass from inflating the data by more than the word length will allow. This test must be made *before* each pass through the transform to prevent overflow during that pass.

```
;SINE LOOK-UP ROUTINE
;ASSUMES 513 POINT SINE LOOK-UP TABLE
;RANGING FROM 0 TO 77777 IN 513 LOCATIONS STARTING AT SINTAB
;MUST BE GENERATED BY USER
;USES R2 AND R3
LOKSIN: CLR SIGN          ;REMEMBERS BIT 15
        TST R2            ;LOOK AT BIT 15 OF ARGUMENT
        BPL SIN01         ;SKIP OVER THIS IF +
        BEQ SIN05         ;SINE OF ZERO IS 0
        INC SIGN          ;REMEMBER IF SIGN WAS -
SIN01:  BIT #40000, R2    ;TEST BIT 14
        BEQ SIN02         ;DO NOTHING IF IT WAS ZERO
        COM R2            ;SIN (-Y) IF BIT 14 SET, QUADS 2 & 4
SIN02:  MOV R2, R3
        BIC #177740, R3   ;SAVE INTERPOLATION BITS
        ASH #-4, R2       ;NOW SHIFT TO REPRESENT TABLE LENGTH(512)
        BIC #176001, R2   ;CLEAR OUT EXTRANEOUS SIGN BITS
        MOV SINTAB(R2), SINLO    ;GET THIS SINE
        MOV SINTAB+2(R2), SINHI  ;GET NEXT SINE FOR INTERPOLATION
        SUB SINLO, SINHI         ;GET DIFFERENCE
        CLR R2            ;ADD IN INTERPOLATION HERE
        TST R3            ;CHECK FO ZERO CASE
        BEQ SIN04         ;NO INTERPOLATION
SIN03:  MOV R3, R2        ;MOVE BACK TO MULTIPLY
        MUL SINHI, R2     ;MULTIPLY DIFFERENCE BY INTERP BITS
        ASHC #-5, R2      ;SHIFT INTO POSITION
        MOV R3, R2        ;THIS IS THE INTERPLATION AMOUNT
SIN04:  ADD SINLO, R2     ;ADD SINE TO IT
        TST SIGN          ;NOW RESTORE SIGN
        BEQ SIN05
        NEG R2
SIN05:  RTS PC
```

Figure 20.11   Listing of sine look-up routine.

Fourier Transform of a Real Function

Thus far we have dealt exclusively with the FFT of complex data. Such complex data might be obtained when the input data came from some quadrature detection method,[9,10] but in general only a single input of real data is used for the transform. The output, however, will always be complex, consisting of a real and an imaginary part. To perform the transform on real data, we make use of a modification of the $2N$-point function transform algorithm discussed by Cooley, Lewis, and Welch[11] and by Brigham.[12]

We consider the function X_k for $k = 0$ to $2N - 1$. We then divide X_k into two functions:

$$\left.\begin{array}{l} h(k) = X(2k) \\ \\ g(k) = X(2k + 1) \end{array}\right\}, \quad k = 0, 1, \ldots, N - 1$$

We then treat the real data as if they were the complex function

$$Y(k) = h(k) + ig(k), \quad k = 0, 1, \ldots, N - 1 \qquad (20.27)$$

After computing the FFT of $Y(k)$ as though the even points were real and the odd points imaginary:

$$Y(n) = \sum_{k=0}^{N-1} Y(k) \exp\left(\frac{-2\pi i n k}{N}\right) \qquad (20.28)$$

we have the transformed function $Y(n)$, consisting of a real and an imaginary part:

$$Y(n) = R(n) + iI(n)$$

To perform the postprocessing point shuffle, it is necessary to have complex points $Y(n)$, where n varies from 0 to N. There are thus $N + 1$ data points in $Y(n)$, and the transform produces only N of them. The last point, $Y(N)$, may be calculated and inserted when calculating equations 20.30 and 20.31 by

$$Y(N) = R(0) - iI(0) \qquad (20.29)$$

In other words, $Y(N)$ is the *complex conjugate* of $Y(0)$.

Then, letting

$$R_p = \frac{R(n)}{2} + \frac{R(N-n)}{2}$$

$$I_p = \frac{I(n)}{2} + \frac{I(N-n)}{2}$$

$$R_m = \frac{R(n)}{2} - \frac{R(N-n)}{2}$$

$$I_m = \frac{I(n)}{2} - \frac{I(N-n)}{2}$$

we perform the following point shuffle:

$$A_r(n) = R_p + \cos\left(\frac{\pi n}{N}\right) I_p - \sin\left(\frac{\pi n}{N}\right) R_m \qquad (20.30)$$

$$n = 0, 1, \ldots, N - 1$$

$$A_i(n) = I_m - \sin\left(\frac{\pi n}{N}\right) I_p - \cos\left(\frac{\pi n}{N}\right) R_m \qquad (20.31)$$

where $A_r(n)$ and $A_i(n)$ are the real and imaginary parts of the transformed real array. Thus the peaks in the imaginary will always seem to be 90° out of phase with those in the real, as shown in Figure 20.4, where the phase-corrected real and imaginary parts of the transform of Figure 20.1 are illustrated.

Baseline Correction. Before Fourier transformation to the frequency domain, it is customary to *baseline-correct* the time domain data so that their average baseline is 0. This is necessary because the inclusion of a spurious dc offset in the time domain will cause a spike in the first address in the frequency domain, corresponding to a 0-hertz peak.

This correction is usually performed by taking the integral of the entire spectrum, dividing it by the number of points, and subtracting the quotient from each data point:

$$X_i' = X_i - \frac{\left(\sum_{i=1}^{N} X_i\right)}{N} \tag{20.32}$$

Occasionally, there will really be a 0-hertz peak, in which case some account must be taken of it. In complex data methods such as quadrature nmr, the 0-hertz peak will have decayed by the end of the free-induction decay, and such quadrature data can be baseline-corrected by calculating the correction from the last quarter of the free-induction decay.

Inverse Real Transforms

To take an inverse transform of complex data we simply reverse equation 20.1:

$$X_k = \sum_{k=0}^{N-1} A_r \exp\left(\frac{+2\pi irk}{N}\right), \qquad k = 0, 1, \ldots, N - 1 \tag{20.33}$$

On expanding the exponential term, we have

$$R1' = R1 + R2 \cos y - I2 \sin y \tag{20.34}$$
$$R2' = R1 - R2 \cos y + I2 \sin y \tag{20.35}$$
$$I1' = I1 + R2 \sin y + I2 \cos y \tag{20.36}$$
$$I2' = I1 - R2 \sin y - I2 \cos y \tag{20.37}$$

where we note that these differ from equations 20.19–20.22 only in the sign of the sine terms. Thus we can convert a forward transform to an inverse by simply negating the four sine terms as we calculate them, or by changing the sign of the angle by which we look up, since

$$\sin(-y) = -\sin y$$

Now, to perform an inverse transform on real data, we must first unshuffle the points that were postprocessed according to equations 20.30

and 20.31. We do this with

$$A_r(n) = R_p - \cos\left(\frac{\pi n}{N}\right) I_p - \sin\left(\frac{\pi n}{N}\right) R_m \qquad (20.38)$$

$$A_i(n) = I_m - \sin\left(\frac{\pi n}{N}\right) I_p + \cos\left(\frac{\pi n}{N}\right) R_m \qquad (20.39)$$

which are exactly like equations 20.30 and 20.31 except for the sign of the cosine terms. Thus the postprocessing shuffle during a forward transform can be converted to the preprocessing shuffle before an inverse transform by simply negating the *cosine* terms as they are calculated.

Correlation

Auto- and cross-correlation techniques can be used to measure the periodicity of a function. Autocorrelation is used on a time function, $f(t)$, as a measure of the statistical dependence of a later value, $f(t_0 + t)$, of the wave form on the present value, $f(t_0)$. In some heart diseases, for example, the relative times between the different pulses of the cardiac cycle may vary from those of a healthy heart, but in a weak heart these pulses may be obscured by the noise. However, since the wave form has a periodicity buried under the noise, autocorrelation of the function may produce a cleaner signal.

In cross-correlation, on the other hand, the resulting spectrum will be that produced by the correlation of two different spectra. For example, Dadok and Sprecher[13] and Gupta, Ferretti, and Becker[14] have examined the rapid-scan nmr technique, which allows a fast frequency sweep of the nmr spectrum to be recorded in a computer memory and then cross-correlated with a spectrum of a single line swept at the same rate. The result looks like a conventional slow-passage spectrum without the overlapping ringing caused by violating slow-passage conditions.[1] Figure 20.12a shows the rapid-scan pmr spectrum of the quartet of ethylbenzene, and Figure 20.12b the response of a rapid scan of the singlet of chloroform. These two spectra can be cross-correlated to produce the spectrum shown in Figure 20.12c.

Mathematics of Correlation

The process of correlation can be carried out discretely in the following fashion:

$$z(k) = \sum_{i=0}^{N-1} h(i)x(k + i) \qquad (20.40)$$

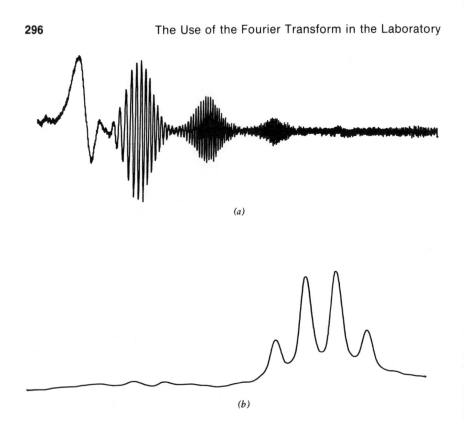

(a)

(b)

Figure 20.12 (a) Rapid scan through quartet of ethylbenzene. (b) Rapid scan through the TMS singlet. (c) Cross-correlation product of (a) and (b).

where both $x(k)$ and $h(k)$ are periodic functions of period N. This essentially means multiplying every point in one spectrum by every point in the other spectrum or a total of N^2 multiplications, as in the Fourier transform. However, the same time savings may become available by using the fact that correlation may be performed through the Fourier transform process. This can be done by first computing the discrete Fourier transform of the data:

$$X(n) = \sum_{k=0}^{N-1} x(k) \exp\left(\frac{-2\pi ink}{N}\right) \tag{20.41}$$

$$H(n) = \sum_{k=0}^{N-1} h(k) \exp\left(\frac{-2\pi ink}{N}\right) \tag{20.42}$$

Then the product is calculated:

$$Y(n) = X(n)H(n) \tag{20.43}$$

and the inverse discrete transform of $Y(n)$ is computed:

$$y(k) = \frac{1}{N} \sum_{n=0}^{N-1} Y(n) \exp\left(\frac{2\pi ink}{N}\right) \qquad (20.44)$$

Thus, whereas equation 20.40 takes N^2 multiplications, we can perform the same operation through the transform process doing only the following multiplications.

Equation	Multiplications
20.41	$N \log_2 (N)$
20.42	$N \log_2 (N)$
20.43	N
20.44	$N \log_2 (N)$
Total	$N + 3N \log_2 (N)$

This can still be a significant saving for larger correlograms. For example, a 1024-point time function could be auto- (or cross-) correlated directly in $(2^{10})^2 = 2^{20}$ multiplications. It can be cross-correlated by this method in $2^{10} + 30 \times 2^{10} = 31,744$ multiplications versus 1,048,576 multiplications by equation 20.40. This is a factor of 33 savings in time.

Noise in the Transform

We saw in Chapter 16 that the highest dynamic range signal that could be detected was limited by the length of the ADC, so that 2^{d+1} represented the absolute upper limit of the dynamic range for an ADC of length d bits. However, the design of longer and longer ADC's will not necessarily increase the dynamic range of an integer Fourier transform because of noise introduced by round-off during the scaling of the transform.

The phenomenon of this round-off has been investigated by Welch[15] and others[16] and the dynamic range observable after a transform has been found to be proportional to the computer's word length, the number of points in the transform and the number of locations that are almost full at the outset of the transform.

For example Table 20.1 shows the different dynamic ranges observed for various transform lengths in a 16- and 20-bit computer depending on whether the large peak occurs at the Nyquist frequency or at low frequency.

Since the dynamic range in 16-bit single precision transforms has been so limited, a number of techniques for enhancing it have also been in-

Table 20.1   Dynamic Range of Transformed
Spectra

| *Large Peak at the Nyquist Frequency* | | |
Size	$w = 20$	$w = 16$
4096	29127	5891
8172	43691	3567
16384	29127	2608
32768	24966	2608
Large Peak at a Low Frequency		
4096	43691	2608
8192	43691	4855
16384	58254	4369
32768	43691	3492

vestigated. The most important changes have been in the scaling methods used in these transforms. In longer word-length computers, it is sufficient to scale the data before each pass so that it can't overflow memory during that pass. If instead, this scaling is only done when overflow actually occurs, a substantial increase in dynamic range is noted.[17] If, in addition, the two-word fractional multiplication is handled so that the upper word is rounded up if the most significant bit of the lower word is a 1, additional dynamic range is noted. This is shown in Table 20.2. These techniques are of much less importance in 24- and 32-bit computers.

Table 20.2   Effect of Scaling and Round-Up Techniques on Dynamic Range
After Fourier Transforms

Size	Scaling Method: Before Pass Rounding Method: Truncation	Before Pass Round-Up	On Overflow Round-Up
	20-bit word		
4096	56,842	93,128	145,512
8192	56,842	116,408	145,512
16384	36,378	116,408	181,896
32768	36,378	116,408	181,896
	16-bit word		
4096	6102	7628	9536
8192	3124	6102	···
16384	2500	6102	···
32768	2000	4882	9536

Speeding Up the Fourier Transform

In this text we have discussed the scaling techniques needed to utilize a short 16-bit word in an integer transform. It is also possible to speed up the integer transform somewhat by reducing the number of repetitions of the sine lookup. In such an algorithm, each similar point in all cells are calculated once the sine and cosine for that point in the cell is determined. This leads to a 10–20% increase in transform speed.

However, using this particular scheme somewhat complicates the scaling-on-overflow techniques suggested here, and it is not utilized in the following program.

The Fourier Transform Calling Program

The program shown in Figure 20.13 is given as a calling program for the FFT routines. It calculates the first available address after the program, calculates the sine table, allows entry of a filename and reads in that file, and then allows a forward or inverse transform on that file. The calling routine thus uses the .CSISPC, .READW, .TTINR and .TTYOUT RT-11 calls, as well as illustrating a single-character command interpreter.

A Fourier Transform Routine

In this section, we present a PDP-11 assembly language program (Figure 20.14) for obtaining forward and inverse real and complex transforms. The method followed is that of Figure 20.9, where bit inversion occurs first. A flowchart and FORTRAN and ALGOL programs are given by Brigham,[8] and a Pascal program has been given in reference 18.

Several important shortcuts are taken in this program. First, since the data are shuffled in bit-inverted order first, the exponents of W are calculated in increasing order. The changes in these exponents as we proceed through the cells are given by an increment that starts at 180° in pass 1 and is half as large in each successive pass. We need only calculate the increment to y for each new point pair in the cell and, starting with $y = 0$, proceed to higher exponents that are in fact simply numbers closer to 2π. We then reset y to 0 for each new cell and recalculate Δy for each new pass.

For the first pass, where no multiplications take place, we have two points per cell and $N/2$ cells. In each succeeding pass, we divide the number of cells by 2 and multiply the number of points per cell by 2. We

```
;FFT DEMO PROGRAM
.TITLE FFT DEMO PROGRAM
;FPMP11 DEFINITIONS
        ;ADDRESSING MODES
        ARM=100
        IMM=200
        RELM=300

;TRAP HANDLER DEFINITIONS IN FPMP11
FPGET=TRAP+71
FPSTR=TRAP+73
FPADD=TRAP+12
FPSUB=TRAP+13
FPMULT=TRAP+21
FPDIV=TRAP+25
FPSIN=TRAP+36
FSQRT=TRAP+46
FCMP=TRAP+17
;LPS-11 DEFINITIONS
DISTAT=170416
XDAC=DISTAT+2
YDAC=XDAC+2

;BIT DEFINITIONS FOR RT-11
JSW=44  ;JOB STATUS WORD
BIT12=10000
FREE=50 ;ADDRESS CONTAINING FIRST FREE ADDRESS ABOVE PROGRAM



;THIS PROGRAM DISPLAYS AND TRANSFORMS A 4K REAL FILE
; IN PLACE, AND DISPLAYS THE RESULT
;SINE TABLE IS GENERATED USING FPMP11 AT STARTUP
.MCALL .EXIT, .TTINR,.TTYOUT
.MCALL .CSISPC, .LOOKUP, .READW

.GLOBL TRAPH,$POLSH,$IR,$RI,FORWD,INVERS,FDATA,FSIZE,SINTAB

.ASECT
.=1000
START:  MOV #., SP
        BIS #BIT12, @#JSW        ;TTY SPECIAL MODE
        MOV @#FREE, FDATA        ;PUT ADDRESS FOR DATA IN DATA WORD
        JSR PC, CRLF     ;INIT CARRIAGE
        MOV #TRAPH, @#34
        MOV #340, @#36  ;NO INTERRUPTS DURING FPMP11
        TST SINFLG      ;WERE SINE ALREADY GENREATED?
        BNE S20         ;YES
;GENERATE 513-POINT SINE LOOK-UP TABLE
        MOV #SINTAB, POINT      ;STARTING HERE
        MOV #511., COUNT        ;CAN'T USE REGISTERS IN POLISH
        CLR @ POINT
        ADD #2, POINT           ;GO ON TO 2ND ENTRY
        MOV #1, SINFUN          ;NUMERATOR OF N/512
SINLOP: MOV SINFUN, -(SP)       ;FLOAT IT
        JSR R4, $POLSH
        $IR             ;FLOAT
        .+2             ;EXIT FROM POLSIH MODE
        FPGET
        FPDIV+IMM       ;POP AND DIVIDE
        .FLT2 512.      ;BY 512
        FPMULT+IMM      ;MULTIPLY BY
        .FLT2 1.570796327       ;PI/2
        FPSIN           ;TAKE THE SINE
        FPMULT+IMM      ;CONVER TO LARGE INTEGER
```

Figure 20.13 The FTCALL driver program.

300

```
            .FLT2 32768.
            FPSTR           ;PUSH FOR FIXING
            JSR R4, $POLSH
            $RI             ;FIX IT
            .+2             ;AND EXIT
            MOV (SP)+, @POINT       ;ENTER IN TABLE
            INC SINFUN      ;ADVANCE SINE FUNCTION
            ADD #2, POINT   ;ADVANCE POINTER
            DEC COUNT                ;DONE?
            BNE SINLOP               ;NO
            MOV #77777, @POINT      ;LAST ENTRY IS 1.0
            INC SINFLG      ;SINE TABLE IS DONE

;READY TO DISPLAY DATA
            JSR PC, CRLF            ;TYPE NEW CR TO INDICATE WE'RE READY
520:        MOV FDATA, R0           ;SET DISPLAY POINTER
            MOV DSIZE, R1   ;SET COUNTER
            MOV #10, DISTAT         ;DISPLAY
            MOV #-1, XDAC
530:        MOV (R0)+, R2           ;GETEACH POINT
            ASH VSHIFT, R2          ;SHIFT
            ADD #4000, R2           ;BIAS FOR DISPLAY
            INC XDAC                ;ADVANCE X
            MOV R2, YDAC            ;AND LOAD Y
            SOB R1, 530            ;AND CONTINUE
;NOW CHECK FOR TTY COMMANDS
            .TTINR          ;KEYBOARD STRUCK?
            BCS 520         ;NO, KEEP DISPLAYING

;COMMANDS ARE
;       C-          CONTRACT
;       X-          EXPAND
;       F-          FORWARD TRANSFORM
;       I-          INVERSE TRANSFORM
;       R-          READ IN FILE
;CTRL/C EXIT TO RT-11

;COMMAND IS IN R0 IF WE GET HERE
            JSR PC, TYPE
            BIC #-200, R0   ;STRIP OFF PARITY
            CMP R0, #'C
            BEQ CONTR               ;CONTRACT
            CMP R0, #'X
            BEQ XPAND               ;EXPAND
            CMP R0, #'F             ;F
            BEQ FORWAR              ;FT
            CMP R0, #'I             ;I
            BEQ INVER               ;INVERSE
            CMP R0, #'R             ;R
            BEQ READIN              ;READ IN SPEC'D FILE
            CMP R0, #3              ;CTRL/C
            BNE 520                 ;GO BACK IF NONE
            EXIT                    ;RETURN TO RT-11


;CONTRACT
CONTR:  DEC VSHIFT
        BR 520

;EXPAND
XPAND:  INC VSHIFT
        BR 520

;FORWARD FT
FORWAR: JSR PC, FORWD
        BR 520
```

Figure 20.13 (*Continued*)

```
; INVERSE FT
INVER:  JSR PC, INVERS
        BR S20
;READ IN SPECIFIED FILE USING CSI IN SPECIAL MODE
READIN: BIC #BIT12, @#JSW       ; CLEAR SPEC MODE BIT
        .CSISPC #BLOCK,#DEFEXT, #0       ;SPECS FROM TTY

; LOOK UP FIRST FILE NAME ENTERED AS INPUT
; WILL BE IN 16TH WORD OF BLOCK

        .LOOKUP #LAREA, #1,#INBLOK
        BCS LKERR       ;ERROR IF CARRY SET
        ASH #8., R0     ;CONVERT BLOCKS TO WORDS
        MOV R0, DSIZE                   ;SAVE SIZE OF FILE

; READ IN THE WHOLE FILE
        .READW #RAREA, #1, FDATA, #4096.,#16.
LKXIT:  BIS #BIT12, @#JSW       ;RESTORE JSW
        BR S20
;LOOKUP ERROR, TYPE ? AND RESET
LKERR:  JSR PC, CRLF
        MOV #'?, R0                     ;?
        JSR PC, TYPE
        JSR PC,CRLF
        BR LKXIT        ;AND RESET KEYBOARD MODE

;CONSTANTS:
POINT:  0
COUNT:  0
SINFUN: 0
VSHIFT: 0
SINFLG: 0
FDATA:  0       ;START OF DATA FILE GOES HERE
DSIZE:  4096.   ;SIZE OF FILE GOES HERE
FSIZE:  2048.   ;COMPLEX SIZE

;CR-LF ROUTINE
CRLF:   MOV #15, R0
        JSR PC, TYPE
        MOV #12, R0
        JSR PC, TYPE
        RTS PC

TYPE:   .TTYOUT ;PRINT THROUGH RT-11
        RTS PC
; ARGUMENT BLOCKS USED BY RT-11 CALLS
DEFEXT: .BLKW 4 ;NO DEFAULT EXTENSIONS
LAREA:  .BLKW 3 ;LOOKUP ARGS GO HERE
RAREA:  .BLKW 5 ;READ-IN ARGS GO HERE
BLOCK:  .BLKW 15.        ;CSISPC OUTPUT FILENAMES START HERE
INBLOK: .BLKW 24.        ;INPUT FILENAMES START HERE

;SINE LOOKUP TABLE CALC'D AND PLACED HERE
SINTAB: .BLKW 513.


    END START
```

Figure 20.13 (*Continued*)

```
;FOURIER TRANSFORM SUBROUTINES FOR FORWARD AND INVERSE
;REAL AND COMPLEX
;IF THE ROUTINES "FORWD" AND "INVERS" ARE USED, IT IS ASSUMED THAT
;LOCATIONS "DATA" AND "SIZE" ARE PRE-LOADED WITH THE STARTING ADDRESS
;OF THE ARRAY TO BE TRANSFORMED AND THE COMPLEX SIZE OF THE ARRAY.
;ALTERNATIVELY YOU CAN MAKE CALLS OF THIS TYPE USING YOUR OWN
;MEMORY LOCATION REFERENCES
.GLOBL FOURIE, FORWD, INVERS, POST,SHUFFL,PFLAG, FSET
.GLOBL SINTAB,FDATA,FSIZE


;FORWARD REAL TRANSFORM
FORWD:  CLR PFLAG        ;INDICATES FORWARD TRANSFORM
        JSR R5, FSET     ;SET UP PARAMETERS
        FDATA                    ;ADDRESS CONTAINING STARTING ADDRESS OF BLOCK
        FSIZE    ;ADDRESS CONTAINING THE COMPLEX SIZE OF THE BLOCK
        JSR PC, FOURIE   ;NOW DO THE FORWARD TRANSFORM
        JSR PC, POST     ;AND POST PROCESSING
        JSR PC, SHUFFL   ;SHUFFLE DATA INTO 1ST HALF 2ND HALF
        RTS PC   ;AND RETURN TO MAIN PROGRAM

;INVERSE REAL TRANSFORM
INVERS: INC PFLAG        ;INDICATES INVERSE
        JSR R5, FSET     ;SET UP THE PARAMETERS
        FDATA            ;ADDRESS OF START OF DATA BLOCK
        FSIZE            ;ADDRESS OF SIZE OF DATA BLOCK
        JSR PC, SHUFFL   ;UN-SHUFFLE
        JSR PC, POST     ;PRE-PROCESS
        JSR PC, FOURIE   ;AND DO THE INVERSE TRANSFORM
        RTS PC           ;AND EXIT

;FOURIER TRANSFORM SET-UP ROUTINE
;CALLED BY
;       JSR R5, FSET
;       ADDRESS CONTAINING START ADDRESS OF BLOCK
;       ADDRESS CONTAINING THE COMPLEX SIZE OF BLOCK
;(A 4096-POINT REAL TRANSFORM THUS HAS A COMPLEX SIZE OF 2048)

FSET:   MOV @(R5)+, DATA          ;STARTING ADDRESS
        MOV @(R5)+, N2            ;NUMBER OF COMPLEX POINTS
        CLR SFACTR               ;SCALING FACTOR
        MOV N2, N                ;SAVE # OF COMPLEX POINTS
        ASL N                    ;NO. OF BYTES IN CPLX OR WORDS IN REAL
;CALCULATE B=LOG2(N-1) BY FINDING FIRST POWER OF TWO LARGER THAN N2
        MOV N2, R0
        MOV #4, R1
        MOV #2, B        ;POWER OF 2 OF DATA
BLOOP:  CMP R1, R0       ;IS THIS ONE > N2?
        BGE B2           ;YES
        ASL R1           ;NEXT GREATER ONE
        INC B            ;NEXT POWER OF 2
        BR BLOOP         ;GO BACK
B2:     RTS R5           ;EXIT FROM SET-UP ROUTINE
```

Figure 20.14   The FFT program, including postprocessing.

arrange y to start at 0 and Δy to be $2\pi/4$. In each succeeding pass, we divide Δy by 2.

We should further note that equations 20.30 and 20.31 are not cast in their most convenient form, since we cannot calculate $R(n)$ and $R(N-n)$ for the first point, replace it, and still have the information necessary to perform this calculation on the last point. Instead, we must start at the outside and work in, calculating $A_r(0)$, $A_i(0)$, $A_r(N)$, and $A_i(N)$

```
;BIT INVERT THE POSITIONS  OF THE DATA
FOURIE: CLR R0           ;FIRST ADDRESS
        CLR R1           ;BIT INVERTED ADDRESS
        MOV N2, R5       ;COUNTER
BIV10:  MOV B, R4        ;BIT SHIFTING COUNTER
        CLR R2           ;SHIFTED INTO R2
BIV20:  CLC              ;BITS WILL BE COPIED INTO C-BIT BY RPTATES
        ROR R1           ;SHIFT OUT OF R1
        ROL R2           ;AND INTO R2
        SOB R4, BIV20    ;B TIMES
        MOV R0, R1       ;COPY
        CMP R2, R1       ;PREVENT DOUBLE SWAP
        BLT BIV30
        ASL R1           ;LEFT SHIFT FOR BYTE ADDRESSING
        ASL R2           ;SAME
        ASL R1           ;X2 FOR REAL ALTERNATE
        ASL R2           ;DITTO
        ADD DATA, R2     ;STARTING ADDRESS
        ADD DATA, R1     ;DESTN ADDRESS
        JSR PC, SWAP     ;SWAP DATA POINTED TO BY R0 & R1
        CMP (R1)+, (R2)+         ;ADVANCE TO IMAGINARY
        JSR PC, SWAP
BIV30:  INC R0   ;GO ON TO NEXT REAL
        MOV R0, R1       ;COPY R0 TO R1 FOR INVERSION
        SOB R5, BIV10    ; DO NEXT SET


;FIRST PASS THROUGH THE DATA HAS NO  MULTIPLICATIONS
;R'M = RM+RN
;I'M = IM+IN
;R'N = RM-RN
;I'N - IM-IN
;SINCE PAIRS ARE ADJACENT ONLY 2 EQUATIONS ARE NECESSARY
        JSR PC, SCALIT   ;ALWAYS SCALE THE DATA FIRST TO PREVENT OVERFLOW
        MOV DATA, R0     ;FIRST POINT = RM
        MOV R0, R1
        CMP (R1)+,(R1)+  ;SECOND POINT = RN
        MOV N2, R5       ;COUNTER
        ASR R5           ;COUNTER/2
PASS1:  MOV (R0), R2     ;SAVE RM
        ADD (R1),(R0)+   ;RM=RM+RN
        SUB (R1), R2     ;RN=RM-RN 0
        MOV R2, (R1)+    ;STORE IT
        MOV (R0), R2
        ADD (R1), (R0)+
        SUB (R1), R2
        MOV R2, (R1)+
        CMP (R0)+, (R1)+         ;GO ON TO  NEXT PAIR
        CMP (R0)+, (R1)+
        SOB R5, PASS1    ;GO BACK N2/2 TIMES

;SET UP DELTAY TO VARY FROM 0 TO 2PI IN EACH CELL
;77777 = 1.0 _ FIRST PASS CELLS HAVE Y/N= 0,1 ONLY
;SECOND PASS DELTAY IS HALF AS GREAT
        MOV #40000, DELTAY        ;SIZE OF INCREMENT OF Y/N
        MOV N2, CELNUM            ;NUMBERS OF CELLS
        ASR CELNUM                ;NUMBER IN FIRST PASS
        ASR CELNUM                ;NUMBER IN 2ND PASS
        MOV #2, PAIRNM            ;NUMBER OF PAIRS IN EACH CELL

        MOV #10, CELDIS           ;DISTANCE TO 2ND POINT IN CELL
```

Figure 20.14   *(Continued)*

```
;EACH NEW PASS AFTER FIRST STARTS HERE
NEWPAS: MOV CELNUM, CELCNT        ;CELL COUNTER
        MOV DATA, POINT           ;RESET ARRAY POINTER TO START

;NEW CELL STARTS HERE
NEWCEL: MOV PAIRNM, PARCNT          ;# OF PAIRS PER CELL
        CLR YN               ;EXPONENT OF W STARTS AT 0 IN EACH CELL
NEWC10: MOV POINT, R0        ;R0 IS REAL POINTER
        MOV R0, R1
        ADD CELDIS, R1            ;ADDRESS OF OTHER NODE OF PAIR

;R1'=R1+R2COS(Y)+I2SIN(Y)
;I1'=I1-R2SIN(Y)+I2COS(Y)
;R2'=-R2COS(Y)-I2SIN(Y)+R1
;I2'=I1+R2SIN(Y)-I2COS(Y)

        MOV YN, R2        ;LOOK UP SINE
        JSR PC, LOKSIN
        TST PFLAG
        BEQ FT10
        NEG R2
FT10:   MOV R2, SINE      ;SAVE SINE
        MOV YN, R2
        ADD #40000, R2    ;BIAS OF 90-DEG GETS COSINE
        JSR PC, LOKSIN
        MOV R2, COS       ;SAVE COSINE
        MOV(R1), R2       ;RE2
        JSR R5, MULT      ;RE*COS
        COS
        R2COSY
        MOV (R1), R2
        JSR R5, MULT        ;R2*SINY
        SINE
        R2SINY
        MOV R1, R4        ;COPY ADDRESS
        TST (R4)+         ;ADDRESS OF IMAGINARY
        MOV (R4), R2      ;I2
        JSR R5, MULT      ;R2*SINY
        SINE
        I2SINY
        MOV (R4), R2      ;I2
        JSR R5, MULT      ;I2COSY
        COS
        I2COSY
```

Figure 20.14 (*Continued*)

and then incrementing the index of the first two and decrementing the index of the last two until we meet in the center. In this case we can take advantage of the symmetry of the sine and cosine functions, realizing that:

$$\sin\left[\frac{(N-n)\pi}{N}\right] = \sin\left[\frac{(n)\pi}{N}\right]$$

and

$$\cos\left[\frac{\pi(N-n)}{N}\right] = -\cos\left[\frac{(n)\pi}{N}\right] \tag{20.45}$$

```
; COMBINE TERMS
        MOV (R0),(R1)    ; R2'=R1+...
        JSR R5, ADDR0    ; RE1+RE2COSY
        R2COSY
        JSR R5, ADDR0    ; RE1'=RE1+R2COSY+I2SINY
        I2SINY
        NEG I2SINY       ; READY FOR SUBTRACT
        JSR R5, ADDR1    ; R2=R1-I2SINY
        I2SINY
        NEG R2COSY       ; READY FOR SUBTRACT
        JSR R5, ADDR1    ; R2=R1+I2SINY-R2COSY
        R2COSY
        MOV R4, R1       ; DO IMAGINARIES
        TST (R0)+
        MOV (R0),(R1)
        JSR R5, ADDR0    ; I1=I1+R2SINY+I2COSY
        I2COSY
        JSR R5, ADDR1    ; I2=I1+R2SINY
        R2SINY
        NEG R2SINY
        NEG I2COSY
        JSR R5, ADDR0    ; I1=I1-R2SINY
        R2SINY
        JSR R5, ADDR1    ; I2=I1-R2SINY-I2COSY
        I2COSY
; ONE PAIR DONE, DO NEXT PAIR
        ADD #4, POINT             ; NEXT ADDRESS IN CELL
        ADD DELTAY, YN   ; INCREMENT EXPOENT OF W
        DEC PARCNT                ; COUNT PAIRS DONE
        BNE NEWC10                ; CONTINUE UNTIL CELL IS DONE

; END OF THIS CELL, RESET ANGLE AND PAIR COUNTER
        ADD CELDIS, POINT         ; GO ON TO NEXT CELL-SKIP OVER 2ND HALF
        DEC CELCNT                ; ALL CELLS DONE?
        BNE NEWCEL                ; NO

; PASS DONE -- CHANGE VALUES
        ASR CELNUM       ; LESS CELLS
        BEQ FTEXIT       ; ALL DONE WHEN 0 CELLS
        ASL PAIRNM       ; MORE PAIRS
        ASL CELDIS       ; MORE DISTANT
        ASR DELTAY       ; LESS INCREMENT OF Y PER CELL
        JMP NEWPAS

FTEXIT: RTS PC
```

Figure 20.14   (*Continued*)

and so forth. This leads to the following equations, which are used in the postprocessing shuffle:

$$A_r(n) = R_p + I_p \cos\left(\frac{n\pi}{N}\right) - R_m \sin\left(\frac{n\pi}{N}\right) \qquad (20.46)$$

$$A_i(n) = I_m - I_p \sin\left(\frac{n\pi}{N}\right) - R_m \cos\left(\frac{n\pi}{N}\right) \qquad (20.47)$$

$$A_r(N - n) = R_p - I_p \cos\left(\frac{n\pi}{N}\right) - R_m \sin\left(\frac{n\pi}{N}\right) \qquad (20.48)$$

$$A_i(N-n) = I_m - I_p \sin\left(\frac{n\pi}{N}\right) + R_m \cos\left(\frac{n\pi}{N}\right) \qquad (20.49)$$

```
; ROUND-UP MULTIPLICATION ROUTINE
MULT:    MUL @(R5)+, R2    ; DO MULTIPLICATION
         ASHC #1, R2       ; AND SHIFT FOR BINARY FRACTION
         TST R3            ; CHECK MSB OF 2ND WORD
         BPL MULT10        ; NOT SET
         INC R2            ; SET, SO INCREMENT TO ROUND UP
MULT10:  MOV R2, @(R5)+    ; PUT IN DESTINATION
         RTS R5            ; AND EXIT


; ROUTINE TO ADD ARGUMENT TO ADDRESS POINTED TO BY R0
ADDR0:   MOV @(R5)+, R2    ; GET ARGUMENT
         ADD (R0), R2      ; ADD TO IT
         BVC ADDR01        ; IF SET, THEN SCALE
         JSR PC, SCALIT    ; ELSE SCALE EVERYTHING
ADDR01:  MOV R2, (R0)      ; STORE RESULT
         RTS R5            ; AND EXIT


; ROUTINE TO ADD ARGUMENT TO ADDRESS POINTED TO BY R1
ADDR1:   MOV @(R5)+, R2    ; GET ARGUMENT
         ADD (R1), R2      ; ADD TO IT
         BVC ADDR02        ; IF NOT SET EXIT
         JSR PC, SCALIT    ; ELSE SCALE EVERYTHING
ADDR02:  MOV R2, (R1)      ; STORE RESULT
         RTS R5            ; AND EXIT


; SCALING ROUTINE - DIVIDES DATA BY 2
SCALIT:  ROR R2   ; REDUCE TEMP
         MOV R0, -(SP)
         MOV R1, -(SP)
         MOV DATA, R0
         MOV N, R1
SC10:    ASR (R0)+
         SOB R1, SC10
         ASR R2COSY        ; REDUCE INTERMEDIATES
         ASR I2SINY
         ASR I2COSY
         ASR R2SINY
         MOV (SP)+, R1
         MOV (SP)+, R0
         RTS PC


; SUBROUTINE TO SWAP (R1) AND (R2)
SWAP:    MOV (R1), R3
         MOV (R2), (R1)
         MOV R3, (R2)
         RTS PC
```

Figure 20.14   (*Continued*)

```
; SHUFFLE ROUTINE- SHUFFLES ALTERNATE POINTS
; TO FIRST HALF-SECOND HALF
; AND THE REVERSE DEPENDING ON PFLAG
SHUFFL: TST PFLAG         ; INVERSE?
        BNE ISORT1        ; YES
        MOV CSIZE, CELDIS        ; DISTANCE TO NEXT CELL
        ASL CELDIS              ; FOR BYTE ADDRESSING
        MOV #1, CELNUM          ; 1 CELL FIRST
        MOV CSIZE, PAIRNM       ; NUMBER OF PAIRS PER CELL
        ASR PAIRNM              ; /4
SRT05:  MOV DATA, R0            ; GET ADDRESS OF 1ST HALF
        MOV R0, R1              ; 2ND HALF
        ADD CELDIS, R1
        TST (R0)+               ; 2ND WORD IN 1ST HALF
        MOV CELNUM, CELCNT      ; NUMBER OF POINTS PER CELL
SRT10:  MOV PAIRNM, PARCNT      ; NUMBER OF PAIRS PER CELL
SRT20:  MOV (R0), R2            ; SWAP (R0) AND (R1)
        MOV (R1), (R0)+
        MOV R2, (R1)+
        CMP (R0)+, (R1)+        ; DOING ONLY ALTERNATE POINTS
        DEC PARCNT              ; DONE WITH CELL
        BNE SRT20               ; NO
; NEW CELL
        ADD CELDIS, R0
        ADD CELDIS, R1          ; TO START OF NEW  CELL
        DEC CELCNT              ; ALL CELLS DONE?
        BNE SRT10               ; NO
; NEW PASS
        TST PFLAG
        BNE ISORT2              ; ADVANCE DIFF IF INVERSE
        ASR CELDIS              ; CELLS ARE CLOSER TOGETHER
        ASL CELNUM              ; MORE OF THEM
        ASR PAIRNM              ; LESS PAIRS PER CELL
        CMP CELDIS, #2          ; EXIT CRITERIA
        BGT SRT05
ISEXIT: RTS PC

; TO UNSORT DURING INVERSE USE THESE CONSTANTS
ISORT1: MOV #4, CELDIS          ;  2 APART AT START
        MOV #1, PAIRNM          ; 1 PAIR PER CELL
        MOV SIZE, CELNUM        ;
        ASR CELNUM              ; SIZE/4 CELLS
        BR SRT05

; TERMINATE CELLS DIFFERENTLY, TOO
ISORT2: ASR CELNUM              ; LESS CELLS
        BEQ ISEXIT              ; EXIT CRITERION
        ASL CELDIS              ; FARTHER APART
        ASL PAIRNM              ; MORE PAIRS
        BR SRT05
```

Figure 20.14 (*Continued*)

```
;POST-PROCESSING
;RP=R(M)+R(N-M)
;RM=R(M)-R(N-M)
;IP=I(M)+I(N-M)
;IM=I(M)-I(N-M)


POST:    JSR PC, SCALIT   ;BE SURE IT IS SMALL ENOUGH TO PREVEMT OVERFLOW
         MOV DATA, R0     ;STARTING POINTER
         MOV N2, R1       ;COUNTER
         ASR R1           ;BOTH ENDS OF ARRAY DONE AT ONCE
         MOV #40000, R3
         MOV B, R4
         DEC R4           ;B-1
         NEG R4
         ASH R4, R3
         MOV R3, INCX              ;AMOUNT TO ADVANCE SINE PINTER
         MOV INCX, INDEX           ;SINE POINT FOR EACH POINT
         MOV N, R4
         ASL R4                    ;X2 FOR R+IMAG
         ADD DATA, R4              ;ADDRESS OF LAST POINT
;SKIP FIRST AND LAST POINT AS THEY DO NOT CHANGE
         CMP(R0)+,-(R4)
         CMP (R0)+, -(R4)
         DEC R1
;GET RP, RM, IP ANS IM FOR ALL BUT FIRST POINT
POST5:   MOV INDEX, R2
         JSR PC, LOKSIN
         MOV R2, SINE             ;SINE(N)
         MOV INDEX, R2
         ADD #40000, R2   ;CONVERT TO COSINE ADDRESS
         JSR PC, LOKSIN
         TST PFLAG        ;INVERSE PREPROCESSING?
         BEQ POST7        ;NO
         NEG R2           ;YES
POST7:   MOV R2, COS              ;AND SAVE COSINE
         MOV (R4), RP     ;R(N-M)
         MOV (R4), RM     ;R(N-M)
         NEG RM           ;-R(N-M) FOR RM

         MOV 2(R4), IP            ;2ND WORD IS PIND2, INDEX ADDRESS
         MOV 2(R4), IM
         NEG IM           ;-I(N-M)
         ADD (R0), RP     ;RP=R(M) + R(N-M)
         ADD (R0), RM     ;RM=R(M)-R(N-M)
         ADD 2(R0), IP    ;IP=I(M)-I(N-M)
         ADD 2(R0), IM    ;IM=I(M)-I(N-M)
```

Figure 20.14 (*Continued*)

```
; AR(M)=RP+IP*COS(N)-RM*SIN(N)
; AI(M)=IM - IP*SIN(N)-RM*COS)N)
; AR(N-M)=RP-IPCOS(N)+RMSIN(N)
; AI(N-M)=-IM-IPSIN(N)-RMCOS(N)

POST10:   MOV  RP,  ARM
          MOV  IM,  AIM        ; ARM AND AIM ARE AR(M) AND AR(N)
          MOV  RP,  ARNM       ; AR(N-M)
          MOV  IM,  AINM       ; AI(N-M)
          NEG  AINM            ; -IM
          MOV  IP,  R2
          MUL  COS, R2         ; IPCOS(N)
          ASHC #1,  R2         ; FOR FRACTONAL
POST20:   ADD  R2,  ARM
          SUB  R2,  ARNM       ; AR(N-M)=RP-IPCOS(N)...
POST25:   MOV  IP,  R2
          MUL  SINE, R2        ; NOW DO IP*SIN(N) TERMS
          ASHC #1,  R2
          SUB  R2,  AIM
          SUB  R2,  AINM       ; AI(N-M)=IM-IPSIN(N)...
          MOV  RM,  R2         ; DO RMCOS AND SINE TERMS
          MUL  SINE, R2
          ASHC #1,  R2         ; RM*SIN(N)
          SUB  R2,  ARM        ; SUBTRACT FROM AR(M)
          ADD  R2,  ARNM       ; AR(N-M)=RP-IPCOS(N)+RMSIN(N)
          MOV  RM,  R2
          MUL  COS, R2
          ASHC #1,  R2         ; RMCOS(M)
POST50:   SUB  R2,  AIM        ; SUBTRACT FROM AI(M)
          SUB  R2,  AINM       ; SUB TO AI(N-M)
POST55:   MOV  ARM,  (R0)
          MOV  ARNM, (R4)
          MOV  AIM,  2(R0)     ; STORE IMAGINARIES
          MOV  AINM, 2(R4)
          ADD  INCX, INDEX ; GET NEXT VALUE TO TAKE SINE
          CMP  (R0)+, -(R4)             ; ADVANCE R0 AND DECREASE R4
          CMP  (R0)+, -(R4)
          DEC  R1
          BEQ  POSEXT                   ; EXIT IF DONE
          JMP  POST5                    ; TOO FAR FOR A BNE
POSEXT:   RTS PC               ; AND EXIT
```

Figure 20.14 (*Continued*)

References

1. J. A. Pople, W. G. Schneider, and H. J. Bernstein, *High Resolution Nuclear Magnetic Resonance*, McGraw-Hill, New York, 1959.

2. T. C. Farrar and E. D. Becker, *Pulse and Fourier Transform Nmr*, Academic, New York, 1971.

3. P. B. Fellget, *J. Phys. Radium* **19**, 187, 237 (1958).

4. P. R. Griffiths, C. T. Foskett, and R. Curbelo, *Appl. Spectrosc. Rev.* **6**, 31–78, (1972).

5. P. R. Griffiths, *Chemical Infrared Fourier Transform Spectroscopy*, Wiley-Interscience, New York, 1975.

6. J. W. Cooley and J. W. Tukey, *Math. Comput.* **19**, 297 (1965).

7. W. T. Cochran et al., *IEEE Trans. Audio. Electroacoust.* **AU-15**, 45–55 (1967). Collected in L. R. Rabiner and C. M. Rader, *Digital Signal Processing*, IEEE Press, 1972.

8. E. Oran Brigham, *The Fast Fourier Transform*, Prentice-Hall, Englewood Cliffs, NJ, 1974, pp. 148ff.

9. J. D. Ellet, M. G. Gibby, U. Haeberlein, L. M. Huber, M. Mehring, A. Pines, and J. S. Waugh, *Adv. Magn. Reson.* **5**, 117 (1971).

10. E. O. Stejskal and J. Schaeffer, *J. Magn. Reson.* **14**, 160 (1974).

11. J. W. Cooley, P. W. Lewis, and P. D. Welch, *J. Sound Vib.* **12**, 315–337 (1970). (Also collected in Rabiner and Rader, see reference 7.

12. E. O. Brigham, *J. Sound Vib.* **12**, 163 (1970).

13. J. Dadok and R. Sprecher, *J. Magn. Reson.* **13**, 243–248 (1974).

14. R. K. Gupta, J. A. Ferretti, and E. D. Becker, *J. Magn. Reson.* **13**, 275–290 (1974).

15. P. D. Welch, *IEEE Trans. Audio Electroacoust.* **AU-17**, 151–570 (1969).

16. J. W. Cooper, *J. Magn. Reson.* **22**, 345 (1976); **28**, 405 (1977).

17. J. W. Cooper, 21st Experimental Nmr Conference, Tallahassee, Florida, 1979.

18. J. W. Cooper, *Introduction to Pascal for Scientists*, Wiley-Interscience, New York, 1981, pp. 211–216.

CHAPTER TWENTY-ONE

WINDOW FUNCTIONS
AND PHASE
CORRECTION METHODS

When we multiply data by any function that changes the characteristics of these data, we refer to the function as a *window function*. The most common such functions in Fourier transform nmr are the increasing and decreasing exponentials. Since the spin-spin relaxation process is in fact exponential in nature, the multiplication of the free-induction decay resulting from the pulsed nmr experiment can be multiplied by an exponential function before the Fourier transform to good effect. The result of this exponential window is that

1. If the exponential is negative, the lines are broadened and the signal-to-noise ratio is enhanced.
2. If the exponential is positive, the lines are narrowed and the signal-to-noise ratio is degraded.

This can be readily understood by examining a free-induction decay such as is shown in Figure 20.8. If we multiply by a decreasing exponential e^{-TC}, we will affect the beginning of the free-induction decay (fid) only minimally, but the end of the fid will be markedly decreased in size. Since the signal decreases across the fid, this will have the effect of decreasing the noise contribution at the end of the fid, where little signal remains. However, since the spin-spin relaxation time, T_2, and the Lorentzian line shape are a Fourier transform pair, the line width will be broadened by this decreasing exponential function. This is illustrated in Figure 21.1, where the data are multiplied by several exponentials before the transformation.

In multiplying an array of data by an exponential function we are really multiplying the points $X(k)$ as follows:

$$X'(k) = X(k) \exp\left(\frac{kTC}{N}\right), \qquad k = 0, 1, \ldots, N - 1 \qquad (21.1)$$

where TC is the exponential time constant which is used.

Although equation 21.1 is fairly straightforward to calculate, the continual calculation of exponential functions by floating point usually involves a Taylor series[1] of some complexity. This can be very time consuming for an array of several thousand points. A good approximation

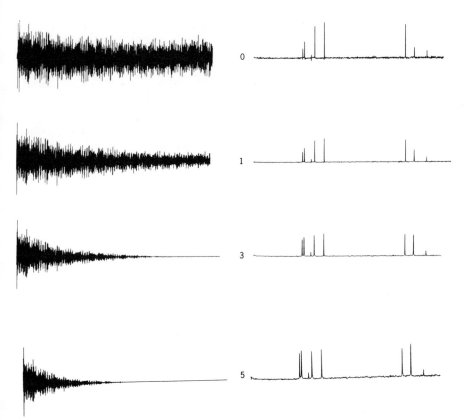

Figure 21.1   Effects of positive and negative exponential window.

of this function can be obtained by rewriting equation 21.1 is

$$X'(k) = X(k) \left[\exp \left(\frac{TC}{N} \right) \right]^k \tag{21.2}$$

We now recognize that

$$e^x \cong 1 + x \tag{21.3}$$

for x small.[2] Then

$$\exp \left(\frac{TC}{N} \right) \cong 1 + \frac{TC}{N}$$

since TC is usually a number between 1 and 10, and N several thousand. Using our representation of 1 as a fixed point fraction, we find for the 16-bit computer that

$$X'(k) = X(k) \left(1 + \frac{TC}{N} \right)^k, \qquad k = 0, 1, \ldots, N - 1 \tag{21.4}$$

where 1 is 77777_8, and TC/N is a small binary fraction added to it. Thus we calculate TC/N once in floating point and scale it to a fixed point fraction by multiplying by 2^{16} and fixing it. Then we multiply the first point, $X(1)$, by 1, the second point by $1 + TC/N$, the third point by $(1 + TC/N)^2$, and so forth. In other words there are only two fixed point multiplications for each new data point, rather than several floating point operations evaluating a Taylor series.

Apodization

Etymologically, *apodization* means to "cut the feet off" some data by appropriate window function. In nmr and ir, we usually mean the multiplication of data by an increasing or decreasing diagonal line. This usually involves the generation of a trapezoidal window having one or two sloping sides. Trapezoidal apodization of m points usually means starting at point zero, multiplying it by 0, and then multiplying each further point by k/m:

$$X'(k) = X(k) \left(\frac{k}{m} \right), \qquad k = 0, 1, \ldots, m - 1 \tag{21.5}$$

This can be done fairly efficiently by determining $1/m$ in advance as a fixed point fraction and adding it to the multiplier before determining each new point.

If a decreasing window function is used, so that point m is multiplied by 1.0 and each successive point by a smaller number up to point $N - 1$, we are performing the operation.

$$X'(k) = X(k) \left(\frac{N - k}{N - m} \right), \qquad k = m, m + 1, \ldots, N - 1 \quad (21.6)$$

In this case it is not advisable to start at point m and subtract the value $1/(N - m)$ from each succeeding point, since round-off error in the determination of $1/(N - m)$ may not make the last point exactly 0. Indeed, in some extreme cases, the expression $(m - k)/m$ may become negative, leading to a small lobe at the point where zero data are intended. This problem can be readily avoided by starting at point $N - 1$ and working backward to point m.

Magnitude Spectra

Once the Fourier transform of time data is calculated, the data may or may not be presented in a meaningful phase relationship. Time data from most biological systems have no meaningful phase information in them, and the peaks of the frequency spectrum are all made positive by calculating the *magnitude spectrum*, also referred to as the *power spectral density* (PSD):

$$M_i = (A_r^2 + A_i^2)^{1/2} \quad (21.7)$$

Nuclear Magnetic Resonance Phase Correction

However, the output of phase detectors of many scientific instruments produce meaningful information, and the real and imaginary parts of the spectrum can be convolved with some simple function to produce a phase-corrected spectrum. For example, Figure 21.2 shows the real and imaginary parts of the un-phase-corrected carbon-13 nmr spectrum of 3-ethylpyridine. Figure 21.3 shows the same spectrum after phase correction.

There are three major causes of mixed phase information, as shown in Figure 21.1:

1. Spectrometer phase detector setting.
2. Delay between pulse and start of data acquisition.
3. Filter settings.

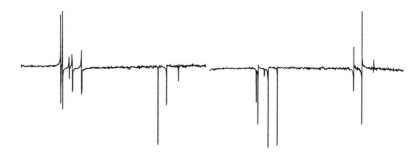

Figure 21.2 Un-phase-corrected spectrum of 3-ethylpyridine.

The spectrometer phase detector is usually optimized at the beginning of a series of spectra using a strong sample containing sharp peaks. Just as in cw spectroscopy, the smallest changes will affect this parameter: Variation in sample tubes, solvents, or even spinning rate may influence the phase of the information entering the detector. This effect is a zero order one, which causes the same shift in phase for each data point, regardless of frequency. Examples of out-of-phase and in-phase cosine waves are shown in Figure 21.4, along with their transforms.

The remaining two factors, delay time and filters, have first order effects on the spectrum; that is, at frequency *zero* the phase shift is 0, and at the highest frequency in the spectrum the phase shift is large. It is customary to refer to this first order phase shift in terms of the phase angle shift of the highest frequency. A 170° first order phase shift is one in which the first frequency domain point has zero phase shift and the last one 170° of phase shift.

The delay time causes a frequency dependent phase shift related to the dwell time. The highest frequency point will be shifted by an amount equal to 180° × delay/dwell. In other words, for each dwell time unit of delay there will be a first order phase shift of 180°. Note that in the case

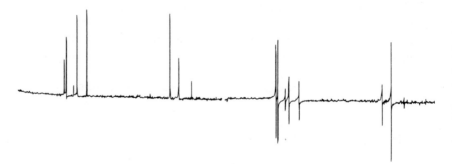

Figure 21.3 Phase-corrected spectrum of 3-ethylpyridine.

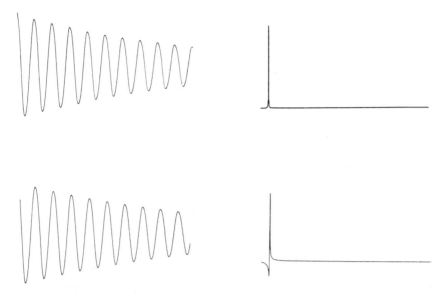

Figure 21.4   Top: Time and frequency domain spectra of an in-phase cosine wave. Bottom: Time and frequency domain spectra of an out-of-phase cosine wave.

of first order shifts a 0° shift and a 360° shift are *not* equivalent. In the 360° shift the highest frequency point is "wrapped" around 360° in phase from the first point, causing the phases of the data points to spiral in the phase plane around one complete circle.

Logically, one can also see that first order phase shifts can be *greater* than 360° and can in fact attain almost any number of degrees. For instance, if the dwell time in a particular experiment is 75 microseconds and the delay time 300 microseconds, then

$$\frac{300 \text{ μsec dwell}}{75 \text{ μsec delay}} \times 180° = 720°$$

This is equivalent to a spiraling of the phase information in two complete circles.

Let us now consider a physical explanation for this phase shift. Although the zero order shift is quite easy to explain in terms of phase detector settings, the first order shift is a little harder to grasp. We will first return to our discussion of sampling. Recall that the sine waves comprising the fid are sampled at a constant interval called the *dwell time*, as shown in Figure 21.5. We have assumed that these waves are sampled starting exactly at 0° as the first point of the sine or cosine wave.

Let us now assume that a delay of dwell time is introduced for instrumental reasons, such as to minimize pulse feedthrough (Figure 21.6). As

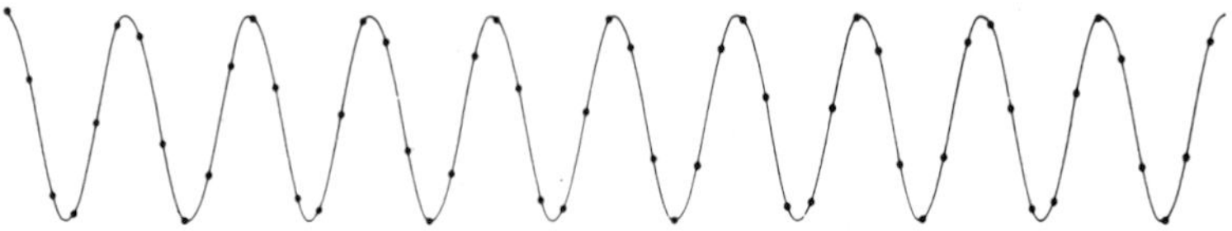

Figure 21.5 Low-frequency sampling without delay.

we can see, the omission of the first two data points causes the phase of this line to begin at about 70° instead of at 0°. This means that this line, when transformed to the frequency domain, will be 70° out of phase.

Let us now consider a line of much higher frequency, one at the *Nyquist frequency*, or at one-half the sampling frequency. At this frequency there are only two data points per cycle of the sine wave (Figure 21.7). If, at the high frequency, we introduce a *one* dwell time delay, we will have wiped out 180° of phase information, so that the wave will start at 180° instead of at 0° (Figure 21.8). Thus, at this frequency, the highest one represented in the fid, the delay of a single dwell time (or of one *address*) will cause a 180° phase shift. It follows from this that a two-address delay will cause a 360° phase shift, and so forth.

A line that is "folded back" or "aliased" from a frequency higher than the Nyquist frequency will have the first order phase dependence that it would have if it were actually observed at its correct frequency. Thus a folded-back line will seem to have an anomalous phase relative to lines around it.

The third cause of phase shifts is the characteristics of the filter used to limit the spectrum bandwidth. In many data systems four-pole But-terworth filters have been chosen because of their extremely linear phase characteristics. The four-pole filter causes a first order shift of −180° if used in the system. This does not vary appreciably with the cutoff fre-quency selected. Thus the total formula for determining the first order phase shift is given by

$$\frac{\text{delay time}}{\text{dwell time}} \times 180° - 180°$$

Phase correction is performed by convolving all points in the real and imaginary spectra with a cosine and sine wave. Zero order phase cor-

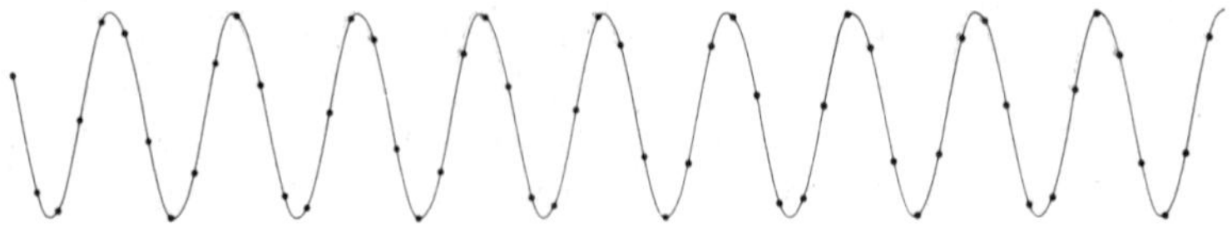

Figure 21.6 Low-frequency sampling with a one sampling interval delay.

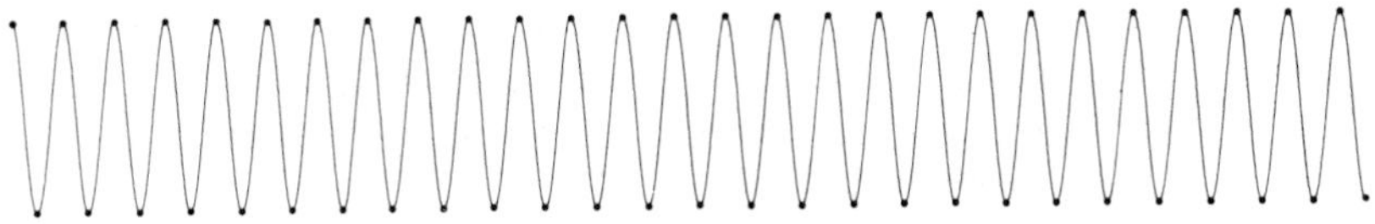

Figure 21.7 Sampling of Nyquist frequency.

rection is performed by

$$R1' = R1 \cos A - I1 \sin A \tag{21.8}$$

$$I1' = I1 \cos A + R1 \sin A \tag{21.9}$$

where A is the zero order phase correction angle.

First order phase correction is performed by varying A linearly as a function of address.

It is clear that, if we have the facility to perform a Fourier transform such as that of equation 20.28, we should have no trouble evaluating equations 21.2 and 21.3. The main trick is finding a method of performing these calculations as rapidly as possible, preferably so that the results can be displayed as the phase is varied by a parameter knob, terminal command, or light pen. This phase correction can, of course, be accomplished entirely in integer mode, using the sine look-up table.

The zero order phase correction parameter is simply a single angle, or position in the look-up table, which is applied to each point in the spectrum. If a first order correction is also to be applied, it is applied simultaneously, by adding a small increment to the angle whose sine is obtained for each new data point across the table. Thus the correction becomes

$$R'_n = R_n \cos(A + B_n) - I_n \sin(A + B_n) \tag{21.10}$$

$$I'_n = R_n \sin(A + B_n) + I_n \cos(A + B_n) \tag{21.11}$$

One rather clever trick reported by Samuelson[3] and by others[4] involves selecting one peak in the nmr spectrum and phasing it perfectly, using a parameter knob, and then holding this phase constant while the phases of all other peaks are tipped around it by varying the first order phase correction parameter B, so that the phase of the selected peak is held

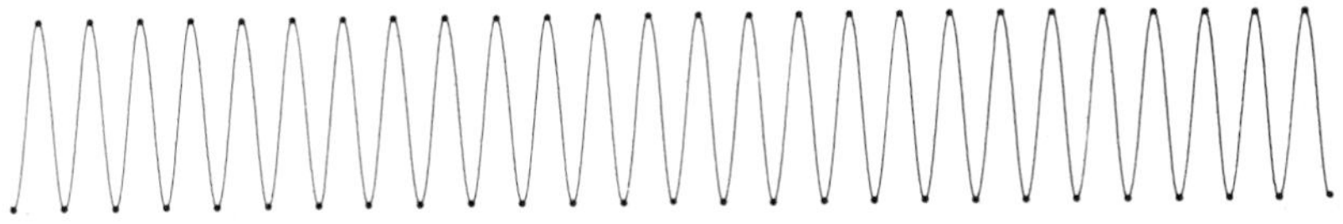

Figure 21.8 Sampling of Nyquist frequency with a one sampling interval delay.

constant. In order that this process can be carried out as rapidly as possible to prevent excessive display flicker, the program is usually combined with a variable display expansion program such as that shown in Chapter 16. The procedure is to

1. Select a peak to be phase corrected, either by finding the largest one automatically or by allowing a cursor routine to select it.
2. Choose the input of a parameter knob to be the phase of this peak, and vary the phase according to the values read from this knob as A, according to equations 21.10 and 21.11.
3. Read another parameter knob and call its value B. Then assuming that the peak being held constant is at address m, apply equations 21.12 and 21.13:

$$R'(n) = R_n \cos[A + B(m - n)] - I_n \sin[A + B(m - n)] \qquad (21.12)$$
$$I'(n) = R_n \sin[A + B(m - n)] + I_n \cos[A + B(m - n)] \qquad (21.13)$$

The subroutine given in Figure 21.9 performs the simplest phase correction in memory by equations 21.10 and 21.11 once A and B have been determined externally. It can easily be expanded and combined with a display routine to implement equations 21.12 and 21.13.

Fourier Transform Infrared Phase Correction and Apodization

In Fourier transform infrared, the phase dependence of various points is not linear, but is still a slowly varying function for a sufficient number of points throughout the spectrum. The following apodization and phase correction techniques follow those given by Mertz,[5] Griffiths, Foskett, and Curbelo[6] and by Codding and Horlick.[7]

When the Fourier transform ir interferogram has been collected, a large zero frequency peak is usually somewhere near the beginning of the spectrum. This is often referred to as the *peak*, since the interference information is much smaller by comparison. Such an interferogram is shown in Figure 21.10.

A linear apodization function is applied so that the first point of the spectrum is multiplied by 0 and a point twice the address of the peak is multiplied by 1. A second decreasing linear apodization from some variable point near the end of the interferogram to the last point is also sometimes used. Then the data are rotated in memory, either actually or by an addressing switch, so that the center of the peak is brought to the first address and the remainder of the peak rotated around to fill up points at the end of the interferogram, as shown in Figure 21.11.

```
;PHASE CORRECTION SUBROUTINE
;ENTER WITH A,B AND SIZE SET

PHASE:   MOV DATA, R0        ;SET POINTER TO REAL
         MOV SIZE, R5        ;LOAD COUNTER
         MOV R0, R1          ;SET IMAGINARY POINTER
         ADD R5, R1          ;2ND HALF OF DATA IS IMAGINARY POINTS
         MOV A, THETA        ;STARTING PHASE

;BEGIN CORRECTION LOOP
PLOOP:   MOV THETA, R2       ;PUT THETA IN R2 FOR SINE LOOKUP
         JSR PC, LOKSIN      ;LOOK UP SINE
         MOV R2, SIN         ;STORE IT
         MOV THETA, R2       ;GET THETA AGAIN
         ADD #40000, R2      ;GO TO NEXT QUADRANT TO GET COSINE
         JSR PC, LOKSIN      ;LOOK IT UP
         MOV R2, COS         ;AND STORE IT

;REAL = REAL*COS -IMAG*SIN
;IMAG=IMAG*COS + REAL * SIN

         MOV (R0), R2        ;GET A REAL POINT
         MOV R2, REAL        ;SAVE A COPY
         MUL COS, R2         ;REAL*COS
         ASHC #1, R2         ;SHIFT FOR FRACTIONAL MULTIPLICATIONS
         MOV R2, R4          ;SAVE
         MOV (R1), R2        ;GET IMAG
         MUL SIN, R2         ;IMAG*SINE
         ASHC #1, R2         ;SHIFT FOR FRACTIONAL MULT
         SUB R2, R4          ;REAL*COS-IMAG*SIN
         MOV R4, (R0)+       ;PUT IN MEMORY
         MOV (R1), R2        ;IMAG
         MUL COS, R2         ;IMAG*COS
         ASHC #1, R2         ;FRACTIONAL MULT SHIFT
         MOV R2, R4          ;SAVE
         MOV REAL, R2        ;GET SAVED OLD REAL
         MUL SIN, R2         ;REAL*SIN
         ASHC #1, R2         ;FRACTIONAL MULT
         ADD R4, R2          ;IMAG*COS+ REAL*SIN
         MOV R2, (R1)+       ;STORE IN MEMORY
         ADD B, THETA        ;ADD TO THETA FOR FREQ DEPENDENT TERM
         SOB R5, PLOOP       ;GO BACK TILL DONE
         RTS PC
DATA:    20000
SIZE:    4096.
SIN:     0
COS:     0
REAL:    0
A:       0
B:       0
THETA:   0
```

Figure 21.9 Listing of phase correction program.

The Fourier transform is calculated, and phase correction is performed by first calculating a *phase array*. This is a small array of a few hundred points obtained by copying the first few points and the last few points of the interferogram into another array and transforming this small array separately.

The magnitude spectrum (equation 21.7) is calculated from this small array and stored in another small array. Then the phase array is nor-

Figure 21.10   Interferogram of air, 4096 points, resolution $= 4$ cm^{-1}.

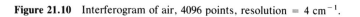

Figure 21.11   Rotated interferogram of air.

malized by dividing the real and imaginary by the magnitude spectrum:

$$RP_n = \frac{R_n}{M_n} \qquad (21.14)$$

$$IP_n = \frac{I_n}{M_n} \qquad (21.15)$$

This normalized array is shown in Figure 21.12.

Actual phase correction is performed by realizing that the entries in the phase array are the cosine and sine of the phase angle at that point. Thus we simply write

$$R'_n = R_n(RP_n) + I_n(IP_n) \qquad (21.16)$$

$$I'_n = R_n(IP_n) + I_n(RP_n) \qquad (21.17)$$

Since the small phase arrays have many fewer points than the data arrays, a large amount of interpolation is usually necessary.

Since the imaginary array does not contain useful phase information once the correction is complete, it is often replaced at that point by the

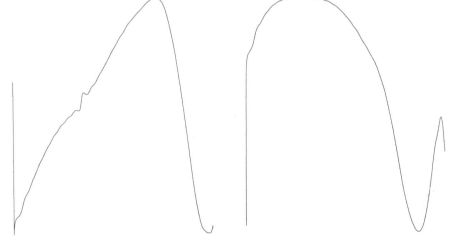

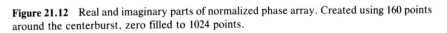

Figure 21.12   Real and imaginary parts of normalized phase array. Created using 160 points around the centerburst, zero filled to 1024 points.

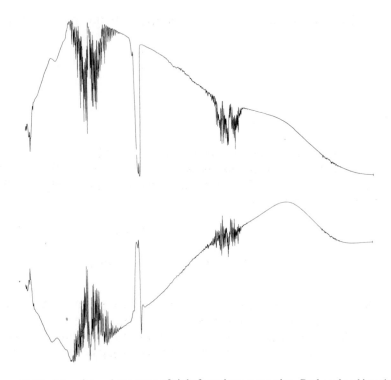

Figure 21.13   Transformed spectrum of air before phase correction. Both real and imaginary parts shown.

magnitude spectrum. Figure 21.13 shows the transformed spectrum before phase correction; Figure 21.14, the spectrum after phase correction.

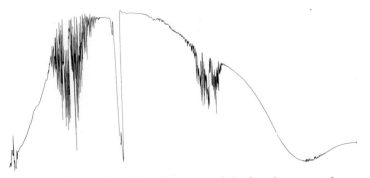

Figure 21.14   Transformed spectrum of air after phase correction.

Smoothing

Frequency domain data can be smoothed in any number of ways. One of the simplest of these is the three-point smooth, where

$$X_i' = (X_{i-1} + 2X_i + x_{i+1})/2$$

for all the points from $i = 2$ to $N - 1$. Points 1 and N are generally not smoothed at all in this simple procedure. A simple smoothing subroutine is given below:

```
;SMOOTHING SUBROUTINE
;CALL THROUGH R5 WITH ADDRESS OF ARRAY IN CALL+2
;AND SIZE OF ARRAY IN CALL+4

SMOOTH: MOV @(R5)+, R0    ;GET ADDRESS OF ARRAY INTO R0
        MOV (R5)+, R4     ;GET SIZE OF ARRAY
        CMP -(R4),-(R4)   ;DECREMENT COUNTER BY 2
;COPY FIRST POINT INTO R3.
        MOV (R0), R3
;R3 ALWAYS HOLDS THE NEXT VALUE TO BE PUT BACK AT END
;OF EACH PAS THROUGH THE SMOOTHING LOOP

LOOP:   MOV (R0)+, R2     ;GET I-1 POINT
        ASH #-2, R2       ;DIVIDE BY 4
        MOV (R0), R1      ;GET MIDDLE POINT, LEAVE R0 POINTING THERE
        ASR R1            ;DIVIDE BY 2
        ADD R1, R2        ;AND ADD INTO SMOOTHED VALUE
        MOV 2(R0), R1     ;GET I+1 POINT
        ASH #-2, R1       ;DIVIDE BY 4
        ADD R1, R2        ;AND ADD INTO SMOOTHED SUM
        MOV R3, -2(R0)    ;PUT LAST SMOOTHED POINT INTO MEMORY
        MOV R2, R3        ;SAVE NEW SMOOTHED POINT IN R3 FOR NEXT PASS
        SOB R4, LOOP      ;GO BACK N-2 TIMES
        MOV R2, (R0)      ;PUT LAST SMOOTHED POINT INTO ARRAY
        RTS R5            ;AND EXIT FROM SMOOTHING ROUTINE
```

This technique of smoothing is termed a "moving average" and can easily be expanded to many more than three points. Tables for performing these moving averages and other related smoothing functions have been given by Savitzky and Golay.[8] Smoothing using larger numbers of points is substantially slower, however, since floating point calculations become necessary for both the coefficients and the normalizing factors.

References

1. C. Hastings, *Approximations for Digital Computers*, Princeton University, Princeton, NJ, 1955.
2. See, for example, R. E. Johnson and F. L. Kiokemeister, *Calculus with Analytic Geometry*, Allyn and Bacon, 1960, or any other calculus textbook.
3. G. Samuelson, JEOL Inc., private communication.
4. J. W. Cooper, *Topics in Carbon-13 Nmr Spectroscopy*, Volume 2, G. Levy, ed., Wiley, New York, 1976, p. 372.
5. L. Mertz, *Infrared Phys.* **7**, 17 (1967).
6. P. R. Griffiths, C. T. Foskett, and R. Curbelo, *Appl. Spectrosc. Rev.* **6**, 31–78 (1972).
7. E. G. Codding and G. Horlick, *Appl. Spectrosc.* **27**, 85–92 (1972).
8. A. Savitzky and M. Golay, *Anal. Chem.* **36**, 1627–1639 (1964).

NUMERICAL OP
CODE LIST

OP Code			Mnemonic	OP Code			Mnemonic	OP Code			Mnemonic
00	00	00	HALT	00	60	DD	ROR	10	40	00	⎫
00	00	01	WAIT	00	61	DD	ROL		↕		⎬ EMT
00	00	02	RTI	00	62	DD	ASR	10	43	77	⎭
00	00	03	BPT	00	63	DD	ASL				
00	00	04	IOT	00	64	NN	MARK				
00	00	05	RESET	00	67	DD	SXT	10	44	00	⎫
00	00	06	RTT						↕		⎬ TRAP
00	00	07	⎫ (unused)	00	70	00	⎫	10	47	77	⎭
00	00	77	⎭		↕		⎬ (unused)				
00	01	DD	JMP	00	77	77	⎭	10	50	DD	CLRB
00	02	0R	RTS					10	51	DD	COMB
				01	SS	DD	MOV	10	52	DD	INCB
00	02	10	⎫	02	SS	DD	CMP	10	53	DD	DECB
	↕		⎬ (reserved)	03	SS	DD	BIT	10	54	DD	NEGB
				04	SS	DD	BIC	10	55	DD	ADCB
00	02	27	⎭	05	SS	DD	BIS	10	56	DD	SBCB
				06	SS	DD	ADD	10	57	DD	TSTB
00	02	40	NOP								
				07	0R	SS	MUL	10	60	DD	RORB
00	02	41	⎫	07	1R	SS	DIV	10	61	DD	ROLB
	↕		⎬ cond	07	2R	SS	ASH	10	62	DD	ASRB
			codes	07	3R	SS	ASHC	10	63	DD	ASLB
00	02	77	⎭	07	4R	DD	XOR	10	64	SS	MTPS
								10	67	DD	MFPS

OP Code	Mnemonic	OP Code	Mnemonic	OP Code	Mnemonic
00 03 DD	SWAB	07 50 0R	FADD		
		07 50 1R	FSUB	11 SS DD	MOVB
00 04 XXX	BR	07 50 2R	FMUL	12 SS DD	CMPB
00 10 XXX	BNE	07 50 3R	FDIV	13 SS DD	BITB
00 14 XXX	BEQ			14 SS DD	BICB
00 20 XXX	BGE	07 50 40		15 SS DD	BISB
00 24 XXX	BLT	↑	(unused)	16 SS DD	SUB
00 30 XXX	BGT	↓			
00 34 XXX	BLE	07 67 77			
00 4R DD	JSR	07 7R NN	SOB		
00 50 DD	CLR	10 00 XXX	BPL		
00 51 DD	COM	10 04 XXX	BMI		
00 52 DD	INC	10 10 XXX	BHI		
00 53 DD	DEC	10 14 XXX	BLOS		
00 54 DD	NEG	10 20 XXX	BVC		
00 55 DD	ADC	10 24 XXX	BVS		
00 56 DD	SBC	10 30 XXX	BCC, BHIS		
00 57 DD	TST	10 34 XXX	BCS, BLO		

POWERS OF TWO

2^n	n	2^{-n}
1	0	1.0
2	1	0.5
4	2	0.25
8	3	0.125
16	4	0.062 5
32	5	0.031 25
64	6	0.015 625
128	7	0.007 812 5
256	8	0.003 906 25
512	9	0.001 953 125
1 024	10	0.000 976 562 5
2 048	11	0.000 488 281 25
4 096	12	0.000 244 140 625
8 192	13	0.000 122 070 312 5
16 384	14	0.000 061 035 156 25
32 768	15	0.000 030 517 578 125
65 536	16	0.000 015 258 789 062 5
131 072	17	0.000 007 629 394 531 25
262 144	18	0.000 003 814 697 265 625
524 288	19	0.000 001 907 348 632 812 5
1 048 576	20	0.000 000 953 674 316 406 25
2 097 152	21	0.000 000 476 837 158 203 125
4 194 304	22	0.000 000 238 418 579 101 562 5
8 388 608	23	0.000 000 119 209 289 550 781 25
16 777 216	24	0.000 000 059 604 644 775 390 625
33 554 432	25	0.000 000 029 802 322 387 695 312 5
67 108 864	26	0.000 000 014 901 161 193 847 656 25
134 217 728	27	0.000 000 007 450 580 596 923 828 125
268 435 456	28	0.000 000 003 725 290 298 461 914 062 5
536 870 912	29	0.000 000 001 862 645 149 230 957 031 25
1 073 741 824	30	0.000 000 000 931 322 574 615 478 515 625
2 147 483 648	31	0.000 000 000 465 661 287 307 739 257 812 5
4 294 967 296	32	0.000 000 000 232 830 643 653 869 628 906 25
8 589 934 592	33	0.000 000 000 116 415 321 826 934 814 453 125
17 179 869 184	34	0.000 000 000 058 207 660 913 467 407 226 562 5
34 359 738 368	35	0.000 000 000 029 103 830 456 733 703 613 281 25
68 719 476 736	36	0.000 000 000 014 551 915 228 366 851 806 640 625
137 438 953 472	37	0.000 000 000 007 275 957 614 183 425 903 320 312 5
274 877 906 944	38	0.000 000 000 003 637 978 807 091 712 951 660 156 25
549 755 813 888	39	0.000 000 000 001 818 989 403 545 856 475 830 078 125
1 099 511 627 776	40	0.000 000 000 000 909 494 701 772 928 237 915 039 062 5

ASCII CHARACTER TABLE

Value	Character	Value	Character	Value	Character
0	NUL or BRK	52	*	125	U
1	SOH	53	+	126	V
2	STX	54	"	127	W
3	ETX	55	-	130	X
4	EOT	56	.	131	Y
5	ENQ	57	/	132	Z
6	ACK	60	0	133	[
7	BEL	61	1	134	\
10	BSP	62	2	135	]
11	TAB	63	3	136	$\wedge$ or $\uparrow$
12	LF	64	4	137	__ or $\leftarrow$
13	VTAB	65	5	140	`
14	FORM	66	6	141	a
15	CR	67	7	142	b
16	SO	70	8	143	c
17	SI	71	9	144	d
20	DLE	72	:	145	e
21	XON	73	;	146	f
22	DC2	74	<	147	g
23	XOF	75	=	150	h
24	DC4	76	>	151	i
25	NAK	77	?	152	j
26	SYN	100	@	153	k
27	ETB	101	A	154	l
30	CAN	102	B	155	m
31	EM	103	C	156	n

Value	Character	Value	Character	Value	Character
32	SUB	104	D	157	o
33	ESC	105	E	160	p
34	FS	106	F	161	q
35	GS	107	G	162	r
36	RS	110	H	163	s
37	US	111	I	164	t
40	SPACE	112	J	165	u
41	!	113	K	166	v
42	"	114	L	167	w
43	#	115	M	170	x
44	$	116	N	171	y
45	%	117	O	172	z
46	&	120	P	173	{
47	'	121	Q	174	|
50	(	122	R	175	} or altmode
51	)	123	S	176	~
52	*	124	T	177	RUB

ANSWERS
TO EXERCISES

Chapter 2

2.1   2, 55, 35, 25670, 75066, 55612

2.2
010	010	011				
001	001	111	000	111		
010	001	000	110	100	110	011
001	010	110	100			
110	101	110	100	011		
011	000	000	110	101	101	111

2.3   4096, 100, 2110, 4096, 174, 5350

2.4   10000, 144, 1000, 454, 2000001, 37777

2.5   3723, 35, 32360, 41440

2.6   100011, 4337347 to 21 bits

2.7   104215, 177207, 25133, 22054

2.8   104216, 177210, 25134, 22055

2.9   56025, 133013, 131102

2.10 220, 21660, 24570

2.11 3, 0, 122, 40230

2.12 7, 7, 377, 176777

2.13 0, 2, 121, 106007

Chapter 4

4.1   The sum 73426 + 31203 produces 124631, and this is a sign change through arithmetic overflow. Thus N = 1, V = 1, Z = 0, C = 0.

4.2   R4 = 0, Z = 1, N = 0, V = 0, C = 1.

4.3   Note that moving a byte into a register causes the sign to be extended into the upper bits. Start with R2 = 000005.

```
NEG R2          ;R2 = 177773
MOVB R2, R3     ;R3 = 177773
NEG R2          ;R2 = 000005
ADD R3, R2      ;R2 = 0
```

4.4   Z will always be set since the contents of R5 are always equal to themselves. N can never be set since the answer will never be negative. If R5 is negative, C and V will be set.

4.5   COM R0; R0 = 001342, DEC R0; R0 = 001341.

4.6   R5 = 176401; the lower byte is 001. DECB R5; lower byte is 000, so N = 0, Z = 1, V = 0, C = 0.

4.7   R5 = 003702; NEGB R5; lower byte is 302, and after negation it will be 76, so R5 = 3476.

```
.TITLE SYMBOL DEFINTIONS USED
;SYMBOL DEFINITIONS USED IN THESE ANSWERS
;BASIC REGISTER DEFINITIONS
R0=%0
R1=%1
R2=%2
R3=%3
R4=%4
R5=%5
SP=%6
PC=%7
PS=177776


;DEVICE DEFINITIONS
TKS=177560
TKB=TKS+2          ;KEYBOARD STATUS AND BUFFER
TPS=TKB+2          ;PRINTER STATUS AND BUFFER
TPB=TPS+2
HSRS=177550        ;HIGH SPEED READER STTATUS
HSRB=HSRS+2
HSPS=HSRB+2        ;HIGH SPEED PUNCH
HSPB=HSPS+2


;LPS-11 DEVICE DEFINITIONS
ADSTAT=170400
ADBUF=ADSTAT+2
CLSTAT=ADBUF+2   ;CLOCK
CLKBUF=CLSTAT+2
IOSTAT=CLKBUF+2  ;DIGITAL I/O
INBUF=IOSTAT+2
OUTBUF=INBUF+2
DISTAT=OUTBUF+2  ;DISPLAY
XDAC=DISTAT+2    ;X-DAC
YDAC=XDAC+2      ;YDAC
```

Chapter 5

5.1.	INST	R3	R4
	MOV R3, R4	4562	4562
	MOV(R3),R4	4562	2616
	MOV (R3)+, R4	4564	2616
	MOV @(R3)+,R4	4564	13
	MOV -(R3),R4	4560	1342
	MOV @-(R3),R4	4560	1515
	MOV -2(R3),R4	4562	1342
	MOV @-2(R3),R4	4562	1515

```
         ;5.2

016374   MOV 2(R3),@-2(R4)          ;ASSEMBLED INTO THREE WO
000002
177776
         ;GIVEN R3=4562 & R4=17630, THE INSTRUCTION WILL
         ;MOVE THE CONTENTS OF ADD. 4564 INTO 13042
         ;THE NUMBER 17760 WILL BE DEPOSITED IN 13042
```

```
5.3
INST                        R3      R4       ADDRESS CHANGED
-----------                 ----    ----     ----------------
ADD (R3),@-(R4)   4562     17626    13042/(5662+2616)
MOV R4, (R3)      4562     17630    4562/1402
MOV R4,(R3)+      4564     17630    4562/17630
DEC -(R4)         4562     17626    17626/13041
CLR (R4)+         4562     17632    17630/0
CLR R3               0     17630    ------
SUB (R4),(R4)+    4562     17632    17630/0
```

```
;5.4

        ADD R0, R1
        SUB R2, R3
        DEC R3
        HALT

;5.5

        MOV R0,R1
        MOV (R0),R2
        MOV (R0)+, R3
        MOV @(R0)+, R4
        MOV -(R0), R5
        HALT


              ;5.6
              ;ASSEMBLES AS FOLLOWS:

012700   MOV #123, R0        ;PUT 123 IN R0
000123
005400   NEG R0              ;R0=-123 = 177655
162700   SUB #3, R0          ;ADD 177775 TO R0
000003
000000   HALT
```

;5.7   THE PC, (R7) ALWAYS CONTAINS THE ADDRESS OF THE NEXT INSTRUCTION
;TO BE EXECUTED. THUS IT POINTS TO THE LOCATION AFTER THE HALT.

```
              ;5.8
      000500  .=500
              ;START WITH      R0=0
              ;                R1=100
000500 005700 TST R0           ;SET COND CODES FROM R0
000502 000000 HALT             ;STOP AND LOOK, PRESS CONT TO GO
000504 012100 MOV (R1)+, R0    ;PUT CONTENTS OF 1000 IN R0
000506 000000 HALT             ;STOP AND LOOK
000510 062100 ADD (R1)+, R0    ;ADD 2ND ENTRY TO IT
000512 000000 HALT             ;STOP AND LOOK
000514 062100 ADD (R1)+, R0    ;ADD THIRD TO THAT
000516 000000 HALT             ;STOP AND LOOK

       001000 .=1000
001000 031726 31726
001002 101241 101241
001004 050132 50132
```

```
;5.9
;START WITH     R2=516
;               R3=3216

        ADD R3, R2      ;ADD 3216 TO 516. R2=3734
        SUB R2,R3       ;R3=3216-3734=-516=177262
        CLR R2          ;R2=0
        SUB R3, R2      ;R2=516
        INC R2          ;R2=517
        COM R2          ;R2=177260
        SUB R2, R3      ;R3=2
        HALT

;5.10

        MOV R5,R4
        CLR R3
        ADD (R4)+, R3
        BNE .-2
        SUB R3,R5
        HALT

;5.11

        MOV #500, SP
        MOV #340, @#36
        MOV #1000, R0
        MOV #10, R1
        CLR R2
        ADD (R0), R2

;5.12
START:  MOV #6, R0      ;PUT 6 IN R0
        SUB R0, TEMP    ;SUBTRACT R0 FROM TEMP
        BEQ START       ;GO BACK IF EQUAL
        COM TEMP        ;IF NOTE EQUAL, COMPLEMENT TEMP
        HALT            ;AND STOP
TEMP:   6               ;INITIALLY CONTAINS 6
;IN THE FIRST TIME THROUGH THE PROGRAM, TEMP WILL BECOME ZERO
;AND THE BRANCH WILL BE SATISFIED. THIS SENDS THE PROGRAM
;BACK TO START. IN THE SECOND TIME THROUGH, TEMP WILL BE 0 AND
;R0 WILL BE 6. SUBTRACTING 6 FROM ZERO GIVES 177772 WHICH WILL BE THE NE

;CONTENTS OF TEMP. THIS IS NON-ZERO AND TEMP IS THEN COMPLEMENTED,
;GIVING A FINAL VALUE OF 5 FOR TEMP. R0 WILL BE 6.
                        ;5.13
        000500
000500  010067  MOV R0, A513
000502  000002
000504  000000  HALT
000506  000000  A513:   0


        000500  .=500
000500  010037  MOV R0, @#506
000502  000506
000504  000000  HA
000506  000000  0

        ;BOTH PROGRAMS PUT THE CONTENTS OF R0 INTO LOCATION 506.
        ;THE DIFFERENCE IS IN THE ADDRESSING MODE. THE FIRST PROGRAM
        ;USES THE PC-INDEXED MODE, AND IS THUS RELOCATABLE, WITH
        ;THE LOCATION OF A ALWAYS 6 BELOW THE MOV INSTRUCTION.
        ;THE SECOND METHOD ALWAYS PUTS R0 IN LOCATION 506
        ;REGARDLESS OF WHERE THE PROGRAM IS LOADED.
```

Chapter 6

```
;6.1
;ADD AND SUBTRACT ALTERNATE NUMBERS UNTIL ZERO IS FOUND
;ASSUME WE WILL CHECK 8K OF MEMORY FROM 0 TO 37776
;INCLUDING THE PROGRAM ITSELF

START:   CLR R0              ;SUM IN R0
ST61:    MOV #4096., R1      ;SET COUNTER TO 4096 PAIRS TO CHECK
         CLR R2              ;POINTER TO DATA STARTS AT ADDRESS 0
LOOP61:  ADD (R2)+, R0       ;ADD EACH NUMBER INTO SUM
         BEQ STOP61          ;QUIT IF SUM NOW 0
         SUB (R2)+, R0       ;SUBTRACT NEXT NUMBER
         BEQ STOP61          ;STOP IF 0 NOW, TOO
         SOB R1, LOOP61      ;GO BACK IF NOT 0 YET
;        BR ST61             ;WE COULD EITHER RESET AND CONTINUE
STOP61:  HALT     ;OR HALT WHEN MEMORY IS FINISHED


;6.2
;PUT A 4K "RAMP" IN MEMORY
         .=20000            ;CAN'T RUN THIS AT ADDRESS 0
START:   CLR R0   /         ;POINTER
         MOV # 4096., R1 ;COUNTER
         CLR R2            ;RAMP VALUE
LOOP62:  MOV R2, (R0)+     ;PUT EACH VALUE IN MEMORY
         INC R2            ;ADVANCE  RAMP VALUE
         SOB R1, LOOP62    ;GO BACK TILL DONE
         HALT              ;THEN STOP


;6.3
;FIBONACCI SERIES
ST63:    MOV #DATA63, R0             ;POINTER TO TABLE
         MOV #1, (R0)               ;FIRST IN SERIES
         MOV (R0)+, (R0)+            ;2ND IN SERIES
;EACH REMAINING TERM IS SUM OF PREVIOUS TWO
         MOV #13., R2               ;DO 13 MORE
LOOP63:  MOV -2(R0), (R0)           ;GET LAST ONE
         ADD -4(R0), (R0)+          ;ADD PREVIOUS AND INCREMENT
         SOB R1, LOOP63             ;GO BACK IF NOT DONE
         HALT                       ;STOP WHEN DONE
DATA63:  0                  ;TABLE STARTS HERE


;6.4
;COUNT NEGATIVE NUMBERS BETWEEN 1200 AND 4216
STR64:   MOV #1200, R1              ;POINTER
         CLR R0                     ;COUNT NUMBERS HERE
LOOP64:  TST (R1)+                  ;IS THIS ONE NEGATIVE?
         BGE A64                    ;NO
         INC R0                     ;YES, COUNT IT
A64:     CMP R0, #4216              ;DONE YET?
         BLE LOOP64                 ;NO, KEEP CHECKING
         HALT                       ;WITH COUNT IN R0
```

```
;6.5
;COUNT LARGEST NUMBER OF ADJACENT ZEROES IN
;THE FIRST 4K OF MEMORY
START:  CLR R0          ;ADDRESS POINTER
        CLR R3          ;SAVED 'LAST COUNT' OF ZEROES
        MOV #4096.., R1 ;SET COUNTER
ZC10:   CLR R2          ;COUNT ZEROES HERE

LOOP:   TST (R0)+       ;IS THIS LOCATION ZERO?
        BNE ZC20        ;NO, THIS IS THE END OF A BLOCK OF ZEROES
        INC R2          ;YES, COUNT IT
        SOB R1, LOOP    ;GO BACK TILL DONE
;DONE WITH 4K, CHECK LAST 2 COUNTERS
        CMP R2, R3      ;IS CURRENT COUNT > THAN SAVED COUNT?
        BLE ZC30        ;NO
        MOV R2, R3      ;YES, COPY IT
ZC30:   MOV R3, R0      ;COPY LARGEST BLOCK INTO R0
        HALT            ;AND STOP WITH # IN R0

;WHEN A ZERO BLOCK ENDS, COMPARE THIS ONE
;WITH THE LARGEST FOUND BEFORE
ZC20:   CMP R2, R3      ;IS THIS ONE BIGGER?
        BLE ZC10        ;NO, ZERO AND COUNT AGAIN
        MOV R2, R3      ;ELSE COPY THIS ONE INTO SAVED COUNTER
        BR ZC10         ;AND GO BACK TO LOOK FOR MORE
```

Chapter 7

7.1 - J - Move the pointer to the top of the current edit buffer.
FSCHA$OLT$$- Replace the first occurence of 'CHA' with the
characters 'OLT' . This will not execute the apparent command 'OLT'
at all. A better command might be JSCHA$OLT$$ which will search
for 'CHA' without changing it and print out the line containing it.

7.2 The directive .ASCII '123' will put the bytes
```
        031061
          063
```
into memory, but since this is an odd number of bytes, this directive
might best be followed by an
```
        .EVEN
```
directive to make the current location counter even again.
     The directive .ASCIZ ATHIS IS A;MESSAGEA will
stop packing characters at the second occurrence of the delimiter, in this
case 'A' and will thus pack the characters

```
        044124  TH
        051511  IS
        044440   I
        020123  S
          000
```

Note that the inclusion of the semicolon was most fortunate

Chapter 8

;8.1
 Clear bits 5-7 and set bits 1,12, and 14 in R0.

```
            BIC #340, R0      ;CLEAR 5-7
            BIS #50002, R0    ;SET 1,12 AND 14
            HALT              ;AND DISPLAY RESULT
```

;8.2
 ;EXCLUSIVE OR OF C AND D WITH RESULT IN E

```
            MOV C, R0         ;C INTO A REGISTER
            MOV D, E          ;COPY ONE INTO DESTINATION LOCATION
            XOR R0, E         ;XOR ONE WITH OTHER ALREADY IN DESTINATION
            HALT
```

;8.3
 ;SOLVE Y=MX + B
```
            MOV  M, R0        ;GET MULTIPLIER
            MUL X, R0         ;MX- SINGLE PRECISION
            ADD B, R1         ;ADD B TO LOW PART
            MOV R1, Y         ;SINGL E PREC ANSWER INTO Y
            HALT
```

;8.4
 ;EXCLUSIVE OR WITHOUT USING XOR INSTRUCTION
 ;XOR = OR - AND
```
STR84:      MOV D, R0         ;D IS ONE NUMBER TO XOR
            BIS C, R0         ;R0 IS  C OR D
            MOV D, R1         ;GET D AGAIN FOR AND
            MOV C, R2         ;GET C FOR AND
            COM R2            ;COMPLEMENT C TO DO NAND
            BIC R2,R1         ;THIS IS C AND D
            SUB R1, R0        ;OR - AND LEFT IN R0
            HALT
```

;8.5
 ;   R2 IS EVEN -- I.E. BIT 0 IS ZERO
 ;   COULD ALSO BE DONE BY
 ;   BIC #1, R2

;8.6
 ;IF R3 IS LARGE AND NEGATIVE SO THAT 15 AND 14 ARE 1 AND 0

Chapter 9

```
;9.1
;PRINT OUT "THIS PROGRAM WORKS!"
.=500
        MOV #., SP        ;BE SURE TO INITIALIZE THE STACK
        MOV #TEXT91, R1   ;SET POINTER TO TEXT
LOOP91: MOVB (R1)+, R0    ;GET EACH BYTE OF MESSAGE
        BEQ STOP91        ;QUIT ON ZERO BYTE
        JSR PC, TYPE      ;PRINT EACH BYTE
        BR LOOP91         ;GO BACK TILL DONE
STOP91: HALT
TEXT91: .BYTE 15, 12      ;CRLF FIRST FOR INITIALIZATION!
        .ASCIZ "THIS PROGRAM WORKS!"
.EVEN


;9.2
;SAME AS 9.1 EXCEPT FOR TEXT AND PUNCH
.=500
        MOV #., SP        ;INITIALIZE SP
A92:    MOV #DOLLS, R1    ;ADDRESS OF LIST
        MOV #11., R2      ;COUNT 11 BYTES PER DOLL
B92:    MOVB (R1)+, R0    ;PUNCH OUT 1 DOLL CHAR
        JSR PC, PUNCH     ;PUNCH IT
        DEC R2            ;DONE?
        BNE B92           ;GO BACK 11 TIMES
        BR A92            ;THEN REPEAT AD NAUSEAM
DOLLS:  .BYTE 10, 10, 10, 110, 253, 274
.BYTE 253, 110, 10, 10, 10


;9.3
;READ IN CHARACTERS UNTIL $-SIGN IS STRUCK
;THEN PRINT THEM OUT AGAIN
.=500
STAR93: MOV #., SP         ;INIT STACK POINTER
        JSR PC, CRLF       ;INIT TTY CARRIAGE
        MOV #DATA93, R1    ;START OF TEXT STORAGE
        CLR R2             ;R2 IS COUNTER
A93:    JSR PC, READ       ;READ AND PRINT CHARS ON INPUT
        MOVB R0, (R1)      ;STORE IN ARRAY
        INC R2             ;COUNT CHARS TYPED IN
        CMPB #'$, (R1)+    ;WAS IT A DOLLAR SIGN?
        BNE A93            ;NO, KEEP GETTING CHARS
        JSR PC, CRLF       ;YES RE-INIT CARRIAGE AND PRINT OUT CHARS
        MOV #DATA93, R1    ;INIT POINTER
B93:    MOVB (R1)+, R0     ;PUT EACH CHAR IN R0
        JSR PC, TYPE       ;AND PRINT IT
        DEC R2             ;DONE?
        BNE B93            ;NO
        BR STAR93          ;YES,WAIT FOR NEW INPUT
DATA93: 0                  ;TEXT STORAGE STARTS HERE
```

```
;9.4
;ACCEPT CHARACTERS FROM TTY UNTIL $-SIGN STRUCK
;MAKE EVERY CR A CRLF AND STOP IF AN '&' IS STRUCK
.=500
STAR94: MOV #.,  SP       ;INIT SP AS USUAL
        JSR PC, CRLF       ;CRLF
        MOV #DATA94, R1   ;SET POINTER
        CLR R2             ;AND COUNTER
A94:    JSR PC, READ       ;GET EACH CHAR
        MOVB R0, (R1)      ;STORE IN ARRAY
        INC R2             ;AND COUNT IT
        CMPB #'$, (R1)+    ;WAS IT A DOLLAR?
        BEQ B94            ;YES-- GO PRINT IT OUT
        CMP #15, R0        ;WAS IT A CR?
        BEQ C94            ;YES, TYPE LF TOO
        CMP R0, #'&        ;AMPERSAND?
        BEQ STOP94         ;YES, STOP EVERYTHING
        BR A94             ;OTHERWISE KEEP GOING

;PRINT OUT TEXT TO DATE
B94:    MOV #DATA94, R1   ;RESET POINTER
        MOVB (R1), R0      ;GET EACH CHAR
        JSR PC, TYPE       ;PRINT IT
        CMPB #15, R0       ;WAS IT CR?
        BNE D94            ;NO, KEEP GOING
        JSR PC, CRLF       ;YES, TYPE CRLF
D94:    DEC R2             ;DONE?
        BNE B94            ;NO
        BR STAR94          ;YES, GET NEW INPUT

;TYPE LF WHEN CR IS TYPED
C94:    JSR PC, CRLF
        BR A94
STOP94: HALT
DATA94: 0

;9.5
.=500
STAR95: MOV #.,  SP       ;INIT SP TO 500
        CLR R0   ;R0=0
        MOV STAR95, R0    ;PUT CONT OF ADDR. "START" IN R0 = 12706
        JSR PC, DUMMY      ;CALL DUMMY SUBROUTINE
        ADD -(SP), R0      ;ADD RETURN POINTER FROM JSR TO R0
        HALT               ;R0=12706+516 = 13424
DUMMY:  RTS PC             ;RETURN WITHOUT DOING ANYTHING


;9.6
        SR =177570         ;DEFINE IT
        MOV SR, R0         ;PUT IN R0
        HALT               ;HALT WITH DISLAY IN R0
```

```
;9.7
;PRINT OUT SIGNED OCTAL NUMBERS
;CALL WITH JSR PC, OCTOUT, WIHT NUMBER IN RO

OCTOUT: CLR R5            ;SIGN FLAG
        TST RO            ;CHECK SIGN
        BPL A97           ;SKIP OVER IF POSITIVE
        NEG RO            ;ABSOLUTE VALUE
        INC R5            ;AND SAVE SIGN
A97:    MOV RO, R4        ;COPY NUMBER
        MOV #OTABL, R1    ;POINTER TO SHIFT TABLE
        MOV #5, R2        ;COUNTER
        MOV #'+, RO       ;INITIALLY GET PLUS SIGN
        TST R5            ;CHANGE IF NUMBER WAS MINUS
        BPL B97
        MOV #'-, RO       ;GET MINUS SIGN
B97:    JSR PC, TYPE      ;PRINT SIGN
C97:    MOV R4, RO        ;GET EACH NUMBER BACK
        ASH (R1)+, RO     ;SHIFT OVER NEXT 3 BITS
        BIC #177770, RO   ;MASK OUT ALL OTHERS
        ADD #60, RO       ;ADD ASCII BIAS
        JSR PC, TYPE
        SOB R2, C97       ;AND DO THIS 5 TIMES
        RTS PC            ;AND EXIT


;TABLE OF SHIFTS; -NEGATIVE SINCE THEY ARE RIGHT SHIFTS
OTABL:  -12., -9., -6, -3, 0

;9.8
;READ IN OCTAL NUMBERS FROM TTY
OCTIN:  CLR R2            ;NUMBER ACCUMULATES HERE
A98:    JSR PC, READ      ;GET A CHARACTER
        SUB #60, RO       ;SUBTRACT ASCII BIAS
        BLT EXIT98        ;EXIT WITH # TO DATE WHEN ILLEGAL CHAR FOUND
        CMP RO, #7        ;>7?
        BGT EXIT98        ;YES, EXIT
        ASH #3, R2        ;SHIFT OVER NUMBER TO DATE
        BCS OERR98        ;ERROR IF NUMBER OVERFLOWS 16 BITS
        ADD RO, R2        ;ADD IN NEW DIGIT
        BR A98            ;GO BACK TILL ILLEGAL CHAR OR OVERFLOW
EXIT98: RTS PC            ;EXIT WITH NUMBER IN R2
OERR98: MOV #'?, RO       ;TYPE ?-MARK
        JSR PC, TYPE
        JSR PC, CRLF
        BR OCTIN          ;GO BACK AND TRY AGAIN


;9.9
;DECIMAL NUMBER OUTPUT
;ENTER WITH NUMBER TO PRINT IN R1

DECOUT: CLR R5 ;THIS WILL BE THE SIGN FLAG
        TST R1 ;SEE IF THE NUMBER IS NEGATIVE
        BLE A99 ;NO
        NEG R1 ;YES, NEGATE AND SET FLAG
        INC R5            ;SET FLAG
A99:    MOV #DTAB99, R2 ;POINTER TO DIVISOR TABLE
        MOV #5, R3        ;5 OF THEM
B99:    CLR RO            ;CLEAR HIGH WORD BEFORE DIVIDING
        DIV (R2)+, RO     ;DIVIDE BY NEXT DIVISOR
        ADD #60, RO       ;ADD ON ASCII BIAS
        JSR PC, TYPE      ;PRINT IT
        SOB R3, B99       ;GO BACK 5 TIMES
        RTS PC            ;THEN EXIT

DTAB99: 10000., 1000., 100., 10., 1
```

Chapter 10

```
;10.1
;INCREMENT A LOCATION CONTINUOUSLY UNTIL A KEY IS STRUCK
.ASECT
.=1000
START:   MOV #.,  SP       ;ALWAYS INIT SP IF INTERRUPTS TO BE USED
         MOV #KBSERV, @#60       ;SET UP KEYBOARD VECTOR
         MOV #200,  @#62  ;AND PRIORITY
         MOV #100,  TKS   ;ENABLE INTERRUPT TO KEYBOARD
         CLR LOCN         ;HERE IS THE LOCATION TO BE INCREMENTED

A101:    INC LOCN         ;INCREMENT LOCATION
         BR A101          ;CONTINUOUSLY

;UNTIL KEYBOARD INTERRUPTS
KBSERV:  MOV LOCN, R0     ;LEAVE COUNT IN R0
         HALT             ;AND HALT WITH COUNT DISPLAYED



;10.2
;INCREMENT COUNTER UNTIL KEYBOARD IS STRUCK
;THEN PRINT OUT THE RESULT ON THE TTY BY INTERRUPT
;KEEP COUNTING, AND PRINT OUT NEW NUMBER WHEN KEYBOARD
;IS STRUCK AGAIN, EVEN IF THE LAST IS NOT FINISHED

.ASECT
.=1000
STR102:  MOV #., SP       ;INITI SP
         MOV #KBSERV, @#60       ;AND KBD VECTOR
         MOV #200, @#62
         MOV #PRINT, @#64       ;PRINTER VECTOR
         MOV #200, @#66
         CLR PS           ;CLEAR PSW TO START
         CLR R1           ;COUNTING GOES ON IN HERE
B102:    INC R1           ;START COUNTING
         BR B102

;KEYBOARD INTERRUPT SERVICE ROUTINE
;CALCULATES ASCII OUTPUT AND STORES FOR OUTPUT BY "PRINT"
KBSERV:  MOV #DT102, R2         ;POINTER TO DIVISOR TABLE
         MOV #5, R3            ;NUMBER OF DIGITS
         MOV #CTAB, R4        ;CHARACTER OUTPUT TABLE
         MOV TKB, R5          ;DUMMY READ OF CHARACTER
;NOW DIVIDE OUT AND PUT DIGITS IN  TABLE
         CLR R0           ;UPPER WORD FOR DIVISION
E102:    DIV (R2)+, R1        ;DIVIDE BY EACH POWER OF 10
         ADD #60, R0          ;ADD ON ASCII BIAS
         MOVB R0, (R4)+       ;PUT CHARACTER IN OUTPUT TABLE
         CLR R0               ;FOR NEXT DIVISION
         SOB R3, E102         ;DO ALL 5 CHARACTERS
;DONE WITH CHARACTERS, NOW ENABLE PRINTER INTERRUPT
         MOV #100, TPS
         CLR R1               ;RESET COUNTER
         MOV #BTAB, R4        ;RESET TABLE PTR TO INCLUDE CR-LF
         MOV #8., R3          ;NUMBER OF CHARS IN TABLE
         RTI                  ;AND AWAY WE GO
;PRINTER INTERRUPT OUTPUT ROUTINE
PRINT:   MOVB (R4)+, TPB      ;PRINT A CHAR
         DEC R3               ;COUNT THEM
         BNE F102             ;EXIT UNTIL ALL DONE
         CLR TPS              ;WHEN DONE SHUT OFF PRINTER INTERRUPT
F102:    RTI                  ;END EXIT
```

```
;TABLES OF CHARACTERS
BTAB:    .BYTE 15, 12      ;CR-LF
CTAB:    .BLKB 6
;TABLE OF POWERS OF 10
DT102:   10000., 1000., 100., 10., 1

;10.3
;CALL ROUTINE AT LOCATION 7006 THROUGH A TRAP INSTRUCTION
;FIRST PUT THE ADDRESS IN TRAP VECTOR 34
;AND NEW PS IN LOCN 36
CLSET:   MOV #ROUTIN, @#34      ;SET TRAP VECTOR
         MOV #340, @#36   ;ALLOW NO INTERRUPTS WHILE HERE

;THEN CALL THE ROUTINE BY
         TRAP             ;WHICH VECTORS TO LOCN PTD TO BY 34

;AT LOCN 7006, THE ROUTINE STARTS

ROUTIN: MOV 2(SP), PS     ;RESTORE WHATEVER THE PS WAS
;           :
;           :             ;AND DO THE ROUTINE
        RTI               ;THEN EXIT FROM THE ROUTINE
```

Chapter 11

```
;SYMBOL DEFINITIONS USED IN FPMP11 PROGRAMS
.GLOBL TRAPH, FLIP, FLOP, $POLSH
.GLOBL $ADR, $MLR, $DVR
.GLOBL $IR, $RI, FLAC, VFLAG, TCHAR

FPADD=TRAP+12
FPMULT=TRAP+21
FPDIV=TRAP+25
FPSUB=TRAP+13
FCMP=TRAP+17
FPGET=TRAP+71
FPSTR=TRAP+73

ARM=100
IMM=200
RELM=300
```

```
;11.1
;CALCUATE Y=MX+B
.=500
S111:     MOV #.» SP        ;INIT SP BEFORE CALLING FPMP11
          MOV #TRAPH» @#34
          MOV #340» @#36    ;INIT TRAP HANDLER
          JSR PC» CRLF      ;INIT CARRIAGE
          MOV #"=M» R0      ;GET M
          JSR PC» DTYPE     ;TYPE "M=" AND GET M
          FPSTR+RELM
          M111-.            ;STORE IN M
          MOV #"=X» R0
          JSR PC» DTYPE     ;GET X
          FPMULT +RELM
          M111-.            ;MULTIPLY MX
          FPADD + RELM
          B111-.            ;MX+B» B ALREADY KNOWN
          JSR PC» FLOP      ;TYPE OUT RESULT
          BR S111           ;AND DO ANOTHER

DTYPE:    JSR PC» TYPE      ;TYPE 1ST CHAR
          SWAB R0           ;GET OTHER CHARACTER
          JSR PC» TYPE      ;TYPE 2ND CHAR
          JSR PC» FLIP      ;GET NUMBER
          JSR PC» CRLF      ;TYPE CRLF
          RTS PC            ;RETURN WITH VALUE IN FLAC

M111:     0»0               ;BE SURE TO RESERVE TWO LOCATIONS FOR FP NUMBERS
X111:     0»0
B111:     .FLT2 344321      ;FOR EXAMPLE


;11.2
;CALCULATE Y=AX^2 +BX +2 USING POLISH MODE
;FACTOR INTO Y= X(AX+B)+C
.=500
          MOV #.» SP
          MOV C+2» -(SP)
          MOV C» -(SP)      ;PUSH C
          MOV X+2» -(SP)
          MOV X» -(SP)      ;PUSH X
          MOV B+2» -(SP)
          MOV B» -(SP)      ;PUSH B
          MOV A+2» -(SP)
          MOV A»-(SP)       ;PUSH A
          MOV X+2» -(SP)
          MOV X» -(SP)      ;PUSH X
          JSR R4» $POLSH    ;ENTER POLISH MODE
          $MLR              ;AX
          $ADR              ;AX+B
          $MLR              ;AX^2 + BX
          $ADR              ;AX^2 +BX + C
          .+2               ;EXIT
          FPGET             ;POP INTO FLAC
          JSR PC» CRLF      ;INIT TTY
          JSR PC» FLOP      ;AND PRINT RESULT
          HALT              ;AND STOP

;FLOATING CONSTANTS
A:        0»0
B:        0»0
C:        0»0
X:        0»0
```

```
;11.3
;CALCULATE F TO C AND C TO F CONVERSION
CF:       MOV #., SP        ;INIT SP
          MOV #TRAPH, @#34
          MOV #340, @#36
CF10:     JSR PC, CRLF
          JSR PC, FLIP      ;GET NUMBER TO CONVERT
          MOV TKB, R0       ;SAVE TERMINATING CHARACTER
          JSR PC, CRLF      ;TYPE CRLF
          BIC #-200, R0     ;CLER PARITY BIT
          CMP R0, #'F       ;WAS IT F?
          BEQ FTOC          ;YES, CONVERT TO CELSIUS
          CMP R0, #'C       ;WAS IT C?
          BNE CF10          ;IF NOT C, IGNORE AND GET NEW VALUE

;CONVERT F=1.8C +32
          FPMULT+RELM
          K1P8-.            ;FLOATING 1.8
          FPADD +RELM
          K32-.             ;32
CFOUT:    JSR PC, FLOP      ;PRINT RESULT
          BR CF10           ;AND GO BACK FOR NEW VALUE

;C=(F-32)/1.8
FTOC:     FPSUB+RELM        ;F-32
          K32-.
          FPDIV+RELM        ;/1.8
          K1P8-.
          BR CFOUT          ;AND PRINT OUT RESULT
K1P8:     .FLT2 1.8         ;FLOATING 1.8
K32:      .FLT2 32          ;FLOATING 32


;11.4
;SUBROUTINE TO PRINT OUT AND ALLOW MODIFICATION
;OF A NUMBER WHOSE ADDRESS IS IN THE LOCATION FOLLOWING THE CALL
;CALL WITH
;         JSR R5, INOUT
;         address of variable

INOUT:    MOV (R5), R0      ;GET ADDRESS OF VARIABLE
          FPGET +ARM        ;GET IN @R0 MODE
          JSR PC, FLOP      ;PRINT IT OUT
          MOV # SPACE, R0   ;PRINT A SPACE AFTER NUMBER
          JSR PC, TYPE
;NOW GET NEW VALUE OR JUST A RETURN
          JSR PC, FLIP
          JSR PC, CRLF      ;PRINT A CR ON EXIT
          TST VFLAG         ;WAS A VALID NUMBER ENTERED?
          BEQ INEXIT        ;NO, LEAVE CURRENT VALUE ALONE
          MOV (R5)+, R0     ;ELSE STORE FLAC IN @R0
          FPSTR+ARM
          RTS R5            ;AND EXIT
;OTHERWISE EXIT WITH NO CHANGE
INEXIT:   TST (R5)+         ;INCREMENT R5
          RTS R5            ;AND THEN EXIT
```

```
;PRINT OUT CHARACTER STRING
ASCOUT: MOV #ASC114, R1  ;INIT POINTER
        MOV #4, R2       ;FOUR CHARS
C114:   MOV (R1)+, R0    ;GET EACH CHAR
        JSR PC, TYPE     ;PRINT IT
        DEC R2
        BNE C114
        RTS PC

CNT114: 0            ;COUNTER

ASC114: 0,0
NUM114: 0
```

Chapter 15

```
;15.1
;DISPLAY FALLING DIAGONAL LINE ON SCOPE
.ASECT
.=1000
S151:   MOV #4, DISTAT   ;INTENS ON Y-LOAD
        MOV #4096., R0   ;COUNT
        MOV #7777, YDAC  ;TOP OF Y
        CLR XDAC         ;LEFT SIDE
;LINE STARTS AT TOP LEFT
B151:   DEC YDAC         ;AND Y DECREASES
        IN XDAC          ;AS X INCREASES
        SOB R0, B151     ;GO BACK TILL AT BOTTOM RIGHT
        BR S151          ;THEN REFRESH
;THE ABOVE PROGRAM COULD DISPLAY WITH LESS FLICKER
;BY MOVING IN INCREMENTS OF 8 OR 16

;15.2
;DISPLAY A CROSS ON THE CRT
;FIRST DO THE HORIZONTAL BAR
S152:   MOV #4, DISTAT   ;INTENS ON X-LOAD
        MOV #4000, YDAC  ;CENTER
        MOV #4096., R0   ;COUNT POINTS ACROSS
        CLR XDAC         ;START X AT LEFT
;AND MOVE X ACROSS TO RIGHT
B152:   INC XDAC
        SOB R0, B152
;NOW DO THE VERTICAL BAR
        MOV #10, DISTAT  ;INTENS ON Y-LOAD
        MOV #4096., R0   ;COUNTER
        MOV #4000, XDAC  ;X IN CENTER
        CLR YDAC         ;Y AT BOTTOM
C152:   INC YDAC         ;MOV Y UP
        SOB R0, C152
        BR A152          ;GO BACK AND REFRESH

;15.3
;DISPLAY HORIZONTAL BAR WHOSE POSITION
;DEPENDS ON ADC READING
A153:   MOV #4096., R1   ;COUNT ACROSS SCOPE
        MOV #4, DISTAT   ;INTENSIFY ON X-LOAD
        MOV #1001, ADSTAT     ;KNOB 1, WAIT FOR ADC
B153:   TSTB ADSTAT      ;AD DONE?
        BPL B153         ;NOT YET
        MOV ADBUF, YDAC  ;LOAD Y DIRECTLY FROM ADC READING
C153:   INC XDAC         ;NOW DISPLAY BAR A LEVEL DETERMINED BY ADC
        SOB R1, C153     ;GO BACK TILL DONE
        BR A153          ;GET ANOTHER READING
```

```
;15.4
;SMALL VARIABLE CROSS
S154:    MOV #.,, SP
         MOV #1, ADSTAT    ;START ADC ON KNOB 0
         JSR PC, ADREAD    ;READ IT
         MOV R0, XPOS      ;SAVE X-POSITION
         MOV #400, ADSTAT       ;KNOB 1
         JSR PC, ADREAD    ;READ THAT
         MOV R0, YPOS      ;SAVE Y-POSITION
         JSR R5, BAR       ;DO BAR
         4                 ;VALUE FOR DISTAT
         XPOS              ;CENTER OF BAR
         XDAC              ;REGISTER TO VARY
;DO Y-BAR
         JSR R5, BAR
         10                ;VALUE OF DISTAT
         YPOS              ;VALUE OF CENTER
         YDAC              ;REGISTER TO  VARY
         BR S154           ;GET NEW VALUES
;SUBROUTINE TO READ ADC
ADREAD:  TSTB ADSTAT       ;AD READY?
         BPL ADREAD        ;NOT YET
         MOV ADBUF, R0     ;READ INTO R0
         RTS PC            ;AND EXIT


;SUBROUTINE TO DRAW A BAR
BAR:     MOV (R5)+, DISTAT      ;LOAD DISTAT FROM 1ST ARGUMENT
         MOV @(R5)+, R0    ;GET CENTER VALUE FROM SECOND ARGUMENT
         SUB #100, R0      ;100 ON EACH SIDE OF CENTER
         MOV #200, R2      ;NUMBER OF POINTS IN BAR
         MOV (R5)+, R1     ;ADDRESS OF REGISTER TO VARY (XDAC OR YDAC)
BARL:    INC (R1)          ;INCREMENT REGISTER
         SOB R2, BARL      ;GO BACK TILL DONE
         RTS R5            ;AND EXIT


;DO XBAR
         MOV #4, DISTAT    ;INTENS ON XLOAD
         MOV R1,


;15.5
;DISPLAY 2K AND HALT ON 'Q'
S155:    MOV #DAT155, R0   ;ADDRESS OF START OF 2K TO DISPLAY
         MOV #2048., R1    ;COUNTER
         MOV #-2, XDAC     ;X-INITIALIZATION
         MOV #4, DISTAT    ;INTENSIFY ON X-LOAD
A155:    MOV (R0)+, YDAC   ;LOAD Y
         ADD #2, XDAC      ;ADVANCE X
         SOB R1, A155      ;LOOP BACK
         TSTB TKS          ;WAS KBD STRUCK?
         BPL S155          ;NO
         MOV TKB, R0       ;YES
         BIC #-200, R0     ;CLEAR PARITY
         CMP #'Q, R0       ;WAS THIS Q?
         BNE S155          ;NO
         HALT              ;YES, HALT
DAT155:  0                 ;ARRAY STARTS HERE
```

```
;15.6
;CALCULATE Y+MX+B AND DISPLAY RESULT IN LED'S
.GLOBL FCO
        LEDBUF=ADBUF      ;DEFINE SYMBOL BOTH WAYS
.ASECT
.=1000
S146:   MOV #., SP
        JSR PC, CRLF
        MOV #TRAPH, @#34
        MOV #340, @#36
        MOV #"=B, R0              ;READY TO TYPE B=
        JSR PC, DTYPE            ;CALLED FROM PROBLEM 11.1
        FPSTR                    ;PUSH B
        MOV #"=M, R0             ;M=
        JSR PC, DTYPE
        FPMULT          ;MX
        FPADD           ;MX+B
        MOV #ASC146, -(SP)        ;PUSH START OF ASCII BUFFER
        MOV #7., -(SP)           ;F7.2
        MOV #2, -(SP)            ;2
        CLR -(SP)               ;P=0
        FPSTR                   ;PUSH FP #
        JSR PC, $FCO            ;CONVERT

;NOW LOAD LED WITH RESULT
        MOV #ASC146, R0 ;START OF ASCII FIELD
        CLR R1                  ;LED NUMBER
A146:   MOVB (R0)+, R2 ;GET EACH CHAR OF OUTPUT
        CMP R2, #'.     ;IF PERIOD, TURN ON DOT
        BEQ PER146
        CMPB R2, #'*    ;NUMBER TOO BIG?
        BNE B146        ;NO
        MOV #12, R2     ;TEST PATTERN CODE
B146:   BIC #360, R0    ;CLEAR OUT EXTRANEOUS BITS
        MOVB R1, LED+1  ;MEMORY LOCN WHER  BITS ARE ASSEMBLED
        MOVB R0, LED
        MOV LED, LEDBUF ;LOAD LED REGISTER
        INC R1          ;NEXT LED
        CMP R1, #6      ;ALL DONE?
        BNE A146        ;NO
        BR S146         ;YES, GET NEW VALUE

;PUT IN PERIOD
PER146: BISB #20, LED
        MOV LED, LEDBUF ;INTENSIFY DOT
        BR A146         ;GET NEXT NUMBER
LED:    0               ;STORE HERE
ASC146: 0,0,0,0         ;ASCII FIELD
```

```
;15.7

;DISPLAY REGION OF MEMORY DEFINED BY SWITCH REGISTER
A147:     MOV SR,R1          ;GET SIZE
          BIC #177400, R1 ;CLEAR EXTRANEOUS BITS
          ASH #8., R1        ;CONVERT TO COUNT
          MOV SR, R0         ;GET SA FROM SR
          BIC #377, R0       ;CLEAR OUT ANY SIZE BITS, SA REMAINS
          MOV #4, DISTAT     ;INTENS ON X-LOAD
;CALCULATE SHIFTED X-INCREMENT
          MOV #1, R3         ;DOUBLE PRECISION NUMBER 1 0 000 000 000 000 000
          CLR R4
          DIV R1, R3         ;SXINC=1;000000/SIZE
          MOV R3, R4         ;COPY SXINC
          NEG R4             ;FIRST X - ONE
B147:     MOV (R0)+, R5      ;GET EACH POINT
          ASH VSHIFT, R5  ;SHIFT
          ADD #4000, R5
          MOV R5, YDAC       ;LOAD
          ADD R3, R4         ;ADVANCE X
          MOV R4, R3         ;COPY BEFORE SHIFTING
          ASH #-4, R3        ;SHIFT INTO POSITION
          MOV R3, XDAC       ;AND LOAD
          SOB R1, B147       ;GO BACK TILL DONE
          BR A147            ;REREAD SR AND REFRESH

VSHIFT:  0          ;VERTICAL SHIFT FACTOR, USED IN MANY OF FOLLOWING PROBS.
```

Chapter 16

```
;16.1

;4096 POINTS IN 120 SECONDS = 292 MSEC/POINT
          MOV #., SP            ;INIT SP
          MOV #AD151, @#300     ;INIT ADC INTERRUPT VECTOR
          MOV #300, @#302       ;AND BR
          MOV #-292.,CLKBUF     ;SET CLOCK COUNTER
          MOV #DAT151, R4       ;ADC POINTER
          MOV #4096., R5        ;ADC COUNTER
          MOV #140, ADSTAT      ;CLOCK STARTS ADC, INTERRUPT WHEN DONE
          MOV #411, CLSTAT      ;START CLOCK GOING AT 1 MSEC TICKS
A151:     MOV #-1, XDAC         ;INIT XDAC
          MOV #DAT151, R0       ;SET DISPLAY POINTER
          MOV #4096., R1        ;AND COUNTER
B151:     MOV (R0)+, R2         ;GET EACH POINT
          ASH VSHIFT, R2     ;SHIFT
          ADD #4000, R2        ;BIAS
          MOV R2, YDAC         ;LOAD Y
          INC XDAC             ;ADVANCE X
          SOB R1, B151         ;GO BACK
          BR A151              ;RESET COUNTER AND POINTER

;ADC INTERRUPT ROUTINE
AD151:    MOV ADBUF, (R4)+     ;PUT READING INTO MEMORY
          DEC R5               ;DECREASE COUNTER
          BNE X151             ;DONE?
          CLR ADSTAT           ;YES TURN OFF ADC AND CLOCK
          CLR CLSTAT
X151:     RTI                  ;AND RETURN FROM INTERRUPT
DAT151:   0                ;DATA STARTS HERE
```

```
;16.2
;TO MODIFY RATE, ASSUME RANGE OF .4096 SEC TO 134213 SEC=37 HOURS
A152:    JSR PC, CRLF              ;INIT CARRIAGE
         JSR PC, FLIP              ;GET ENTRY
         JSR PC, CRLF              ;TYPE CR
         FCMP+RELM                 ;IS IT <.4096?
         F4096-.
         BLT A152                  ;YES, GET ANOTHER
         FCMP+RELM
         F13421-.                  ;IS IT >134213?
         BGT A152                  ;YES, GET ANOTHER
         FPMULT+IMM                ;CONVERT TO MSEC
         .FLT2 1000.
         FPSTR                     ;PUSH FOR FIXING
         FPDIV+IMM
         .FLT2 4096000    ;CONVERT TO TIME PER POINT IN SECONDS
         FPSTR+RELM
         TIMPPT-.                  ;STORE FOR FUTURE REFERENCE
         JSR R4, $POLSH            ;FIX
         $RI
         .+2                       ;EXIT
         NEG (SP)                  ;NEGATE FIXED RESULT
         MOV (SP)+, CLKBUF         ;AND LOAD CLOCK COUNTER
;          :
;          :
F4096:   .FLT2 .4096
F13421:  .FLT2 134213


;16.3, 4
;TO SCALE DOWN, SEE  OUR TEST EXAMPLE IN CHAPTER
;TO EXPAND AND CONTRACT WITH C AND X
;SEE THE FTCALL ROUTINE IN CHAPTER 20

;TO START ON SCHMITT TRIGGER:
         MOV #20411, CLSTAT       ;ST1 WILL START CLOCK
         MOV#140, CLSTAY ;CLOCK WILL START ADC

         MOV #ADSERV, @#300
         MOV #340, @#302          ;NO INTERRUPTS DURING ADC SERVICE

;THEN THE ADC SERVICE ROUTINE MUST TURN OFF
;THE ST1 START BIT IN CLSTAT TO PREVENT RESTARTS
;DURING AVERAGING
ADSERV: BIC #20000, CLSTAT       ;NO MORE ST1 STARTS DURING SCAN
```

Chapter 18

```
;18.1

;PRINT OUT PEAKS
.GLOBL PEASET,PEAPCK,CENTER,PFIRST,PLAST
            JSR PC,CRLF
            JSR R5,PEASET            ;SET UP BLOCK TO PEAK PICK
            DAT171                   ;STARTING ADDRESS
            SIZ171                   ;SIZE ADDRESS
A171:       JSR PC, PEAPCK           ;PICK NEXT PEAK
            BEQ EX171                ;EXIT WHEN DONE
            MOV CENTER, -(SP)        ;PUSH CENTER ADDRESS
            SUB DAT171, (SP)         ;SUBTRACT SA
            ASR (SP)                 ;DIVIDE BY 2 BECAUSE OF BYTE ADDRESSING
            JSR R4, $POLSH           ;NOW FLOAT IT
            $IR
            .+2                      ;EXIT
            FPGET                    ;POP INTO FLAC
            FPMULT+RELM              ;MULT BY TIME PER POINT
            TIMPPT-.
            JSR PC, FLOP             ;PRINT OUT TIME
            JSR PC, CRLF             ;CR
            JSR PC, DISPLA           ;DISPLAY 4K UNTIL CR STRUCK
;NOW INTENSIFY PEAK 3 TIMES
            MOV #3, R4
B171:       MOV PFIRST, R0
            MOV R0, R2               ;COPY
            SUB DAT171, R2           ;CALC DIFFERENCE
            ASR R2                   ;DIVIDE BY 2 FOR BYTE ADDR.
            DEC R2                   ;PREVIOUS XDAC SETTING
C171:       MOV (R0)+, R1            ;GET EACH POINT
            ASH VSHIFT, R1           ;SHIFT BEFORE DISPLAYING
            ADD #4000, R1            ;BIAS
            MOV R1, YDAC             ;LOAD Y
            INC XDAC                 ;ADVANCE X
            CMP R0, PLAST            ;DONE?
            BLE C171                 ;NO
            SOB R3, B171             ;INTENSIFY REGION 3 TIMES
            BR A171
EX171:      JSR PC, DISPLA           ;NOW JUST DISPLAY 4K W/O STOPPING
            BR EX171
;SUBROUTINE TO DISPLAY 4K UNTIL A CR IS STRUCK
DISPLA: MOV DAT171, R0               ;POINTER
            MOV SIZ171, R1           ;COUNTER
            MOV #-1, XDAC            ;INIT X
            MOV #4, DISTAT
DIS171: MOV (R0)+, R2                ;GET EACH POINT
            ASH VSHIFT, R2           ;SHIFT BY VSHIFT
            ADD #4000, R2
            MOV R2, YDAC
            INC XDAC                 ;LOAD X AND Y
            SOB R1, DIS171
            TSTB TKS
            BPL DISPLA               ;GO BACK UNLESS KBD STRUCK
            MOV TKB, R2
            BIC #-200, R2            ;CLEAR PARITY
            CMP R2, #15              ;WAS IT A CR?
            BNE DISPLA               ;NO, KEEP DISPLAYING
            JSR PC, CRLF             ;YES TYPE CRLF
            RTS PC                   ;AND EXIT

DAT171: 20000                       ;SA OF DATA
SIZ171: 4096.                       ;SIZE OF BLOCK
```

```
;18.3
;INTEGRATION BY 3 METHODS

;RETANGULAR INTEGRATION
        MOV PFIRST, R0          ;GET FIRST POINT
        CLR FLAC                ;ZERO FLAC
        CLR FLAC+2
A173:   MOV (R0)+, -(SP)        ;PUSH EACH VALUE
        JSR R4, $POLSH          ;FLOAT IT
        $IR
        .+2                     ;EXIT
        FPADD                   ;ADD INTO FLAC
        CMP R0, PLAST           ;DONE?
        BLE A173                ;NO
        FPSTR+RELM
        RECT-.                  ;YES, STORE RESULT

        CLR ALTFLG              ;CLR ALTERNATE POINT FLAG
        JSR PC, TRAPNT          ;TRAPEZOIDAL INTEGRATION
        FPSTR+RELM
        TRAPEZ-.                ;STORE INTEGRAL
        INC ALTFLG              ;SET ALTERNATE POINT FLAG
        JSR PC, TRAPNT          ;TRAPEZOIDAL INTEGRAL OF ALTERNATE POINT

;ROMBERG INTEGRATION
;CALCULATE IR = IN + (IN-I2N)/3
        FPSUB +RELM
        TRAPEZ-.                ;I2N-IN
        ADD #100000, FLAC       ;NEGATE =IN-I2N
        FPDIV+IMM
        .FLT2 3.0               ;(IN-I2N)/3
        FPADD + RELM
        TRAPEZ-.                ;IR=IN+(IN-I2N)/3
;NOW PRINT OUT 3 RESULTS
        JSR PC, FLOP            ;ROMBERG
        JSR PC, CRLF
        FPGET+RELM
        TRAPEZ-.
        JSR PC, FLOP            ;TRAPEZOIDAL
        JSR PC, CRLF
        FPGET+RELM
        RECT-.
        JSR PC, FLOP            ;RECTANGULAR
        JSR PC, CRLF
;       :
;       :
```

```
;TRAPEZOIDAL INTEGRATION SUBROUTINE
;TAKES EVERY POINT IF ALTFLG=0
;ALTERNATE POINTS OF ALTFLG=1
TRAPNT: CLR FLAC
        CLR FLAC+2              ;FLAC=0
        MOV PFIRST, R0         ;SET POINTER
B173:   MOV (R0)+, -(SP)       ;GET EACH POINT
        CMP R0, PFIRST         ;FIRST POINT?
        BNE C173               ;NO
        ASR (SP)               ;YES, DIVIDE BY 2
C173:   CMP R0, PLAST          ;LAST POINT?
        BNE D173               ;NO
        ASR (SP)               ;YES, DIVIDE BY 2
D173:   JSR R4, $POLSH         ;FLOAT
        $IR
        .+2                    ;EXIT
        FPADD                  ;POP AND ADD
        TST ALTFLG             ;WAS THIS ALTERNATE POINTS
        BEQ E173               ;NO
        TST (R0)+              ;YES, ADVANCE POINTER AGAIN
E173:   CMP R0, PLAST
        BLE B173               ;STILL MORE POINTS
        RTS PC                 ;EXIT


;CONSTANTS
RECT:   0,0                    ;SAVE 2 SPACES FOR FLTG NUMBERS
TRAPEZ: 0,0
ALTFLG: 0


;18.4

.GLOBL PLOT, DELAY, VSA, VSIZE,YMEM
;TO PLOT A PORTION OF A DISPLAY BUFFER INTO MEMORY
;MODIFY THE PLOT ROUTINE SO THAT ALL DELAYS ARE ELIMINATED BY
;PUTTING AN RTS PC AT LOCATION DELAY:
        MOV #207, DELAY

;THEN ADD INSTRUCTIONS TO PUT THE POINT IN MEMORY EACH TIME
;ADD
        MOV YMEM, @POINT2
;AT PLT37: AND PLT57:
;AND ADD
        ADD #2, POINT2
;IN THE STEPS: SUBROUTINE
;THEN
        MOV DISBUF, POINT2     ;INITIALIZE POINT2
        JSR R3, PLOT           ;CALL PLOT
        VSA                    ;SA FROM VDIS
        VSIZE                  ;SIZE FROM VDIS
        4096.                  ;NUMBER OF POINTS IN DISPLAY BUFFER

;THEN DISPLAY THE RESULT, WITHOUT ANY BIASING OR SHIFTING AT DISBUF
A174:   MOV DISBUF, R0
        MOV #4096., R1         ;COUNTER
        MOV #-1, XDAC
B174:   MOV (R0)+, YDAC
        INC XDAC
        SOB R1, B174
        BR A174

POINT2: 0
DISBUF: 40000

;GOOD LUCK WITH YOUR PROGRAMS!
```

□□□□□□□□□ **APPENDIX V** □

DECIMAL–OCTAL
CONVERSION TABLE

	0	1	2	3	4	5	6	7	8	9
0	0	1	2	3	4	5	6	7	10	11
10	12	13	14	15	16	17	20	21	22	23
20	24	25	26	27	30	31	32	33	34	35
30	36	37	40	41	42	43	44	45	46	47
40	50	51	52	53	54	55	56	57	60	61
50	62	63	64	65	66	67	70	71	72	73
60	74	75	76	77	100	101	102	103	104	105
70	106	107	110	111	112	113	114	115	116	117
80	120	121	122	123	124	125	126	127	130	131
90	132	133	134	135	136	137	140	141	142	143
100	144	145	146	147	150	151	152	153	154	155
110	156	157	160	161	162	163	164	165	166	167
120	170	171	172	173	174	175	176	177	200	201
130	202	203	204	205	206	207	210	211	212	213
140	214	215	216	217	220	221	222	223	224	225
150	226	227	230	231	232	233	234	235	236	237
160	240	241	242	243	244	245	246	247	250	251
170	252	253	254	255	256	257	260	261	262	263
180	264	265	266	267	270	271	272	273	274	275
190	276	277	300	301	302	303	304	305	306	307
200	310	311	312	313	314	315	316	317	320	321
210	322	323	324	325	326	327	330	331	332	333
220	334	335	336	337	340	341	342	343	344	345
230	346	347	350	351	352	353	354	355	356	357
240	360	361	362	363	364	365	366	367	370	371
250	372	373	374	375	376	377	400	401	402	403
260	404	405	406	407	410	411	412	413	414	415
270	416	417	420	421	422	423	424	425	426	427
280	430	431	432	433	434	435	436	437	440	441
290	442	443	444	445	446	447	450	451	452	453
300	454	455	456	457	460	461	462	463	464	465
310	466	467	470	471	472	473	474	475	476	477
320	500	501	502	503	504	505	506	507	510	511
330	512	513	514	515	516	517	520	521	522	523
340	524	525	526	527	530	531	532	533	534	535
350	536	537	540	541	542	543	544	545	546	547
360	550	551	552	553	554	555	556	557	560	561
370	562	563	564	565	566	567	570	571	572	573
380	574	575	576	577	600	601	602	603	604	605
390	606	607	610	611	612	613	614	615	616	617
400	620	621	622	623	624	625	626	627	630	631
410	632	633	634	635	636	637	640	641	642	643
420	644	645	646	647	650	651	652	653	654	655
430	656	657	660	661	662	663	664	665	666	667
440	670	671	672	673	674	675	676	677	700	701
450	702	703	704	705	706	707	710	711	712	713
460	714	715	716	717	720	721	722	723	724	725
470	726	727	730	731	732	733	734	735	736	737
480	740	741	742	743	744	745	746	747	750	751
490	752	753	754	755	756	757	760	761	762	763

	0	1	2	3	4	5	6	7	8	9
500	764	765	766	767	770	771	772	773	774	775
510	776	777	1000	1001	1002	1003	1004	1005	1006	1007
520	1010	1011	1012	1013	1014	1015	1016	1017	1020	1021
530	1022	1023	1024	1025	1026	1027	1030	1031	1032	1033
540	1034	1035	1036	1037	1040	1041	1042	1043	1044	1045
550	1046	1047	1050	1051	1052	1053	1054	1055	1056	1057
560	1060	1061	1062	1063	1064	1065	1066	1067	1070	1071
570	1072	1073	1074	1075	1076	1077	1100	1101	1102	1103
580	1104	1105	1106	1107	1110	1111	1112	1113	1114	1115
590	1116	1117	1120	1121	1122	1123	1124	1125	1126	1127
600	1130	1131	1132	1133	1134	1135	1136	1137	1140	1141
610	1142	1143	1144	1145	1146	1147	1150	1151	1152	1153
620	1154	1155	1156	1157	1160	1161	1162	1163	1164	1165
630	1166	1167	1170	1171	1172	1173	1174	1175	1176	1177
640	1200	1201	1202	1203	1204	1205	1206	1207	1210	1211
650	1212	1213	1214	1215	1216	1217	1220	1221	1222	1223
660	1224	1225	1226	1227	1230	1231	1232	1233	1234	1235
670	1236	1237	1240	1241	1242	1243	1244	1245	1246	1247
680	1250	1251	1252	1253	1254	1255	1256	1257	1260	1261
690	1262	1263	1264	1265	1266	1267	1270	1271	1272	1273
700	1274	1275	1276	1277	1300	1301	1302	1303	1304	1305
710	1306	1307	1310	1311	1312	1313	1314	1315	1316	1317
720	1320	1321	1322	1323	1324	1325	1326	1327	1330	1331
730	1332	1333	1334	1335	1336	1337	1340	1341	1342	1343
740	1344	1345	1346	1347	1350	1351	1352	1353	1354	1355
750	1356	1357	1360	1361	1362	1363	1364	1365	1366	1367
760	1370	1371	1372	1373	1374	1375	1376	1377	1400	1401
770	1402	1403	1404	1405	1406	1407	1410	1411	1412	1413
780	1414	1415	1416	1417	1420	1421	1422	1423	1424	1425
790	1426	1427	1430	1431	1432	1433	1434	1435	1436	1437
800	1440	1441	1442	1443	1444	1445	1446	1447	1450	1451
810	1452	1453	1454	1455	1456	1457	1460	1461	1462	1463
820	1464	1465	1466	1467	1470	1471	1472	1473	1474	1475
830	1476	1477	1500	1501	1502	1503	1504	1505	1506	1507
840	1510	1511	1512	1513	1514	1515	1516	1517	1520	1521
850	1522	1523	1524	1525	1526	1527	1530	1531	1532	1533
860	1534	1535	1536	1537	1540	1541	1542	1543	1544	1545
870	1546	1547	1550	1551	1552	1553	1554	1555	1556	1557
880	1560	1561	1562	1563	1564	1565	1566	1567	1570	1571
890	1572	1573	1574	1575	1576	1577	1600	1601	1602	1603
900	1604	1605	1606	1607	1610	1611	1612	1613	1614	1615
910	1616	1617	1620	1621	1622	1623	1624	1625	1626	1627
920	1630	1631	1632	1633	1634	1635	1636	1637	1640	1641
930	1642	1643	1644	1645	1646	1647	1650	1651	1652	1653
940	1654	1655	1656	1657	1660	1661	1662	1663	1664	1665
950	1666	1667	1670	1671	1672	1673	1674	1675	1676	1677
960	1700	1701	1702	1703	1704	1705	1706	1707	1710	1711
970	1712	1713	1714	1715	1716	1717	1720	1721	1722	1723
980	1724	1725	1726	1727	1730	1731	1732	1733	1734	1735
990	1736	1737	1740	1741	1742	1743	1744	1745	1746	1747

	0	1	2	3	4	5	6	7	8	9
1000	1750	1751	1752	1753	1754	1755	1756	1757	1760	1761
1010	1762	1763	1764	1765	1766	1767	1770	1771	1772	1773
1020	1774	1775	1776	1777	2000	2001	2002	2003	2004	2005
1030	2006	2007	2010	2011	2012	2013	2014	2015	2016	2017
1040	2020	2021	2022	2023	2024	2025	2026	2027	2030	2031
1050	2032	2033	2034	2035	2036	2037	2040	2041	2042	2043
1060	2044	2045	2046	2047	2050	2051	2052	2053	2054	2055
1070	2056	2057	2060	2061	2062	2063	2064	2065	2066	2067
1080	2070	2071	2072	2073	2074	2075	2076	2077	2100	2101
1090	2102	2103	2104	2105	2106	2107	2110	2111	2112	2113
1100	2114	2115	2116	2117	2120	2121	2122	2123	2124	2125
1110	2126	2127	2130	2131	2132	2133	2134	2135	2136	2137
1120	2140	2141	2142	2143	2144	2145	2146	2147	2150	2151
1130	2152	2153	2154	2155	2156	2157	2160	2161	2162	2163
1140	2164	2165	2166	2167	2170	2171	2172	2173	2174	2175
1150	2176	2177	2200	2201	2202	2203	2204	2205	2206	2207
1160	2210	2211	2212	2213	2214	2215	2216	2217	2220	2221
1170	2222	2223	2224	2225	2226	2227	2230	2231	2232	2233
1180	2234	2235	2236	2237	2240	2241	2242	2243	2244	2245
1190	2246	2247	2250	2251	2252	2253	2254	2255	2256	2257
1200	2260	2261	2262	2263	2264	2265	2266	2267	2270	2271
1210	2272	2273	2274	2275	2276	2277	2300	2301	2302	2303
1220	2304	2305	2306	2307	2310	2311	2312	2313	2314	2315
1230	2316	2317	2320	2321	2322	2323	2324	2325	2326	2327
1240	2330	2331	2332	2333	2334	2335	2336	2337	2340	2341
1250	2342	2343	2344	2345	2346	2347	2350	2351	2352	2353
1260	2354	2355	2356	2357	2360	2361	2362	2363	2364	2365
1270	2366	2367	2370	2371	2372	2373	2374	2375	2376	2377
1280	2400	2401	2402	2403	2404	2405	2406	2407	2410	2411
1290	2412	2413	2414	2415	2416	2417	2420	2421	2422	2423
1300	2424	2425	2426	2427	2430	2431	2432	2433	2434	2435
1310	2436	2437	2440	2441	2442	2443	2444	2445	2446	2447
1320	2450	2451	2452	2453	2454	2455	2456	2457	2460	2461
1330	2462	2463	2464	2465	2466	2467	2470	2471	2472	2473
1340	2474	2475	2476	2477	2500	2501	2502	2503	2504	2505
1350	2506	2507	2510	2511	2512	2513	2514	2515	2516	2517
1360	2520	2521	2522	2523	2524	2525	2526	2527	2530	2531
1370	2532	2533	2534	2535	2536	2537	2540	2541	2542	2543
1380	2544	2545	2546	2547	2550	2551	2552	2553	2554	2555
1390	2556	2557	2560	2561	2562	2563	2564	2565	2566	2567
1400	2570	2571	2572	2573	2574	2575	2576	2577	2600	2601
1410	2602	2603	2604	2605	2606	2607	2610	2611	2612	2613
1420	2614	2615	2616	2617	2620	2621	2622	2623	2624	2625
1430	2626	2627	2630	2631	2632	2633	2634	2635	2636	2637
1440	2640	2641	2642	2643	2644	2645	2646	2647	2650	2651
1450	2652	2653	2654	2655	2656	2657	2660	2661	2662	2663
1460	2664	2665	2666	2667	2670	2671	2672	2673	2674	2675
1470	2676	2677	2700	2701	2702	2703	2704	2705	2706	2707
1480	2710	2711	2712	2713	2714	2715	2716	2717	2720	2721
1490	2722	2723	2724	2725	2726	2727	2730	2731	2732	2733

	0	1	2	3	4	5	6	7	8	9
1500	2734	2735	2736	2737	2740	2741	2742	2743	2744	2745
1510	2746	2747	2750	2751	2752	2753	2754	2755	2756	2757
1520	2760	2761	2762	2763	2764	2765	2766	2767	2770	2771
1530	2772	2773	2774	2775	2776	2777	3000	3001	3002	3003
1540	3004	3005	3006	3007	3010	3011	3012	3013	3014	3015
1550	3016	3017	3020	3021	3022	3023	3024	3025	3026	3027
1560	3030	3031	3032	3033	3034	3035	3036	3037	3040	3041
1570	3042	3043	3044	3045	3046	3047	3050	3051	3052	3053
1580	3054	3055	3056	3057	3060	3061	3062	3063	3064	3065
1590	3066	3067	3070	3071	3072	3073	3074	3075	3076	3077
1600	3100	3101	3102	3103	3104	3105	3106	3107	3110	3111
1610	3112	3113	3114	3115	3116	3117	3120	3121	3122	3123
1620	3124	3125	3126	3127	3130	3131	3132	3133	3134	3135
1630	3136	3137	3140	3141	3142	3143	3144	3145	3146	3147
1640	3150	3151	3152	3153	3154	3155	3156	3157	3160	3161
1650	3162	3163	3164	3165	3166	3167	3170	3171	3172	3173
1660	3174	3175	3176	3177	3200	3201	3202	3203	3204	3205
1670	3206	3207	3210	3211	3212	3213	3214	3215	3216	3217
1680	3220	3221	3222	3223	3224	3225	3226	3227	3230	3231
1690	3232	3233	3234	3235	3236	3237	3240	3241	3242	3243
1700	3244	3245	3246	3247	3250	3251	3252	3253	3254	3255
1710	3256	3257	3260	3261	3262	3263	3264	3265	3266	3267
1720	3270	3271	3272	3273	3274	3275	3276	3277	3300	3301
1730	3302	3303	3304	3305	3306	3307	3310	3311	3312	3313
1740	3314	3315	3316	3317	3320	3321	3322	3323	3324	3325
1750	3326	3327	3330	3331	3332	3333	3334	3335	3336	3337
1760	3340	3341	3342	3343	3344	3345	3346	3347	3350	3351
1770	3352	3353	3354	3355	3356	3357	3360	3361	3362	3363
1780	3364	3365	3366	3367	3370	3371	3372	3373	3374	3375
1790	3376	3377	3400	3401	3402	3403	3404	3405	3406	3407
1800	3410	3411	3412	3413	3414	3415	3416	3417	3420	3421
1810	3422	3423	3424	3425	3426	3427	3430	3431	3432	3433
1820	3434	3435	3436	3437	3440	3441	3442	3443	3444	3445
1830	3446	3447	3450	3451	3452	3453	3454	3455	3456	3457
1840	3460	3461	3462	3463	3464	3465	3466	3467	3470	3471
1850	3472	3473	3474	3475	3476	3477	3500	3501	3502	3503
1860	3504	3505	3506	3507	3510	3511	3512	3513	3514	3515
1870	3516	3517	3520	3521	3522	3523	3524	3525	3526	3527
1880	3530	3531	3532	3533	3534	3535	3536	3537	3540	3541
1890	3542	3543	3544	3545	3546	3547	3550	3551	3552	3553
1900	3554	3555	3556	3557	3560	3561	3562	3563	3564	3565
1910	3566	3567	3570	3571	3572	3573	3574	3575	3576	3577
1920	3600	3601	3602	3603	3604	3605	3606	3607	3610	3611
1930	3612	3613	3614	3615	3616	3617	3620	3621	3622	3623
1940	3624	3625	3626	3627	3630	3631	3632	3633	3634	3635
1950	3636	3637	3640	3641	3642	3643	3644	3645	3646	3647
1960	3650	3651	3652	3653	3654	3655	3656	3657	3660	3661
1970	3662	3663	3664	3665	3666	3667	3670	3671	3672	3673
1980	3674	3675	3676	3677	3700	3701	3702	3703	3704	3705
1990	3706	3707	3710	3711	3712	3713	3714	3715	3716	3717

	0	1	2	3	4	5	6	7	8	9
2000	3720	3721	3722	3723	3724	3725	3726	3727	3730	3731
2010	3732	3733	3734	3735	3736	3737	3740	3741	3742	3743
2020	3744	3745	3746	3747	3750	3751	3752	3753	3754	3755
2030	3756	3757	3760	3761	3762	3763	3764	3765	3766	3767
2040	3770	3771	3772	3773	3774	3775	3776	3777	4000	4001
2050	4002	4003	4004	4005	4006	4007	4010	4011	4012	4013
2060	4014	4015	4016	4017	4020	4021	4022	4023	4024	4025
2070	4026	4027	4030	4031	4032	4033	4034	4035	4036	4037
2080	4040	4041	4042	4043	4044	4045	4046	4047	4050	4051
2090	4052	4053	4054	4055	4056	4057	4060	4061	4062	4063
2100	4064	4065	4066	4067	4070	4071	4072	4073	4074	4075
2110	4076	4077	4100	4101	4102	4103	4104	4105	4106	4107
2120	4110	4111	4112	4113	4114	4115	4116	4117	4120	4121
2130	4122	4123	4124	4125	4126	4127	4130	4131	4132	4133
2140	4134	4135	4136	4137	4140	4141	4142	4143	4144	4145
2150	4146	4147	4150	4151	4152	4153	4154	4155	4156	4157
2160	4160	4161	4162	4163	4164	4165	4166	4167	4170	4171
2170	4172	4173	4174	4175	4176	4177	4200	4201	4202	4203
2180	4204	4205	4206	4207	4210	4211	4212	4213	4214	4215
2190	4216	4217	4220	4221	4222	4223	4224	4225	4226	4227
2200	4230	4231	4232	4233	4234	4235	4236	4237	4240	4241
2210	4242	4243	4244	4245	4246	4247	4250	4251	4252	4253
2220	4254	4255	4256	4257	4260	4261	4262	4263	4264	4265
2230	4266	4267	4270	4271	4272	4273	4274	4275	4276	4277
2240	4300	4301	4302	4303	4304	4305	4306	4307	4310	4311
2250	4312	4313	4314	4315	4316	4317	4320	4321	4322	4323
2260	4324	4325	4326	4327	4330	4331	4332	4333	4334	4335
2270	4336	4337	4340	4341	4342	4343	4344	4345	4346	4347
2280	4350	4351	4352	4353	4354	4355	4356	4357	4360	4361
2290	4362	4363	4364	4365	4366	4367	4370	4371	4372	4373
2300	4374	4375	4376	4377	4400	4401	4402	4403	4404	4405
2310	4406	4407	4410	4411	4412	4413	4414	4415	4416	4417
2320	4420	4421	4422	4423	4424	4425	4426	4427	4430	4431
2330	4432	4433	4434	4435	4436	4437	4440	4441	4442	4443
2340	4444	4445	4446	4447	4450	4451	4452	4453	4454	4455
2350	4456	4457	4460	4461	4462	4463	4464	4465	4466	4467
2360	4470	4471	4472	4473	4474	4475	4476	4477	4500	4501
2370	4502	4503	4504	4505	4506	4507	4510	4511	4512	4513
2380	4514	4515	4516	4517	4520	4521	4522	4523	4524	4525
2390	4526	4527	4530	4531	4532	4533	4534	4535	4536	4537
2400	4540	4541	4542	4543	4544	4545	4546	4547	4550	4551
2410	4552	4553	4554	4555	4556	4557	4560	4561	4562	4563
2420	4564	4565	4566	4567	4570	4571	4572	4573	4574	4575
2430	4576	4577	4600	4601	4602	4603	4604	4605	4606	4607
2440	4610	4611	4612	4613	4614	4615	4616	4617	4620	4621
2450	4622	4623	4624	4625	4626	4627	4630	4631	4632	4633
2460	4634	4635	4636	4637	4640	4641	4642	4643	4644	4645
2470	4646	4647	4650	4651	4652	4653	4654	4655	4656	4657
2480	4660	4661	4662	4663	4664	4665	4666	4667	4670	4671
2490	4672	4673	4674	4675	4676	4677	4700	4701	4702	4703

	0	1	2	3	4	5	6	7	8	9
2500	4704	4705	4706	4707	4710	4711	4712	4713	4714	4715
2510	4716	4717	4720	4721	4722	4723	4724	4725	4726	4727
2520	4730	4731	4732	4733	4734	4735	4736	4737	4740	4741
2530	4742	4743	4744	4745	4746	4747	4750	4751	4752	4753
2540	4754	4755	4756	4757	4760	4761	4762	4763	4764	4765
2550	4766	4767	4770	4771	4772	4773	4774	4775	4776	4777
2560	5000	5001	5002	5003	5004	5005	5006	5007	5010	5011
2570	5012	5013	5014	5015	5016	5017	5020	5021	5022	5023
2580	5024	5025	5026	5027	5030	5031	5032	5033	5034	5035
2590	5036	5037	5040	5041	5042	5043	5044	5045	5046	5047
2600	5050	5051	5052	5053	5054	5055	5056	5057	5060	5061
2610	5062	5063	5064	5065	5066	5067	5070	5071	5072	5073
2620	5074	5075	5076	5077	5100	5101	5102	5103	5104	5105
2630	5106	5107	5110	5111	5112	5113	5114	5115	5116	5117
2640	5120	5121	5122	5123	5124	5125	5126	5127	5130	5131
2650	5132	5133	5134	5135	5136	5137	5140	5141	5142	5143
2660	5144	5145	5146	5147	5150	5151	5152	5153	5154	5155
2670	5156	5157	5160	5161	5162	5163	5164	5165	5166	5167
2680	5170	5171	5172	5173	5174	5175	5176	5177	5200	5201
2690	5202	5203	5204	5205	5206	5207	5210	5211	5212	5213
2700	5214	5215	5216	5217	5220	5221	5222	5223	5224	5225
2710	5226	5227	5230	5231	5232	5233	5234	5235	5236	5237
2720	5240	5241	5242	5243	5244	5245	5246	5247	5250	5251
2730	5252	5253	5254	5255	5256	5257	5260	5261	5262	5263
2740	5264	5265	5266	5267	5270	5271	5272	5273	5274	5275
2750	5276	5277	5300	5301	5302	5303	5304	5305	5306	5307
2760	5310	5311	5312	5313	5314	5315	5316	5317	5320	5321
2770	5322	5323	5324	5325	5326	5327	5330	5331	5332	5333
2780	5334	5335	5336	5337	5340	5341	5342	5343	5344	5345
2790	5346	5347	5350	5351	5352	5353	5354	5355	5356	5357
2800	5360	5361	5362	5363	5364	5365	5366	5367	5370	5371
2810	5372	5373	5374	5375	5376	5377	5400	5401	5402	5403
2820	5404	5405	5406	5407	5410	5411	5412	5413	5414	5415
2830	5416	5417	5420	5421	5422	5423	5424	5425	5426	5427
2840	5430	5431	5432	5433	5434	5435	5436	5437	5440	5441
2850	5442	5443	5444	5445	5446	5447	5450	5451	5452	5453
2860	5454	5455	5456	5457	5460	5461	5462	5463	5464	5465
2870	5466	5467	5470	5471	5472	5473	5474	5475	5476	5477
2880	5500	5501	5502	5503	5504	5505	5506	5507	5510	5511
2890	5512	5513	5514	5515	5516	5517	5520	5521	5522	5523
2900	5524	5525	5526	5527	5530	5531	5532	5533	5534	5535
2910	5536	5537	5540	5541	5542	5543	5544	5545	5546	5547
2920	5550	5551	5552	5553	5554	5555	5556	5557	5560	5561
2930	5562	5563	5564	5565	5566	5567	5570	5571	5572	5573
2940	5574	5575	5576	5577	5600	5601	5602	5603	5604	5605
2950	5606	5607	5610	5611	5612	5613	5614	5615	5616	5617
2960	5620	5621	5622	5623	5624	5625	5626	5627	5630	5631
2970	5632	5633	5634	5635	5636	5637	5640	5641	5642	5643
2980	5644	5645	5646	5647	5650	5651	5652	5653	5654	5655
2990	5656	5657	5660	5661	5662	5663	5664	5665	5666	5667

	0	1	2	3	4	5	6	7	8	9
3000	5670	5671	5672	5673	5674	5675	5676	5677	5700	5701
3010	5702	5703	5704	5705	5706	5707	5710	5711	5712	5713
3020	5714	5715	5716	5717	5720	5721	5722	5723	5724	5725
3030	5726	5727	5730	5731	5732	5733	5734	5735	5736	5737
3040	5740	5741	5742	5743	5744	5745	5746	5747	5750	5751
3050	5752	5753	5754	5755	5756	5757	5760	5761	5762	5763
3060	5764	5765	5766	5767	5770	5771	5772	5773	5774	5775
3070	5776	5777	6000	6001	6002	6003	6004	6005	6006	6007
3080	6010	6011	6012	6013	6014	6015	6016	6017	6020	6021
3090	6022	6023	6024	6025	6026	6027	6030	6031	6032	6033
3100	6034	6035	6036	6037	6040	6041	6042	6043	6044	6045
3110	6046	6047	6050	6051	6052	6053	6054	6055	6056	6057
3120	6060	6061	6062	6063	6064	6065	6066	6067	6070	6071
3130	6072	6073	6074	6075	6076	6077	6100	6101	6102	6103
3140	6104	6105	6106	6107	6110	6111	6112	6113	6114	6115
3150	6116	6117	6120	6121	6122	6123	6124	6125	6126	6127
3160	6130	6131	6132	6133	6134	6135	6136	6137	6140	6141
3170	6142	6143	6144	6145	6146	6147	6150	6151	6152	6153
3180	6154	6155	6156	6157	6160	6161	6162	6163	6164	6165
3190	6166	6167	6170	6171	6172	6173	6174	6175	6176	6177
3200	6200	6201	6202	6203	6204	6205	6206	6207	6210	6211
3210	6212	6213	6214	6215	6216	6217	6220	6221	6222	6223
3220	6224	6225	6226	6227	6230	6231	6232	6233	6234	6235
3230	6236	6237	6240	6241	6242	6243	6244	6245	6246	6247
3240	6250	6251	6252	6253	6254	6255	6256	6257	6260	6261
3250	6262	6263	6264	6265	6266	6267	6270	6271	6272	6273
3260	6274	6275	6276	6277	6300	6301	6302	6303	6304	6305
3270	6306	6307	6310	6311	6312	6313	6314	6315	6316	6317
3280	6320	6321	6322	6323	6324	6325	6326	6327	6330	6331
3290	6332	6333	6334	6335	6336	6337	6340	6341	6342	6343
3300	6344	6345	6346	6347	6350	6351	6352	6353	6354	6355
3310	6356	6357	6360	6361	6362	6363	6364	6365	6366	6367
3320	6370	6371	6372	6373	6374	6375	6376	6377	6400	6401
3330	6402	6403	6404	6405	6406	6407	6410	6411	6412	6413
3340	6414	6415	6416	6417	6420	6421	6422	6423	6424	6425
3350	6426	6427	6430	6431	6432	6433	6434	6435	6436	6437
3360	6440	6441	6442	6443	6444	6445	6446	6447	6450	6451
3370	6452	6453	6454	6455	6456	6457	6460	6461	6462	6463
3380	6464	6465	6466	6467	6470	6471	6472	6473	6474	6475
3390	6476	6477	6500	6501	6502	6503	6504	6505	6506	6507
3400	6510	6511	6512	6513	6514	6515	6516	6517	6520	6521
3410	6522	6523	6524	6525	6526	6527	6530	6531	6532	6533
3420	6534	6535	6536	6537	6540	6541	6542	6543	6544	6545
3430	6546	6547	6550	6551	6552	6553	6554	6555	6556	6557
3440	6560	6561	6562	6563	6564	6565	6566	6567	6570	6571
3450	6572	6573	6574	6575	6576	6577	6600	6601	6602	6603
3460	6604	6605	6606	6607	6610	6611	6612	6613	6614	6615
3470	6616	6617	6620	6621	6622	6623	6624	6625	6626	6627
3480	6630	6631	6632	6633	6634	6635	6636	6637	6640	6641
3490	6642	6643	6644	6645	6646	6647	6650	6651	6652	6653

	0	1	2	3	4	5	6	7	8	9
3500	6654	6655	6656	6657	6660	6661	6662	6663	6664	6665
3510	6666	6667	6670	6671	6672	6673	6674	6675	6676	6677
3520	6700	6701	6702	6703	6704	6705	6706	6707	6710	6711
3530	6712	6713	6714	6715	6716	6717	6720	6721	6722	6723
3540	6724	6725	6726	6727	6730	6731	6732	6733	6734	6735
3550	6736	6737	6740	6741	6742	6743	6744	6745	6746	6747
3560	6750	6751	6752	6753	6754	6755	6756	6757	6760	6761
3570	6762	6763	6764	6765	6766	6767	6770	6771	6772	6773
3580	6774	6775	6776	6777	7000	7001	7002	7003	7004	7005
3590	7006	7007	7010	7011	7012	7013	7014	7015	7016	7017
3600	7020	7021	7022	7023	7024	7025	7026	7027	7030	7031
3610	7032	7033	7034	7035	7036	7037	7040	7041	7042	7043
3620	7044	7045	7046	7047	7050	7051	7052	7053	7054	7055
3630	7056	7057	7060	7061	7062	7063	7064	7065	7066	7067
3640	7070	7071	7072	7073	7074	7075	7076	7077	7100	7101
3650	7102	7103	7104	7105	7106	7107	7110	7111	7112	7113
3660	7114	7115	7116	7117	7120	7121	7122	7123	7124	7125
3670	7126	7127	7130	7131	7132	7133	7134	7135	7136	7137
3680	7140	7141	7142	7143	7144	7145	7146	7147	7150	7151
3690	7152	7153	7154	7155	7156	7157	7160	7161	7162	7163
3700	7164	7165	7166	7167	7170	7171	7172	7173	7174	7175
3710	7176	7177	7200	7201	7202	7203	7204	7205	7206	7207
3720	7210	7211	7212	7213	7214	7215	7216	7217	7220	7221
3730	7222	7223	7224	7225	7226	7227	7230	7231	7232	7233
3740	7234	7235	7236	7237	7240	7241	7242	7243	7244	7245
3750	7246	7247	7250	7251	7252	7253	7254	7255	7256	7257
3760	7260	7261	7262	7263	7264	7265	7266	7267	7270	7271
3770	7272	7273	7274	7275	7276	7277	7300	7301	7302	7303
3780	7304	7305	7306	7307	7310	7311	7312	7313	7314	7315
3790	7316	7317	7320	7321	7322	7323	7324	7325	7326	7327
3800	7330	7331	7332	7333	7334	7335	7336	7337	7340	7341
3810	7342	7343	7344	7345	7346	7347	7350	7351	7352	7353
3820	7354	7355	7356	7357	7360	7361	7362	7363	7364	7365
3830	7366	7367	7370	7371	7372	7373	7374	7375	7376	7377
3840	7400	7401	7402	7403	7404	7405	7406	7407	7410	7411
3850	7412	7413	7414	7415	7416	7417	7420	7421	7422	7423
3860	7424	7425	7426	7427	7430	7431	7432	7433	7434	7435
3870	7436	7437	7440	7441	7442	7443	7444	7445	7446	7447
3880	7450	7451	7452	7453	7454	7455	7456	7457	7460	7461
3890	7462	7463	7464	7465	7466	7467	7470	7471	7472	7473
3900	7474	7475	7476	7477	7500	7501	7502	7503	7504	7505
3910	7506	7507	7510	7511	7512	7513	7514	7515	7516	7517
3920	7520	7521	7522	7523	7524	7525	7526	7527	7530	7531
3930	7532	7533	7534	7535	7536	7537	7540	7541	7542	7543
3940	7544	7545	7546	7547	7550	7551	7552	7553	7554	7555
3950	7556	7557	7560	7561	7562	7563	7564	7565	7566	7567
3960	7570	7571	7572	7573	7574	7575	7576	7577	7600	7601
3970	7602	7603	7604	7605	7606	7607	7610	7611	7612	7613
3980	7614	7615	7616	7617	7620	7621	7622	7623	7624	7625
3990	7626	7627	7630	7631	7632	7633	7634	7635	7636	7637

	0	1	2	3	4	5	6	7	8	9
4000	7640	7641	7642	7643	7644	7645	7646	7647	7650	7651
4010	7652	7653	7654	7655	7656	7657	7660	7661	7662	7663
4020	7664	7665	7666	7667	7670	7671	7672	7673	7674	7675
4030	7676	7677	7700	7701	7702	7703	7704	7705	7706	7707
4040	7710	7711	7712	7713	7714	7715	7716	7717	7720	7721
4050	7722	7723	7724	7725	7726	7727	7730	7731	7732	7733
4060	7734	7735	7736	7737	7740	7741	7742	7743	7744	7745
4070	7746	7747	7750	7751	7752	7753	7754	7755	7756	7757
4080	7760	7761	7762	7763	7764	7765	7766	7767	7770	7771
4090	7772	7773	7774	7775	7776	7777	10000	10001	10002	10003
4100	10004	10005	10006	10007	10010	10011	10012	10013	10014	10015
4110	10016	10017	10020	10021	10022	10023	10024	10025	10026	10027
4120	10030	10031	10032	10033	10034	10035	10036	10037	10040	10041
4130	10042	10043	10044	10045	10046	10047	10050	10051	10052	10053
4140	10054	10055	10056	10057	10060	10061	10062	10063	10064	10065
4150	10066	10067	10070	10071	10072	10073	10074	10075	10076	10077
4160	10100	10101	10102	10103	10104	10105	10106	10107	10110	10111
4170	10112	10113	10114	10115	10116	10117	10120	10121	10122	10123
4180	10124	10125	10126	10127	10130	10131	10132	10133	10134	10135
4190	10136	10137	10140	10141	10142	10143	10144	10145	10146	10147
4200	10150	10151	10152	10153	10154	10155	10156	10157	10160	10161
4210	10162	10163	10164	10165	10166	10167	10170	10171	10172	10173
4220	10174	10175	10176	10177	10200	10201	10202	10203	10204	10205
4230	10206	10207	10210	10211	10212	10213	10214	10215	10216	10217
4240	10220	10221	10222	10223	10224	10225	10226	10227	10230	10231
4250	10232	10233	10234	10235	10236	10237	10240	10241	10242	10243
4260	10244	10245	10246	10247	10250	10251	10252	10253	10254	10255
4270	10256	10257	10260	10261	10262	10263	10264	10265	10266	10267
4280	10270	10271	10272	10273	10274	10275	10276	10277	10300	10301
4290	10302	10303	10304	10305	10306	10307	10310	10311	10312	10313
4300	10314	10315	10316	10317	10320	10321	10322	10323	10324	10325
4310	10326	10327	10330	10331	10332	10333	10334	10335	10336	10337
4320	10340	10341	10342	10343	10344	10345	10346	10347	10350	10351
4330	10352	10353	10354	10355	10356	10357	10360	10361	10362	10363
4340	10364	10365	10366	10367	10370	10371	10372	10373	10374	10375
4350	10376	10377	10400	10401	10402	10403	10404	10405	10406	10407
4360	10410	10411	10412	10413	10414	10415	10416	10417	10420	10421
4370	10422	10423	10424	10425	10426	10427	10430	10431	10432	10433
4380	10434	10435	10436	10437	10440	10441	10442	10443	10444	10445
4390	10446	10447	10450	10451	10452	10453	10454	10455	10456	10457
4400	10460	10461	10462	10463	10464	10465	10466	10467	10470	10471
4410	10472	10473	10474	10475	10476	10477	10500	10501	10502	10503
4420	10504	10505	10506	10507	10510	10511	10512	10513	10514	10515
4430	10516	10517	10520	10521	10522	10523	10524	10525	10526	10527
4440	10530	10531	10532	10533	10534	10535	10536	10537	10540	10541
4450	10542	10543	10544	10545	10546	10547	10550	10551	10552	10553
4460	10554	10555	10556	10557	10560	10561	10562	10563	10564	10565
4470	10566	10567	10570	10571	10572	10573	10574	10575	10576	10577
4480	10600	10601	10602	10603	10604	10605	10606	10607	10610	10611
4490	10612	10613	10614	10615	10616	10617	10620	10621	10622	10623

	0	1	2	3	4	5	6	7	8	9
4500	10624	10625	10626	10627	10630	10631	10632	10633	10634	10635
4510	10636	10637	10640	10641	10642	10643	10644	10645	10646	10647
4520	10650	10651	10652	10653	10654	10655	10656	10657	10660	10661
4530	10662	10663	10664	10665	10666	10667	10670	10671	10672	10673
4540	10674	10675	10676	10677	10700	10701	10702	10703	10704	10705
4550	10706	10707	10710	10711	10712	10713	10714	10715	10716	10717
4560	10720	10721	10722	10723	10724	10725	10726	10727	10730	10731
4570	10732	10733	10734	10735	10736	10737	10740	10741	10742	10743
4580	10744	10745	10746	10747	10750	10751	10752	10753	10754	10755
4590	10756	10757	10760	10761	10762	10763	10764	10765	10766	10767
4600	10770	10771	10772	10773	10774	10775	10776	10777	11000	11001
4610	11002	11003	11004	11005	11006	11007	11010	11011	11012	11013
4620	11014	11015	11016	11017	11020	11021	11022	11023	11024	11025
4630	11026	11027	11030	11031	11032	11033	11034	11035	11036	11037
4640	11040	11041	11042	11043	11044	11045	11046	11047	11050	11051
4650	11052	11053	11054	11055	11056	11057	11060	11061	11062	11063
4660	11064	11065	11066	11067	11070	11071	11072	11073	11074	11075
4670	11076	11077	11100	11101	11102	11103	11104	11105	11106	11107
4680	11110	11111	11112	11113	11114	11115	11116	11117	11120	11121
4690	11122	11123	11124	11125	11126	11127	11130	11131	11132	11133
4700	11134	11135	11136	11137	11140	11141	11142	11143	11144	11145
4710	11146	11147	11150	11151	11152	11153	11154	11155	11156	11157
4720	11160	11161	11162	11163	11164	11165	11166	11167	11170	11171
4730	11172	11173	11174	11175	11176	11177	11200	11201	11202	11203
4740	11204	11205	11206	11207	11210	11211	11212	11213	11214	11215
4750	11216	11217	11220	11221	11222	11223	11224	11225	11226	11227
4760	11230	11231	11232	11233	11234	11235	11236	11237	11240	11241
4770	11242	11243	11244	11245	11246	11247	11250	11251	11252	11253
4780	11254	11255	11256	11257	11260	11261	11262	11263	11264	11265
4790	11266	11267	11270	11271	11272	11273	11274	11275	11276	11277
4800	11300	11301	11302	11303	11304	11305	11306	11307	11310	11311
4810	11312	11313	11314	11315	11316	11317	11320	11321	11322	11323
4820	11324	11325	11326	11327	11330	11331	11332	11333	11334	11335
4830	11336	11337	11340	11341	11342	11343	11344	11345	11346	11347
4840	11350	11351	11352	11353	11354	11355	11356	11357	11360	11361
4850	11362	11363	11364	11365	11366	11367	11370	11371	11372	11373
4860	11374	11375	11376	11377	11400	11401	11402	11403	11404	11405
4870	11406	11407	11410	11411	11412	11413	11414	11415	11416	11417
4880	11420	11421	11422	11423	11424	11425	11426	11427	11430	11431
4890	11432	11433	11434	11435	11436	11437	11440	11441	11442	11443
4900	11444	11445	11446	11447	11450	11451	11452	11453	11454	11455
4910	11456	11457	11460	11461	11462	11463	11464	11465	11466	11467
4920	11470	11471	11472	11473	11474	11475	11476	11477	11500	11501
4930	11502	11503	11504	11505	11506	11507	11510	11511	11512	11513
4940	11514	11515	11516	11517	11520	11521	11522	11523	11524	11525
4950	11526	11527	11530	11531	11532	11533	11534	11535	11536	11537
4960	11540	11541	11542	11543	11544	11545	11546	11547	11550	11551
4970	11552	11553	11554	11555	11556	11557	11560	11561	11562	11563
4980	11564	11565	11566	11567	11570	11571	11572	11573	11574	11575
4990	11576	11577	11600	11601	11602	11603	11604	11605	11606	11607

INDEX TO TECO
COMMANDS AND
SPECIAL CHARACTERS

Command Character	Function
CTRL/C	Stop execution.
CTRL/D	Set radix to decimal.
CTRL/E	Form feed flag.
CTRL/EA	(Match control character) Match alpha.
CTRL/EC	(Match control character) Match RAD50.
CTRL/ED	(Match control character) Match digit.
CTRL/EL	(Match control character) Match line terminator.
CTRL/EQq	(String build character) Use contents of Q-reg q.
CTRL/ER	(Match control character) Match alphanumeric.
CTRL/ES	(Match control character) Match spaces or tabs
CTRL/EX	(Match control character) Match any character
CTRL/F	Contents of console switch register.
CTRL/G CTRL/G	Kill command string.
CTRL/G[space]	Retype current command line.

Command Character	Function
CTRL/G*	Retype current command input.
TAB	Insert tab and text
LF	Line terminator, ignored in commands.
VT	Ignored in commands
FF	Page terminator, ignored in commands.
CR	Ignored in commands.
CTRL/N	End of file flag.
CTRL/Nx	(Match control character) Match all but *x*.
CTRL/O	Set radix to octal.
CTRL/O	Stop terminal typeout.
CTRL/Q	Convert line arg into character arg.
CTRL/Q	(String build character) Use next character literally.
nCTRL/R	Identical to nFS command.
CTRL/S	-(length) of last string inserted or found.
CTRL/S	(Match control character) Match separator character.
CTRL/T	ASCII code of next character typed.
nCTRL/T	Output ASCII character of value *n*.
CTRL/U	Kill command line.
CTRL/Uq	Put string into Q-register q.
CTRL/V	Version of TECO.
CTRL/W	Redisplay text buffer immediately.
CTRL/X	Search mode flag.
CTRL/X	(Match control character) Match any character.
CTRL/Y	Equivalent to ".+CTRL/S,.".
CTRL/Z	Number of characters in Q-register area.
ESC	String and command terminator.
SP	Ignored in commands.
!	Define label.
"	Start conditional.
n"A	Test for alphabetic.
n"C	Test for RAD50.
n"D	Test for digit.
n"E	Test for equal to zero.
n"F	Test for false.
n"G	Test for greater than zero.
n"L	Test for less than zero.

Command Character	Function
n"N	Test for not equal to zero.
n"R	Test for alphanumeric.
n"S	Test for successful.
n"T	Test for true.
n"U	Test for unsuccessful.
n"V	Test for uppercase alphabetic.
n"W	Test for lowercase alphabetic.
n">	Test for greater than 0.
n"<	Test for less than 0.
#	Logical OR.
n%q	Add n to Q-register q.
&	Logical AND.
'	End conditional
(	Expression grouping.
)	Expression grouping.
*	Multiplication
*q	Put last command in Q-register q.
+	Addition
,	Argument separator.
−	Subtraction or negation.
.	Current pointer position.
/	Division.
0–9	Digit.
:	Make next command return a value.
::	Make next search a compare.
:=	Type in decimal, no carriage return.
:==	Type in octal, no carriage return.
:Gq	Type Q-register q on terminal.
Qq	Size of text in Q-register q.
n;	Exit iteration if $n \geq 0$.
n<	Iterate n times.
=	Type in decimal.
==	Type in octal.
>	End iteration.
?	Toggle trace mode.
?	Type command string up to error.
@	Use alternate string delimiter.
A	Append to buffer.
nA	ASCII value of character in buffer.

Command Character	Function
B	0 or beginning of buffer.
nC	Advance n characters.
nD	Delete n characters.
EA	Select secondary output stream.
EB	Open input and output.
EC	Close out (copy input to output and close).
ED	Edit level flag.
EF	Close output file.
EG	Close out and exit.
EH	Edit help level.
EI	Open indirect command file.
EK	Kill output file.
EN	Wild card lookup.
EP	Select secondary input stream.
ER	Open input file.
ES	Search verification flag.
ET	Type out control flag.
EU	Case flagging flag.
EV	Edit verify flag
EW	Open output file.
EX	Close out and exit.
nFN	Global string replace.
FR	Replace last string.
nFS	Local string replace.
m,nFS	Bounded local string replace.
Gq	Get string in Q-register q into buffer.
G*	Get last filespec string into buffer.
G_	Get last search string into buffer.
H	Equivalent to "B,Z".
I	Insert text.
nI	Insert ASCII character "n".
nJ	Move pointer to "n".
nK	Kill n lines.
m,nK	Delete between m and n.
nL	Advance n lines.
Mq	Execute string in Q-register q.
nN	Global search
O	Go to label.
nP	Advance n pages.

**Command
Character**   **Function**

m,nP	Write out characters *m* to *n*.
nPW	Write buffer *n* times.
m,nPW	Write out characters *m* to *n*.
Qq	Number in Q-register q.
nR	Back up *n* characters.
nS	Local search.
m,nS	Bounded local search.
nT	Type *n* lines.
m,nT	Type from *m* to *n*.
nUq	Put number n in Q-register q.
nV	Type *n* current lines.
W	Not a TECO command.
nXq	Put *n* lines into Q-register q.
m,nXq	Put characters *m* to *n* into Q-register q.
Y	Read into buffer.
Z	End of buffer value.
]q	Q-register push.
\	Value of digit string in buffer.
n\	Convert *n* to digits in buffer.
]q	Q-register pop.
^	Interpret next command character as a control character.
^	Interpret next search character as a control character.
n	Global search without output.
a–z	Treated the same as uppercase A–Z.
DEL	Delete last character typed in.

INDEX□